Multiscale Modeling and Uncertainty Quantification of Materials and Structures

Manolis Papadrakakis • George Stefanou
Editors

Multiscale Modeling and Uncertainty Quantification of Materials and Structures

Proceedings of the IUTAM Symposium held at Santorini, Greece, September 9–11, 2013

 Springer

 International Union of Theoretical and Applied Mechanics

Editors
Manolis Papadrakakis
Institute of Structural Analysis
  & Antiseismic Research
National Technical University of Athens
Athens, Greece

George Stefanou
Institute of Structural Analysis
  & Antiseismic Research
National Technical University of Athens
Athens, Greece

Institute of Structural Analysis & Dynamics
  of Structures
Aristotle University of Thessaloniki
Thessaloniki, Greece

ISBN 978-3-319-06330-0   ISBN 978-3-319-06331-7 (eBook)
DOI 10.1007/978-3-319-06331-7
Springer Cham Heidelberg New York Dordrecht London

Library of Congress Control Number: 2014943257

Printed on acid-free paper

Springer is part of Springer Science+Business Media (www.springer.com)

Preface

Over the last few years, the intense research activity at microscale and nanoscale reflected the need to account for disparate levels of uncertainty from various sources and across scales. As even over-refined deterministic approaches are not able to account for this issue, an efficient blending of stochastic and multiscale methodologies is required to provide a rational framework for the analysis and design of materials and structures. The purpose of the Symposium was to promote achievements in uncertainty quantification combined with multiscale modeling and to encourage research and development in this growing field with the aim of improving the safety and reliability of engineered materials and structures.

The Symposium took place from September 9 to September 11, 2013 in Santorini Island, Greece and has been attended by 39 participants from 12 countries. Special emphasis was placed on multiscale material modeling and simulation as well as on the multiscale analysis and uncertainty quantification of fracture mechanics of heterogeneous media. The homogenization of two-phase random media was also thoroughly examined in several presentations. Various topics of multiscale stochastic mechanics, such as identification of material models, scale coupling, modeling of random microstructures, analysis of CNT-reinforced composites and stochastic finite elements, have been analyzed and discussed. A large number of papers were finally devoted to innovative methods in stochastic dynamics.

This book consists of 20 chapters which are extended versions of selected papers presented at the Symposium. The chapters are grouped into the following five thematic topics: Damage and fracture, homogenization, inverse problems–identification, multi-scale stochastic mechanics and stochastic dynamics.

The editors would like to express their deep appreciation to all contributors for their active participation in the Symposium and for the time and effort devoted to the completion of their contributions to this volume. Special thanks are also due to the reviewers for their constructive comments and suggestions which enhanced the quality of the book. Finally, the editors would like to thank the Scientific Committee of the Symposium and the personnel of Springer for their most valuable support during the publication process.

Athens, Greece Manolis Papadrakakis
March 2014 George Stefanou

Contents

Part I
Damage and Fracture

Fracture Simulations of Concrete Using Discrete Meso-level Model with Random Fluctuations of Material Parameters

Jan Eliáš, Miroslav Vořechovský, and Jia-Liang Le

Abstract The paper presents numerical simulations of concrete fracture performed on beams of variable size and notch depth using the stochastic meso-level discrete model. The model includes a substantial part of randomness in concrete heterogeneity by accounting for the largest grains when assembling the lattice geometry. The remaining randomness, caused by finer particles and the non-uniformity of the mixing process, is introduced by random fluctuations of material parameters represented by a random field. The results of the stochastic meso-level discrete model are compared with published fracture experiments performed on concrete beams loaded in three-point bending. The effects of randomness in connection with different beam size and notch depth are discussed, as well as observed differences in dissipated energy.

Keywords Lattice-particle model • Concrete • Random field • Strength • Fracture energy • Size effect

1 Introduction

The reliability of reinforced concrete components is crucial for modern engineering structures. The need to understand the phenomenon of concrete failure has resulted in the development of complex numerical models that can predict the strength and post-critical behavior of concrete components. It is generally agreed that

J. Eliáš (✉) • M. Vořechovský
Faculty of Civil Engineering, Institute of Structural Mechanics, Brno University of Technology, Veveří 331/95, Brno 60200, Czech Republic
e-mail: elias.j@fce.vutbr.cz; vorechovsky.m@fce.vutbr.cz

J.-L. Le
Department of Civil Engineering, University of Minnesota, 500 Pilsbury Drive SE, Minneapolis, MN 55455 USA

M. Papadrakakis and G. Stefanou (eds.), *Multiscale Modeling and Uncertainty Quantification of Materials and Structures*, DOI 10.1007/978-3-319-06331-7__1,
© Springer International Publishing Switzerland 2014

the failure process in concrete and similar quasibrittle materials (ceramics, ice, etc.) is characterized by the gradual release of stress within a fracture process zone (FPZ) ahead of the crack tip. This gradual release is understood to be a consequence of concrete heterogeneity (van Mier 1997). This belief has led to attempts to include heterogeneity in the developed model. Although this can be achieved using continuous material description (Caballero et al. 2006), meso-level modeling of concrete fracture is usually performed using discrete models. The least phenomenological of them are probably the classical lattice models (Herrmann et al. 1989; Schlangen and Garboczi 1997; van Mier et al. 1997; Man and van Mier 2008; van Mier 2013) with elasto-brittle lattice elements and lattice geometry that is independent of the considered material heterogeneity. However, such fine-resolution models are, due to their high computational cost, suitable for small specimens only. A reasonable compromise seems to be to use a less dense lattice with each node corresponding to one mineral grain (Bažant et al. 1990; Jirásek and Bažant 1994; Cusatis et al. 2003; Cusatis and Cedolin 2007; Cusatis et al. 2011). These models are referred to as the lattice-particle models. Their disadvantage is more complicated and phenomenological constitutive law; however, one can use them for analysing substantially larger volumes of concrete.

Gaining an understanding of the fracture process is further complicated by the presence of random fluctuations in the material. This randomness comes from several sources, such as the randomness of the concrete constituents themselves (material properties, geometric properties), from mixing the constituents together (grain locations, non-homogeneous distribution of water, cement, finer grains and additives), from non-uniform drying, etc. Material randomness is quite often ignored. Meso-level models have the advantage that a substantial part of randomness is included through the consideration of the meso-level structure of material. To improve the stochastic description of concrete, further random fluctuations of model parameters are typically used. Fluctuations are then usually included in the form of a stationary autocorrelated random field (Vořechovský and Sadílek 2008; Grassl and Bažant 2009; Grassl et al. 2013). Each source of randomness naturally has its own characteristics such as correlation length, distribution type and coefficient of variation.

The lattice-particle model, enhanced by random fluctuations, was used here to simulate an extensive experimental series of three-point-bended beam tests carried out at Northwestern University by Hoover et al. (2013). The experimental series included four different beam sizes (with a size ratio of 1:12.5) and variable notch depths (from no notch at all up to a notch extending to 30 % of beam depth). The experiments were controlled by crack mouth opening displacement (CMOD), so it was also possible to measure the post-peak behavior.

The deterministic model parameters were identified from the peak loads and the dissipated energies of the beams with the deepest notch only. Stochastic parameters were unfortunately not identified but only fabricated. The main objective was to demonstrate the ability of the model to closely match and (after identification of its parameters) also predict experimental results. Furthermore, the influence of size and boundary conditions on energy dissipation in the model was studied as well as the effect of randomness.

The paper is structured as follows. A brief introduction of the deterministic model, followed by a description of the introduced randomness, is presented in Sect. 2. A description of the published experimental series as well as the identification of model parameters can be found in Sect. 3. The results of the deterministic (and stochastic) model are compared with the results of experiments in Sect. 4 (and Sect. 5, respectively). Finally, the energy dissipation is discussed in Sect. 6.

2 Model Description

The modeling of material by using an assembly of discrete units has become a well established approach with several advantages, such as the relative simplicity of constitutive law formulation, the ability of representing material inhomogeneity, the automatic weakening of the modeled material only in directions perpendicular to cracks, etc. On the other hand, extensive computer resources are often needed to use such models. The present study is based on the meso-level discrete lattice-particle model developed by Cusatis and Cedolin (2007), which is an extension of Cusatis et al. (2003, 2006). Information regarding the further development of the model can be found in Cusatis et al. (2011).

2.1 Deterministic Model

The material is represented by a discrete three-dimensional assembly of ideally rigid cells. The cells are created by a tessellation according to the pseudo-random locations and radii of computer generated spheres – virtual mineral aggregates of concrete. Every cell contains one aggregate (Fig. 1a, b). Rigid cells are connected through their common facets, where nonlinear (cohesive) constitutive laws are applied. Damage done to the facets then represents cracking in the matrix and interfacial transition zone between two aggregates. The single damage variable ω controls the loss of material integrity both in the normal and the tangential directions of the facet. It depends on facet strains and previous loading history. For a detailed description of the model's features, see Cusatis and Cedolin (2007). The confinement effect (present in the full version of the model) is not implemented here.

To save computer time, the lattice-particle model only covered the regions in which cracking was expected. The other regions of the beams were assumed to remain linear elastic and were therefore modeled by standard 8-node isoparametric finite elements and connected to the discrete model via auxiliary zero-diameter particles (Eliáš and Bažant 2011). Elastic constants for these elements might be found via the fitting of a displacement field of a continuous homogeneous model to displacements of the particle system when subjected to low-level uniaxial compression.

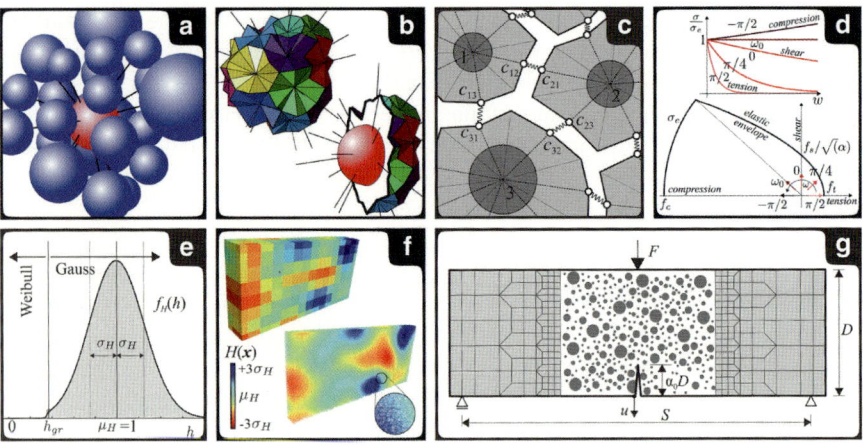

Fig. 1 (**a**) Concrete meso-structure simulated by the random placement of grains; (**b**) tessellation providing rigid cells around every grain; (**c**) elements between adjacent cells assumed in the common facet center; (**d**) constitutive relations assigned to the interparticle elements – elastic envelope (*bottom*) and exponential softening (*top*), both dependent on the direction of straining; (**e**) Weibull-Gauss probability distribution used for randomization of element parameters; (**f**) a random field of model parameters generated on a regular grid and projected to the model elements; (**g**) specimen shape and boundary conditions

2.2   Stochastic Extension of the Model

Since the stochastic extension of the model was not included in Cusatis and Cedolin (2007), it is elucidated here in greater depth. Material parameters are assigned at each inter-particle connection according to a stationary autocorrelated random field. The value of the c-th realization of the discretized field at a spatial coordinate x is denoted $H^c(x)$. For a given coordinate x_0, $H(x_0)$ is a random variable H of cdf $F_H(h)$. Since the random field is stationary, cdf $F_H(h)$ is identical for any position x_0.

The strength of components made of quasibrittle material is typically governed by the material's strength and fracture energy. Realistic fracture models should therefore incorporate the random spatial variability of at least these two variables. We consider the material strength fully correlated (with a correlation coefficient of one) with the fracture energy (Grassl and Bažant 2009). Furthermore, the lattice-particle model also includes the shear strength f_s and fracture energy in shear G_s, which are again assumed to be fully correlated with the tensile strength f_t and fracture energy in tension G_t. Assuming identical distribution type and identical variance (and also higher statistical moments), we can use the same random field to generate values for both material strengths and both fracture energies. When any of the four aforementioned mechanical properties are substituted for X, one can write

$$X(x) = \bar{X} H(x), \tag{1}$$

where $\bar{X}$ stands for the mean value of the particular property. The mean value of the (field) random variable H equals 1.

Papers by Bažant and Pang (2007), Le et al. (2011), and Le and Bažant (2011) have suggested approximating the strength distribution of a quasibrittle material using a Gaussian cdf onto which a Weibullian tail is grafted from the left hand side. We use this distribution for our H variable.

$$
F_H(h) = \begin{cases} r_f \left(1 - e^{-\langle h/s_1 \rangle^m}\right) & h \leq h_{\text{gr}} \qquad\qquad (2) \\[2em] p_{\text{gr}} + \dfrac{r_f \displaystyle\int_{h_{\text{gr}}}^{h} e^{-\frac{(h-\mu_G)^2}{2\delta_G^2}}\, dh}{\delta_G \sqrt{2\pi}} & h > h_{\text{gr}} \qquad\qquad (3) \end{cases}
$$

where $\langle\cdot\rangle = \max(\cdot, 0)$, $s_1 = s_0 r_f^{1/m}$, m is the Weibull modulus (shape parameter) and s_0 is the scale parameter of the Weibull tail, μ_G and δ_G are the mean value and standard deviation of the Gaussian distribution that describes the Gaussian core and $p_{\text{gr}} = F_H(h_{\text{gr}})$ is the grafting probability. The Weibull-Gauss juncture at point at h_{gr} requires equality in the probability density: $(dF_H/dh)|_{h_{\text{gr}}^+} = (dF_H/dh)|_{h_{\text{gr}}^-}$; here, r_f is a scaling parameter normalizing the distribution to satisfy the condition $F_H(\infty) = 1$. The distribution has four independent parameters in total.

The spatial fluctuations of the field are characterized through an autocorrelation function. It determines the spatial dependence pattern between the random variables at any pair of nodes. The correlation coefficient ρ_{ij} between two field variables at coordinates x_i and x_j can be assumed to obey the squared exponential function

$$
\rho_{ij} = \exp\left[-\left(\frac{\|x_i - x_j\|}{d}\right)^2\right]. \qquad\qquad (4)
$$

It introduces a new parameter, d, called the autocorrelation length.

To produce the random field $\boldsymbol{H}$ of a non-Gaussian variable H, the most frequent procedure is to generate the Gaussian field $\hat{H}$ and then transform it via the isoprobabilistic (memoryless) transformation

$$
\boldsymbol{H}(x) = F_H^{-1}(\Phi(\hat{H}(x))), \qquad\qquad (5)
$$

where Φ stands for the cdf of the Gaussian field. Such a transformation distorts the correlation structure of the field $\boldsymbol{H}$. Thus, when generating the underlying Gaussian field $\hat{H}$, the correlation coefficients must be modified in order to fulfil the desired pairwise correlations of the non-Gaussian field $\boldsymbol{H}$. This is performed using an approximation to the Nataf model (Li et al. 2008).

There are several methods of generating a Gaussian random field. One simple way is to use the Karhunen–Loève expansion based on the spectral decomposition of the covariance matrix $\boldsymbol{C}$ of mutual correlation coefficients (principal component analysis). The realizations of the random field need to be evaluated at every shared facet center of the lattice-particle model. This can be computationally demanding for a large number of facets, because the covariance matrix is then large as well. In our simulation, we might have about 200,000 inter-particle connections. To overcome this computational cost, the expansion optimal linear estimation method – EOLE (Li and Der Kiureghian 1993), is adopted. This method can significantly reduce the time required for random field generation. The field is initially generated on a regular grid of nodes with spacing $d/3$ (see Fig. 1) instead of at the facet centers. Therefore only about 800 grid nodes are needed. The values of the random field at the model facets are then obtained from the expression

$$\hat{H}^c(x) = \sum_{k=1}^{K} \frac{\xi_k^c}{\sqrt{\lambda_k}} \boldsymbol{\psi}_k^T \boldsymbol{C}_{xg}, \tag{6}$$

where λ and $\boldsymbol{\psi}$ are the eigenvalues and eigenvectors of the covariance matrix of the grid nodes, and $\boldsymbol{C}_{xg}$ is a covariance matrix between the facet center at coordinates x and the grid nodes. ξ_k are independent standard normal variables. After the Gaussian random field values at facet centers are obtained by EOLE (Eq. 6), they need to be transformed to the non-Gaussian space by Eq. 5.

Besides significant time savings, another advantage of using EOLE is that one can simply use the same field realization for several different granular positions. By keeping the c-th realization of the decomposed independent variables $\boldsymbol{\xi}^c$ unchanged, the field realization can be adapted for any configuration of the facets in the discrete model.

3   Experimental Series and Identification of Model Parameters

The experimental series was tested at Northwestern University by Hoover et al. (2013) and Hoover and Bažant (2013). Three-point-bended beams with or without a notch were loaded by a prescribed displacement. The tests were controlled via CMOD, which also allowed the measurement of the descending branch of load-CMOD response. The series contains beams with four different depths D of 500, 215, 93 and 40 mm, denoted by the upper-case letters A, B, C and D, respectively. The thickness $t = 40$ mm was the same for all the specimens and the span was $2.176 \times D$. Notch depth varied from no notch to a notch cut to 30 % of specimen depth. Five notch depths were tested: $\alpha_0 = a_0/D0 = 0, 0.025, 0.075, 0.15$ and 0.3. These are denoted by the lower-case letters a ($\alpha_0 = 0.3$), b, c, d and e ($\alpha_0 = 0$) in test descriptions. All the size-notch depth combinations were tested except the shallowest notch b and the two smallest beam sizes, C and D.

Identification of material parameters is based on simple minimization of the difference between the experimentally measured and simulated responses. There is no attempt to specify model parameters from information about the material composition. The only important parameter for modeling is the maximum grain size, which is 10 mm. Instead of a real sieve curve, which was not available, the aggregate diameters were considered to be distributed according to the Fuller curve.

3.1 Identification of Deterministic Parameters

In the first step of identification, only deterministic parameters were found. The deterministic model has several adjustable parameters; however, most of them were kept unchanged since they are either related to a different mode of failure (compressive strength, compressive hardening) or are hard to identify from the available limited set of experiments (shear strength and fracture energy, asymptote of the hyperbolic elastic envelope). Only four free parameters were considered to be identified: elastic modulus of matrix E_c, parameter α determining the macroscopic Poisson ratio, tensile strength f_t and fracture energy in tension G_t. The remaining parameters were considered (based on Cusatis and Cedolin 2007) to be as follows: elastic modulus of grains $E_a = 3E_c$, shear strength $f_s = 3f_t$, shear fracture energy $G_s = 16G_t$, compressive strength $f_c = 16f_t$, initial slope of compressive hardening $K_c = 0.26E_c$, slope of elastic envelope hyperbola asymptote $\mu = 0.2$, parameter of compressive hardening $n_c = 2$, and parameter of compressive elastic envelope $\beta = 1$.

For the identification of deterministic parameters, only the responses of beams with the deepest notch (Aa, Ba, Ca, Da) were used. There are two reasons for this: (i) the presence of a strong stress concentrator such as a deep notch minimizes the effect of spatial randomness on the mean response (Eliáš et al. 2013; Eliáš and Vořechovský 2013); (ii) to test whether the model can provide reasonable predictions by simulating remaining beam geometries and comparing the results to experimental data that were not used in the identification process.

The macroscopic Poisson's ratio was considered to be 0.19, which roughly corresponds to parameter $\alpha = 0.29$. The elastic modulus of matrix E_c was found by trial-error fitting of the elastic part of the load-CMOD curves to be approximately 25 GPa. The corresponding macroscopic modulus of finite elements was found via fitting of the displacement of particles under low-strain uniaxial compression via the use of linear elasticity theory of continuum.

The tensile strength and fracture energy were identified through the simple automatic minimization of objective function F_o. The function was formulated as the maximum relative difference between measured and simulated peak loads and areas under load-opening curves. The mean values of the experimentally measured maximal load $\bar{P}^{exp}$ and areas under the load-opening curve up to opening 0.15 mm

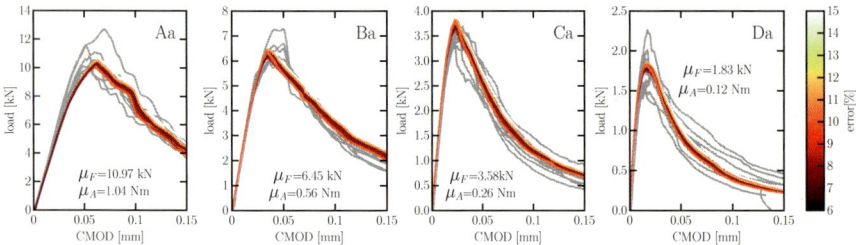

Fig. 2 Comparison of experimental load-CMOD curves and simulated responses as obtained by automatic optimization

$\bar{A}^{\text{exp}}$ were calculated, and represent the values we would like to exactly reproduce. The corresponding simulated values (P^{sim}, A^{sim}) were evaluated for every iteration of the optimization algorithm. The error represented by the objective function is calculated as

$$F_o = \max \left(\frac{|\bar{A}_c^{\text{exp}} - A_c^{\text{sim}}|}{\bar{A}_c^{\text{exp}}}, \frac{|\bar{P}_c^{\text{exp}} - P_c^{\text{sim}}|}{\bar{P}_c^{\text{exp}}} \right) \text{ for } c \in \{\text{Aa, Ba, Ca, Da}\} . \quad (7)$$

For the sake of saving computational time, the simulated results A^{sim} and P^{sim} were calculated using the deterministic model with one (constant) grain position only. For more reliable identification, one should perform several simulations with different grain positions for every evaluation of the objective function. The diagrams resulting from the optimization process are shown in Fig. 2. The minimum objective function value found was 0.067 (6.7 %).

3.2   Identification of Stochastic Parameters

In the second step, at least some of the stochastic parameters were expected to be identified. (i) It was assumed that it would be possible to identify the coefficient of variation of H from the deep notch results. As was shown in Eliáš et al. (2013) and Eliáš and Vořechovský (2013), the spatial fluctuations of local strengths and fracture energies have a negligible effect on the response if the crack initiates from a deep notch. One can separate the local properties of the randomness (distribution F_H) from the spatial properties (correlation length d) by introducing a strong stress concentrator. By matching the variability of experimental responses for a deep notch, it should be theoretically possible to estimate the coefficient of variation of the random field. (ii) Once this was done, one could identify the correlation length by matching the peak loads of the unnotched beams. As was shown in the aforementioned papers (Eliáš et al. 2013; Eliáš and Vořechovský 2013), the mean value of the peak load strongly depends on the correlation length.

Unfortunately, this theoretical procedure was not applied in the current study for two reasons. First, the experimental scatter for deep notch beams was already very close to the statistical scatter of the deterministic model, where the randomness is only present via the location of grains. The coefficient of variation of H should therefore be considered to be close to zero. Second, introducing the randomness into no-notch simulation can only lead to a decrease in mean peak loads. However, the deterministic model already exhibited lower peak loads for no-notch beams than was experimentally measured. It again suggests that any randomness other than that caused by the location of the largest grains was negligible in these tests. We therefore decided to consider the variability present in the deterministic model to be sufficient for reproduction of the variability in the experimental series. Instead of using the random field to get closer to the measured data, we performed the numerical study with an artificially excessive coefficient of variation (0.25) in order to further study the effect of randomness on the model.

The following local parameters of the random field were used: Weibull modulus $m = 24$; $s_1 = 0.486\,\text{MPa}$; grafting point $h_{\text{gr}} = 0.364\,\text{MPa}$; standard deviation of the Gaussian core $\delta_G = 0.25\,\text{MPa}$. These parameters provided the overall mean value $\mu_H = 1$; standard deviation $\delta_H \approx 0.25$, and grafting probability $F_H(h_{\text{gr}}) \approx 10^{-3}$. The probability density function of this distribution is shown in Fig. 1e.

For the stochastic study, two correlation lengths d were considered: the shorter length $d_4 = 40\,\text{mm}$ (as found in Grassl and Bažant 2009) and the longer length $d_8 = 80\,\text{mm}$ (as found in Vořechovský and Sadílek 2008).

4 Deterministic Modeling

The simulations of all the experimental beam geometries were performed using the parameters identified for the deep notch beams, and without randomness. Ten simulations with differing in the random location of grains were calculated for each geometry. CMOD control is pointless for shallow to no-notch beams, since the crack may initiate outside the midspan. However, in the simulation one can simply measure the opening at several short intervals along the beam span and control the simulation using the largest of these openings. If there is no gap between the intervals, the crack must initiate inside one of them and the CMOD is obtained. This is, however, only rarely possible in real experiments. Therefore, the opening was measured over only one longer interval by one gauge with a the hope that a crack would initiate inside it. This gauge opening was extracted from the simulations as well to make comparison possible.

The experiment-model comparison of responses is shown in Fig. 3. Each sub-graph has the mean value and standard deviation of peak load plotted in its top right corner, the number at the same position gives the relative difference between the mean peak loads of the model and the experiment. Identification was performed for the leftmost column only; all the other columns are model predictions. The correspondence seems to be sufficient. However, some differences are present:

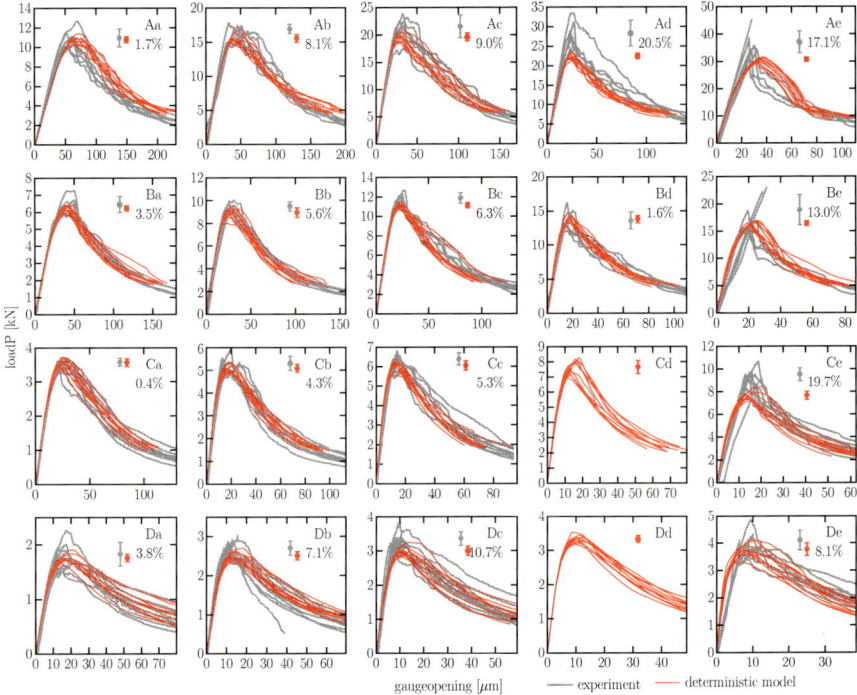

load P [kN]

gaugeopening [μm]   —— experiment   —— deterministic model

Fig. 3 Responses obtained by the deterministic model compared to the responses recorded during experiments

- The model systematically underestimates the peak loads for almost all the geometries. This shows that the identification is not perfect. Consideration of other beam geometries for identification would improve the model's responses, but evaluation of the model's predictive capabilities would be lost.
- The elastic parts of the experimental and model response differ for the smallest unnotched geometry De. One possible explanation is that the length of the measuring interval differs between the model and experiment as well.
- Two of the largest unnotched geometries (Ae and Be) had convergence problems right after the peak due to the large snap back present in the load-deflection record.

Figure 4 shows some damage patterns obtained by the deterministic model. One can see that no-notch simulations predict a wide zone of distributed cracking (which occurs prior to reaching the peak load). However, after the localization that occurs when reaching the peak load, the crack looks more or less the same as in the case of the deep notch geometries. Another interesting point is the clear dependence of post-peak macro-crack width on the specimen size. The larger the specimen, the wider the damage area. This was previously reported in Eliáš and Bažant (2011).

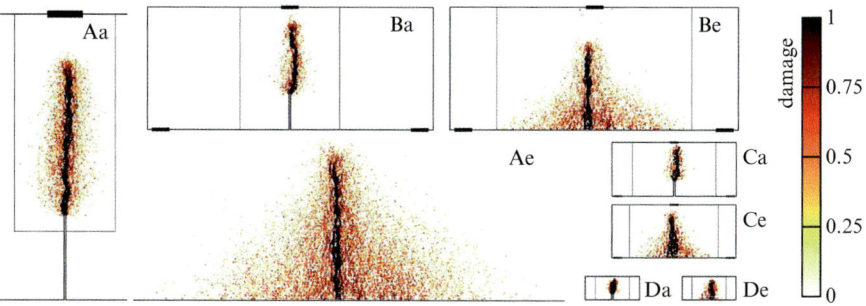

Fig. 4 Some damage patterns obtained by the deterministic lattice-particle model

5 Effects of Spatial Randomness

As stated in Sect. 3, the application of spatial randomness is not meant to bring the model behavior closer to that of the experiments. It is performed here in order to study the effects of randomness on the model's behavior.

For every geometry and every correlation length, 24 simulations were performed differing in both random grain positions and random field realizations. However, the 24 random field realizations for each geometry were obtained from the same 24 grid realizations. The resulting load-gauge opening curves are plotted in Fig. 5, as well as the means and standard deviations of the peak loads in the upper right corner of every subplot.

For the deeply notched specimens, the application of additional randomness leads only to an increase in response variance. The average peak load does not change compared to the deterministic model. Moreover, the observed increase in variance looks more or less independent of the correlation length. It was planned that this expected behavior would be used to identify the coefficient of variation of H.

A different situation appears for unnotched geometries. In contrast to the deeply-notched specimens, in which the crack always starts to propagate from the notch tip, the unnotched specimens are free of any stress concentrator, allowing the crack to initiate anywhere along the bottom surface. Therefore, the region with the worst combination of stress and local strength will serve as an initiation point. Since the crack will systematically start in weak regions, the peak value must decrease compared to the value in the deterministic simulation. The larger the area where the crack may initiate and the shorter the correlation length, the weaker the region that may appear. One can therefore see that the difference between deterministic and stochastic peak load increases with increasing size and decreasing correlation length.

The shallow notches induce weak stress concentrations which, in most cases, suffice to force cracks to initiate from the notch tip. However, it may happen that, due to randomness, a crack initiates outside the shallow notch. Such a case is documented in Fig. 6 for geometry Bd. Also, as the crack may start far from the

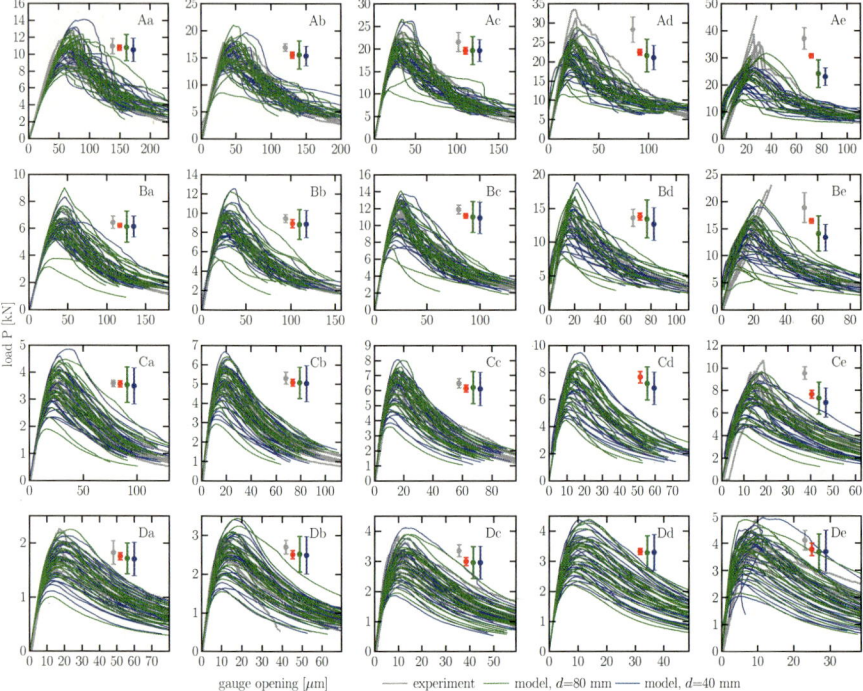

Fig. 5 Responses obtained by the stochastic model (for two different correlation lengths) compared to the responses recorded during experiments

midspan, the gauge length may not cover the crack position. This is why some of the responses for notches d and e exhibit decreasing gauge opening after reaching the peak load.

6   Analysis of Energy Dissipation

It is interesting to study energy dissipation during the simulation. This is done here via variable g, which represents the energy needed to propagate a crack by unit surface. It is calculated by summing the energies G_i dissipated at individual elements i within a horizontal strip of width 2τ. For any vertical coordinate y, one finds all the elements i with vertical coordinate y_i within interval $\langle y - \tau, y + \tau \rangle$ and sums their released energies. This number is then normalized by specimen thickness t and strip width 2τ.

$$g(y) = \sum_{i\,:\,|y_i - y| \leq \tau} \frac{G_i}{2t\tau} \qquad (8)$$

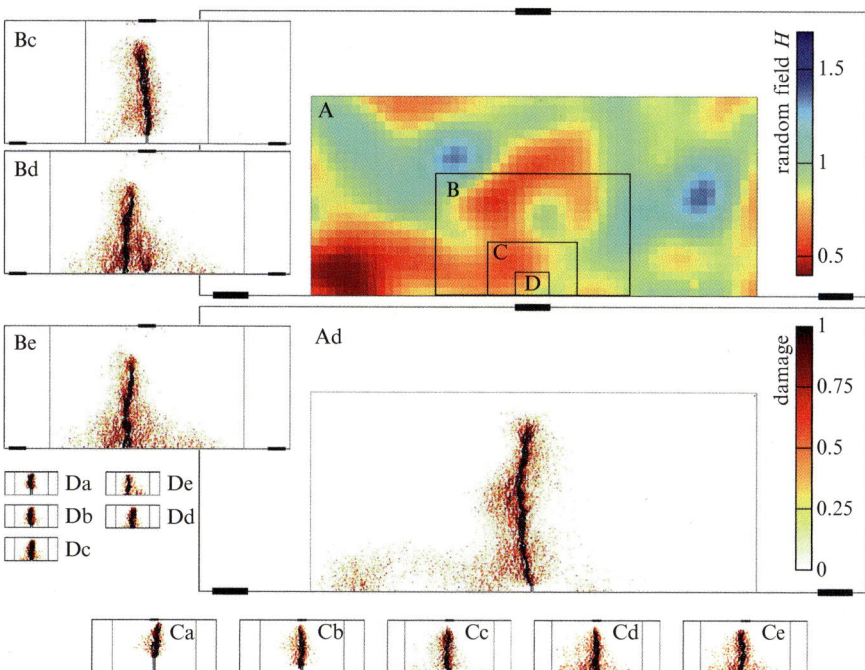

Fig. 6 Some damage patterns obtained by the stochastic lattice-particle model for different geometries but the same realization of the random field

Figure 7 shows the energy variable g along the specimen depth for all the geometries. Deterministic simulations are shown on the left hand side, whereas the right hand side displays the stochastic results. The mean value (bold line) and standard deviation (shaded area) of g is evaluated from 10 deterministic or 24 stochastic realizations.

- Note that besides increased standard deviation, there is no difference between the stochastic and deterministic results or between stochastic results with different correlation length. Only in the deterministic case do unnotched specimens exhibit a large area of distributed cracking prior to reaching peak load, which is visible in the graph as increased energy dissipation close to the bottom surface. The stochastic model lacks the distributed cracking because the pre-peak cracking is already localized into weak regions only. Therefore, no increase in g can be seen.
- The maximal values of g are located close to the notch tip. As we proceed to the upper parts of the specimen, g initially slightly decreases. Then, it decreases rapidly because the simulation was stopped before the stress-free crack reached this depth; g had not reached its *final* value yet. Nevertheless, the *final* value of g is about the same irrespective of notch depth. However, it was shown in Eliáš and Bažant (2011) that the *final* value of g may significantly decrease for extreme notch depths of $\alpha_0 \geq 0.7$.

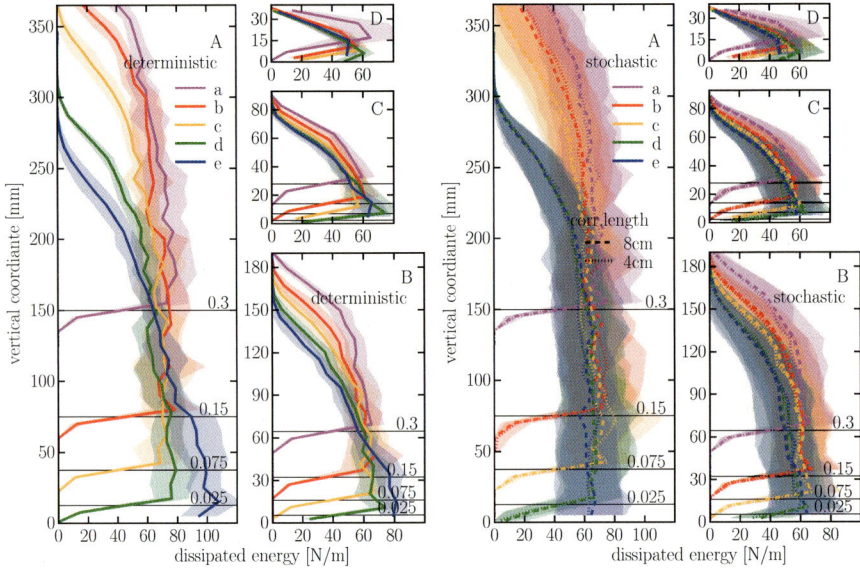

Fig. 7 Energies g dissipated at specific beam depths until the end of the simulations. The *thick line* refers to the average value and the *shaded area* shows standard deviation. *Left*: deterministic model; *right*: stochastic models with correlation length $d = 40$ and 80 mm

- One can see that smaller beams also have smaller *final* values of g. This is attributed to the increasing stress gradient that constrains the development of the fracture process zone for decreasing size. The same effect is responsible for the slight decrease in g with increasing depth which is described in the previous item.

7 Conclusions

The discrete lattice-particle model with identified parameters was employed to reproduce an extensive series of experiments. The comparison confirmed the robustness of the model by showing reasonable agreement between simulated and experimental responses, verifying the predictive capabilities of the model. The deterministic version of the model (which contains randomness due to the random locations of the largest concrete grains) seems to be already sufficient to produce the variations measured in the experiment.

The stochastic study with artificially chosen parameters confirmed the previously observed effects of randomness. These effects strongly depend on the initial notch depth.

Acknowledgements The financial support received from the Ministry of Education, Youth and Sports of the Czech Republic under Project No. LH12062 and the Czech Science Foundation under Project No. GC13-19416J is gratefully acknowledged.

References

Bažant ZP, Pang SD (2007) Activation energy based extreme value statistics and size effect in brittle and quasibrittle fracture. J Mech Phys Solids 55(1):91–131

Bažant Z, Tabbara M, Kazemi M, Pijaudier-Cabot G (1990) Random particle model for fracture of aggregate or fiber composites. J Eng Mech ASCE 116(8):1686–1705

Caballero A, López C, Carol I (2006) 3d meso-structural analysis of concrete specimens under uniaxial tension. Comput Method Appl Mech 195(52):7182–7195

Cusatis G, Cedolin L (2007) Two-scale study of concrete fracturing behavior. Eng Fract Mech 74(1–2):3–17

Cusatis G, Bažant ZP, Cedolin L (2003) Confinement-shear lattice model for concrete damage in tension and compression: I. Theory. J Eng Mech ASCE 129(12):1439–1448

Cusatis G, Bažant ZP, Cedolin L (2006) Confinement-shear lattice CSL model for fracture propagation in concrete. Comput Method Appl Mech 195(52):7154–7171

Cusatis G, Pelessone D, Mencarelli A (2011) Lattice discrete particle model (LDPM) for failure behavior of concrete. I: theory. Cement Concr Compos 33(9):881–890

Eliáš J, Bažant ZP (2011) Fracturing in concrete via lattice-particle model. In: Onate E, Owen D (eds) 2nd international conference on particle-based methods – fundamentals and applications, Barcelona

Eliáš J, Vořechovský M (2013) Lattice modeling of concrete fracture including material spatial randomness. Eng Mech 20:413–426

Eliáš J, Vořechovský M, Bažant ZP (2013) Stochastic lattice simulations of flexural failure in concrete beams. In: 8th international conference on fracture mechanics of concrete and concrete structures, Toledo, pp 1340–1351

Grassl P, Bažant ZP (2009) Random lattice-particle simulation of statistical size effect in quasi-brittle structures failing at crack initiation. J Eng Mech ASCE 135:85–92

Grassl P, Gregoire D, Solano LR, Pijaudier-Cabot G (2013) Meso-scale modelling of the size effect on the fracture process zone of concrete. Int J Solids Struct 49(43):1818–1827

Herrmann H, Hansen A, Roux S (1989) Fracture of disordered, elastic lattices in two dimensions. Phys Rev B 39(1):637–648

Hoover CG, Bažant ZP (2013) Comprehensive concrete fracture tests: size effects of types 1 & 2, crack length effect and postpeak. Eng Fract Mech 110:281–289

Hoover CG, Bažant ZP, Vorel J, Wendner R, Hubler MH (2013) Comprehensive concrete fracture tests: description and results. Eng Fract Mech 114:92–103

Jirásek M, Bažant Z (1994) Macroscopic fracture characteristics of random particle systems. Int J Fract 69(3):201–228

Le JL, Bažant ZP (2011) Unified nano-mechanics based probabilistic theory of quasibrittle and brittle structures: II. Fatigue crack growth, lifetime and scaling. J Mech Phys Solids 59:1322–1337

Le JL, Bažant ZP, Bazant MZ (2011) Unified nano-mechanics based probabilistic theory of quasibrittle and brittle structures: I. Strength, static crack growth, lifetime and scaling. J Mech Phys Solids 59(7):1291–1321

Li C, Der Kiureghian A (1993) Optimal discretization of random fields. J Eng Mech ASCE 119(6):1136–1154

Li H, Lèu Z, Yuan X (2008) Nataf transformation based point estimate method. Chin Sci Bull 53(17):2586–2592

Man HK, van Mier J (2008) Influence of particle density on 3D size effects in the fracture of (numerical) concrete. Mech Mater 40(6):470–486

Schlangen E, Garboczi E (1997) Fracture simulations of concrete using lattice models: computational aspects. Eng Fract Mech 57(2–3):319–332

van Mier JGM (1997) Fracture processes of concrete: assesment of material parameters for fracture models. CRC, Boca Raton

van Mier J (2013) Concrete fracture: a multiscale approach. CRC, Boca Raton

van Mier J, Chiaia B, Vervuurt A (1997) Numerical simulation of chaotic and self-organizing damage in brittle disordered materials. Comput Method Appl Mech 142(1–2):189–201

Vořechovský M, Sadílek V (2008) Computational modeling of size effects in concrete specimens under uniaxial tension. Int J Fract 154(1–2):27–49

Sequentially Linear Analysis of Structures with Stochastic Material Properties

George Stefanou, Manolis Georgioudakis, and Manolis Papadrakakis

Abstract This paper investigates the influence of uncertain spatially varying material properties on the fracture behavior of structures with softening materials. Structural failure is modeled using the sequentially linear analysis (SLA) proposed by Rots (Sequentially linear continuum model for concrete fracture. In: de Borst R, Mazars J, Pijaudier-Cabot G, van Mier J (eds) Fracture mechanics of concrete structures. Balkema, Lisse, 2001, pp 831–839), which replaces the incremental nonlinear finite element analysis by a series of scaled linear analyses and the nonlinear stress-strain law by a saw-tooth curve. In this work, SLA is implemented in the framework of a stochastic setting. The proposed approach constitutes an efficient procedure avoiding the convergence problems encountered in regular nonlinear FE analysis. The effect of uncertain material properties (Young's modulus, tensile strength, fracture energy) on the variability of the load-displacement curves and crack paths is examined. The uncertain properties are described by homogeneous stochastic fields using the spectral representation method in conjunction with translation field theory. The response variability is computed by means of direct Monte Carlo simulation. The influence of the variation of each random parameter as well as of the coefficient of variation and correlation length of the stochastic fields is quantified in a numerical

G. Stefanou (✉)
Institute of Structural Analysis & Antiseismic Research, National Technical University of Athens, 9 Iroon Polytechneiou, Zografou Campus, Athens 15780, Greece

Institute of Structural Analysis & Dynamics of Structures, Aristotle University of Thessaloniki, 54124 Thessaloniki, Greece
e-mail: stegesa@mail.ntua.gr

M. Georgioudakis • M. Papadrakakis
School of Civil Engineering, Institute of Structural Analysis & Antiseismic Research, National Technical University of Athens, Zografou Campus, 15780 Athens, Greece
e-mail: geoem@mail.ntua.gr; mpapadra@central.ntua.gr

M. Papadrakakis and G. Stefanou (eds.), *Multiscale Modeling and Uncertainty Quantification of Materials and Structures*, DOI 10.1007/978-3-319-06331-7_2,
© Springer International Publishing Switzerland 2014

example. It is shown that the load-displacement curves, the crack paths and the failure probability are affected by the statistical characteristics of the stochastic fields.

Keywords Softening materials • Sequentially linear analysis • Stochastic field • Response variability • Monte Carlo simulation

1 Introduction

Several numerical techniques have been recently developed to model the failure of structures in the framework of the finite element method (FEM). For structures made of softening materials, a realistic representation of the softening behavior requires the accurate description of stiffness degeneration due to damage. This description can be achieved in a unified manner using damage mechanics approaches that have been proven advantageous for modeling failure phenomena due to their numerical efficiency. The damage mechanics approach permits the incorporation of the description of damage into the constitutive equations, as well as the combination with different more specific simulation methods, such as the embedded finite element method (Oliver 1996; Oliver et al. 2012), extended finite element method (Moës et al. 1999; Mariani and Perego 2003) and non-local theories (Jirásek 1998).

It is also known in failure mechanics that material softening is often responsible for unstable structural behavior (Bažant and Cedolin 2010). This instability can lead to secondary equilibrium states or bifurcation of the equilibrium path, which require more elaborate incremental/iterative solution schemes (De Borst et al. 2012). As a consequence, the robustness of the numerical procedure used for solving the nonlinear problem is strongly affected. In order to overcome these problems, an alternative method, called sequentially linear analysis (SLA), has been introduced by Rots (2001). This method replaces the incremental nonlinear FE analysis by a series of scaled linear analyses and the nonlinear stress-strain law by a saw-tooth curve. The advantage of this replacement is that the secant linear (saw-tooth) stiffness is always positive and the analysis does always converge. The method is generally applicable for materials with nonlinear softening behavior, but it is particularly beneficial when brittle fracture causes convergence issues. In this paper, SLA is implemented in the framework of a stochastic setting. The proposed approach constitutes an efficient procedure for investigating the influence of uncertain spatially varying material properties on the fracture behavior of structures with softening materials (Georgioudakis et al. 2014).

A benchmark structure (double-edge notched specimen) is analyzed and comparisons with nonlinear analysis results are provided. The effect of uncertain variables such as Young's modulus, tensile strength and fracture energy on the variability of the load–displacement curves and crack paths is examined. The uncertain properties are described by homogeneous stochastic fields using the spectral representation method in conjunction with the translation field theory (Shinozuka and Deodatis

1996; Grigoriu 1998). The response variability is computed by means of direct Monte Carlo simulation (MCS). The influence of the variation of each random parameter as well as of the coefficient of variation and correlation length of the stochastic fields is quantified. It is shown that the response statistics and the failure probability of the structure are affected by the statistical characteristics of the stochastic fields.

2 Sequentially Linear Analysis (SLA)

When analyzing material failure with standard nonlinear FE analysis, problems are often encountered such as bifurcation and divergence of the solution. In particular, cases subjected to tension softening tend to encourage the emergence of multiple equilibrium paths. In order to overcome these problems, the SLA has been developed to address specifically the difficulty of modeling snap-back behavior (Rots 2001), typical in full-scale concrete and masonry structures (DeJong et al. 2009). While generally applicable for materials with nonlinear softening branches, it is particularly beneficial when brittle fracture causes convergence issues.

In SLA, a series of linear analyses are used to model the nonlinear behavior of the structure while the modeling proceeds by directly capturing brittle events, rather than trying to iterate around these critical points in a Newton-Raphson scheme. Hence extensive iterations within the load or displacement increment can be avoided. Furthermore, in this approach, a tensile softening curve of negative slope is replaced by a saw-tooth curve which maintains a positive tangent stiffness (see Fig. 1). The incremental/iterative Newton-Raphson method is no longer required since a series of linear analyses are performed, each with a reduced positive stiffness,

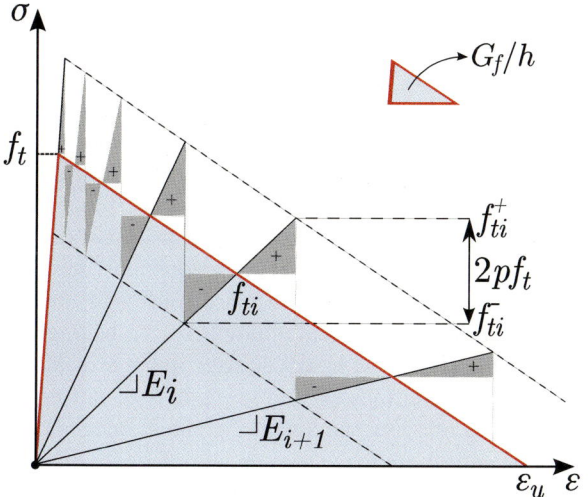

Fig. 1 Stress-strain curve for linear softening and saw-tooth model definitions

---

Algorithm 1 Sequentially linear analysis

1: **repeat**
2:      $\mathsf{K}_N \mathbf{d}_N = \mathbf{P} \rightarrow \sigma_{t_{pr}}$
3:      **for** *element* $= 1, ..., TotalElements$ **do**
4:          **for** *GaussPoint* $= 1, ..., TotalGPperElement$ **do**
5:              Calculate $\frac{f_{t_i}^+}{\sigma_{tpr}}$ and find $\lambda_{cr_N} = max\{\frac{f_{t_i}^+}{\sigma_{tpr}}\}$
6:          **end for**
7:      **end for**
8:      Scale displacements and stress resultants $(\sigma_N, \mathbf{d}_N)$ by factor λ_{cr_N}
9:      **for** *element* $= 1, ..., TotalElements$ **do**
10:          **for** *GaussPoint* $= 1, ..., TotalGPperElement$ **do**
11:              Find new $f_{t_{i+1}}^+, E_{i+1}$ according to Section 2.2
12:          **end for**
13:      **end for**
14:      Update structure stiffness matrix $\mathsf{K}_{N+1} \leftarrow \mathsf{K}_N$
15: **until** damage has spread sufficiently into the structure

---

until the global equilibrium position is achieved. It has been shown that this *event-by-event* strategy is robust and reliable (Rots and Invernizzi 2004), and circumvents bifurcation problems, in contrast to standard nonlinear FE analysis. A more detailed description of the SLA procedure is provided in the following subsections.

2.1 General Procedure

The structure is discretized in the framework of FEM, using standard continuum elements and all material properties (Young's modulus E, Poisson's ratio v, initial strength f_t, as well as fracture energy G_f) are assigned to them. Subsequently, the following steps are carried out sequentially without the need of changing the initial mesh (see Algorithm 1).

Initially, a linear elastic FE analysis is performed (K_N: stiffness matrix of the structure and $\mathbf{d}_N$: vector of unknown displacements in analysis step N) with a reference proportional load $\mathbf{P}$ (line 2). After the calculation of maximum principal tensile stresses ($\sigma_{t_{pr}}$) through the linear elastic analysis, a loop over all integration points for all elements is performed in order to find the *critical element* for which its current strength $f_{t_i}^+$ divided by the maximum principal tensile stress is the highest in the whole structure (lines 3–7). Subsequently (line 8), the reference load $\mathbf{P}$ (along with the corresponding displacements and stress resultants) is scaled proportionally by the critical load multiplier λ_{cr_N} belonging to the critical integration point. Finally (lines 9–13), the damage in the critical integration point is increased by reducing the stiffness E and strength f_t according to the saw-tooth tensile-based constitutive relation (see Sect. 2.2). The aforementioned procedure is repeated sequentially, until the damage has spread sufficiently into the structure.

In this way, the nonlinear response is extracted by linking consecutively the results of each cycle. The smoothness of $P - \delta$ curves depends on the smoothness (number N_t of teeth) of the saw-tooth model (see Sect. 2.2). The SLA procedure allows only one integration point to change its status from elastic to softening at each time, while in nonlinear FE analysis, the use of load increments implies that multiple integration points may crack simultaneously and the local stiffnesses at these points switch from positive to negative, following the softening constitutive laws for quasi-brittle materials.

2.2 Saw-Tooth Model

Several saw-tooth approximations could be specified by adjusting the stiffness, maximum strain and strength of each consecutive saw-tooth. However, the approximation must yield results that are mesh independent. Rots and Invernizzi (2004) investigated several saw-tooth approximations and concluded that any approximation must conserve the dissipated energy, i.e. the area under the softening curve, G_f/h (see Fig. 1).

In this work, the generalized tooth size approach (MODEL C) (Rots et al. 2008) is adopted, which does not require special techniques to handle mesh-size objectivity in order to obtain objective results with respect to the mesh as well as to overcome the lack of consistency. The way in which the stiffness and strength of the critical elements are progressively reduced at each "event", is shown schematically in Fig. 1 where the softening curve of negative slope in the constitutive stress-strain relation is replaced by a discretized, saw-tooth diagram of positive slopes which provides the correct energy dissipation. The linear tensile softening stress-strain curve is defined by the Young's modulus E, the tensile strength f_t and the area under the saw-tooth diagram. This area (see Fig. 1) is always equal to the fracture energy G_f, which is considered here as a material property, divided by the crack bandwidth h, which is associated with the size, orientation and integration scheme of the finite element.

In case of linear softening, ultimate strain ϵ_u is given by:

$$\epsilon_u = \frac{2G_f}{f_t h} \tag{1}$$

Both Young's modulus E and strength f_t can be reduced at the same time in the sequentially linear strategy by a factor a, according to:

$$E_i = \frac{E_{i-1}}{a}, \quad for \quad i = 1, 2, \ldots, N \tag{2}$$

where i and $i-1$ denote the current and previous step, respectively, in the saw-tooth diagram. To find the rule of reducing Young's modulus E as well as strength f_t, by ratio a_i in step i according to Fig. 1, we have:

$$f_{ti}^- = f_{ti}^+ - 2pf_t \tag{3}$$

$$E_{i+1} = \frac{f_{ti}^-}{\epsilon_i} \tag{4}$$

$$a_{i+1} = \frac{E_i}{f_{ti}^-}\epsilon_i = \frac{f_{ti}^+}{f_{ti}^-} = \frac{f_{ti}^+}{f_{ti}^+ - 2pf_t} \tag{5}$$

Thus, for the case of linear softening (Fig. 1) the value of f_{ti}^+ can be easily defined as:

$$f_{ti}^+ = \epsilon_u^+ E_i \frac{D}{E_i + D} \tag{6}$$

where,

$$\epsilon_u^+ = \epsilon_u + p\frac{f_t}{D} \tag{7}$$

and D is the tangent to the tensile stress-strain softening curve. The number N_t of teeth is automatically evaluated, depending on the user specified parameter p. For smaller values of p, a higher N_t is needed to cover the softening branch, leading to more exact results. The procedure ends, regarding the corresponding Gauss point, when the difference between the sum of *positive* triangles above the real curve and the sum of *negative* triangles below the real curve vanishes, as shown in Fig. 1.

3 Representation of Uncertain Material Properties

3.1 Non-Gaussian Translation Fields

As the Gaussian assumption for variables bounded by physical constraints (e.g. material properties that should be strictly positive) may lead to a non-zero probability of violation of these constraints, the simulation of non-Gaussian stochastic processes and fields has received considerable attention in the field of computational stochastic mechanics.

Since all the joint multi-dimensional density functions are needed to fully characterize a non-Gaussian stochastic field, a number of studies have been focused on producing a more realistic (approximate) definition of a non-Gaussian sample function from a simple transformation of some underlying Gaussian field with known second-order statistics. Thus, if $g(\mathbf{x})$ is a homogeneous zero-mean Gaussian field with unit variance and spectral density function (SDF) $S_{gg}(\kappa)$, or equivalently autocorrelation function $R_{gg}(\xi)$, a homogeneous non-Gaussian stochastic field $f(\mathbf{x})$ with power spectrum $S_{ff}^T(\kappa)$ can be defined as:

$$f(\mathbf{x}) = F^{-1} \cdot \Phi[g(\mathbf{x})] \tag{8}$$

where Φ is the standard Gaussian cumulative distribution function and F is the non-Gaussian marginal cumulative distribution function of $f(\mathbf{x})$. The transform $F^{-1} \cdot \Phi$ is a memory-less translation since the value of $f(\mathbf{x})$ at an arbitrary point $\mathbf{x} = (x, y)$ depends on the value of $g(\mathbf{x})$ at the same point only and the resulting non-Gaussian field is called a translation field (Grigoriu 1998).

Translation fields can be used to represent various non-Gaussian phenomena and have a number of useful properties such as the analytical calculation of crossing rates and extreme value distributions. They also have some limitations, the most important one from a practical point of view is that the choice of the marginal distribution of $f(\mathbf{x})$ imposes constraints to its correlation structure. In other words, F and $S^T_{ff}(\kappa)$, or $R^T_{ff}(\xi)$, have to satisfy a specific compatibility condition derived directly from the definition of the autocorrelation function of the translation field (Grigoriu 1998). If F and $S^T_{ff}(\kappa)$ are proven to be incompatible, there is no translation field with the prescribed characteristics. In this case, one has to resort to translation fields that match the target SDF approximately (Shields et al. 2011).

3.2 The Spectral Representation Method

In this paper, Eq. (8) is used for the generation of non-Gaussian translation sample functions representing the uncertain material properties of the problem. Sample functions of the underlying homogeneous Gaussian field $g(\mathbf{x})$ are generated using the spectral representation method (Shinozuka and Deodatis 1996). For a two-dimensional stochastic field, the i-th sample function is given by:

$$g^{(i)}(x, y) = \sqrt{2} \sum_{n_1=0}^{N_1-1} \sum_{n_2=0}^{N_2-1} [A^{(1)}_{n_1 n_2} cos(\kappa_{1n_1} x + \kappa_{2n_2} y + \phi^{(1)(i)}_{n_1 n_2}) +$$

$$+ A^{(2)}_{n_1 n_2} cos(\kappa_{1n_1} x - \kappa_{2n_2} y + \phi^{(2)(i)}_{n_1 n_2})] \tag{9}$$

where $\phi^{(j)(i)}_{n_1 n_2}$, $j = 1, 2$ represent the realization for the i-th simulation of the independent random phase angles uniformly distributed in the range $[0, 2\pi]$. $A^{(1)}_{n_1 n_2}$, $A^{(2)}_{n_1 n_2}$ have the following expressions

$$A^{(1)}_{n_1 n_2} = \sqrt{2 S_{gg}(\kappa_{1n_1}, \kappa_{2n_2}) \Delta \kappa_1 \Delta \kappa_2} \tag{10a}$$

$$A^{(2)}_{n_1 n_2} = \sqrt{2 S_{gg}(\kappa_{1n_1}, -\kappa_{2n_2}) \Delta \kappa_1 \Delta \kappa_2} \tag{10b}$$

where

$$\kappa_{1n_1} = n_1 \Delta \kappa_1 \quad \kappa_{2n_2} = n_2 \Delta \kappa_2 \tag{11}$$

$$\Delta \kappa_1 = \frac{\kappa_{1u}}{N_1} \quad \Delta \kappa_2 = \frac{\kappa_{2u}}{N_2} \tag{12}$$

$$n_1 = 0, 1, \ldots, \quad N_1 - 1 \quad and \quad n_2 = 0, 1, \ldots, N_2 - 1 \tag{13}$$

$N_j, j = 1, 2$, represent the number of intervals in which the wave number axes are subdivided and $\kappa_{ju}, j = 1, 2$, are the upper cut-off wave numbers which define the active region of the power spectrum $S_{gg}(\kappa_1, \kappa_2)$ of the stochastic field. The last means that S_{gg} is assumed to be zero outside the region defined by

$$-\kappa_{1u} \leq \kappa_1 \leq \kappa_{1u} \quad and \quad -\kappa_{2u} \leq \kappa_2 \leq \kappa_{2u} \tag{14}$$

The SDF used in the numerical example (see Sect. 5) is of square exponential type:

$$S_{gg}(\kappa_1, \kappa_2) = \sigma_g^2 \frac{b_1 b_2}{4\pi} exp\left[-\frac{1}{4}(b_1^2 \kappa_1^2 + b_2^2 \kappa_2^2) \right] \tag{15}$$

where σ_g denotes the standard deviation of the stochastic field and b_1, b_2 denote the parameters that influence the shape of the spectrum, which are proportional to the correlation lengths of the stochastic field along the x, y axes, respectively. The squared exponential model is a realistic correlation model for softening materials (e.g. concrete) suggested by the Joint Committee on Structural Safety (JCSS 2001) and used in several publications, e.g. Vořechovský (2008), Yang and Xu (2008), and Eliáš et al. (2014). The SDF of the translation field will be slightly different from S_{gg} due to the spectral distortion caused by the transform of Eq. (8) (Papadopoulos et al. 2009).

4 Stochastic Finite Element Analysis

It is assumed that the Young's modulus E, tensile strength f_t and fracture energy G_f of the material are represented by two dimensional uni-variate (2D-1V) homogeneous stochastic fields. The variation of E is described as follows:

$$E(x, y) = E_0[1 + f(x, y)] \tag{16}$$

where E_0 is the mean value of the elastic modulus and $f(x, y)$ is a zero-mean homogeneous stochastic field. The two other properties are varying in a similar way. The stochastic stiffness matrix is derived using the midpoint method, i.e. one integration point at the centroid of each finite element is used for the computation of the stiffness matrix. This approach gives accurate results for relatively coarse meshes keeping the computational cost at reasonable levels (Stefanou 2009).

Using the procedure described in Sect. 3, a large number N_{SAMP} of sample functions are produced, leading to the generation of a set of stochastic stiffness matrices. The associated structural problem is solved N_{SAMP} times and the response variability can finally be calculated by obtaining the response statistics of the N_{SAMP} simulations.

5 Numerical Example

The double-edge notched specimen under tension (Shi et al. 2000; Nguyen 2008) shown in Fig. 2 is used as a numerical example. The specimen is fixed in both directions at the bottom edge, and in horizontal direction at the top edge. Four-node linear quadrilateral elements under plane stress conditions and a 2×2 Gaussian integration rule are used in the numerical analyses. The uncertain parameters of the problem are the Young's modulus E, tensile strength f_t and fracture energy G_f of the material with mean values equal to 24 GPa, 2.4 MPa and 0.059 N/mm, respectively.

The spatial fluctuation of the uncertain parameters is described by 2D-1V homogeneous lognormal translation fields, sample functions of which are generated using Eqs. (8) and (9). A Weibull distribution could be adopted for the tensile strength f_t and fracture energy G_f, as in Vořechovský (2008) and Yang and Xu (2008), without leading to significant differences in the results. Three different values ($b = 1.2, 12, 120$) of the correlation length parameter b proportional to the dimensions of the structure are used, corresponding to stochastic fields of low, moderate and strong correlation (all values of b are in mm). Sample functions of

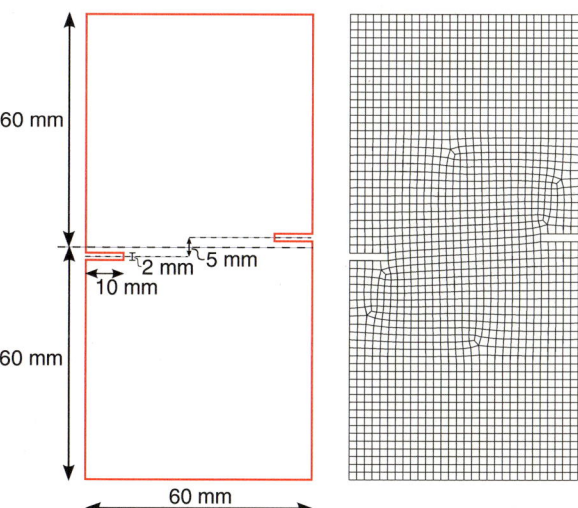

Fig. 2 Double-edge notched specimen (Geometry and FE mesh with 1,950 nodes and 1,850 elements)

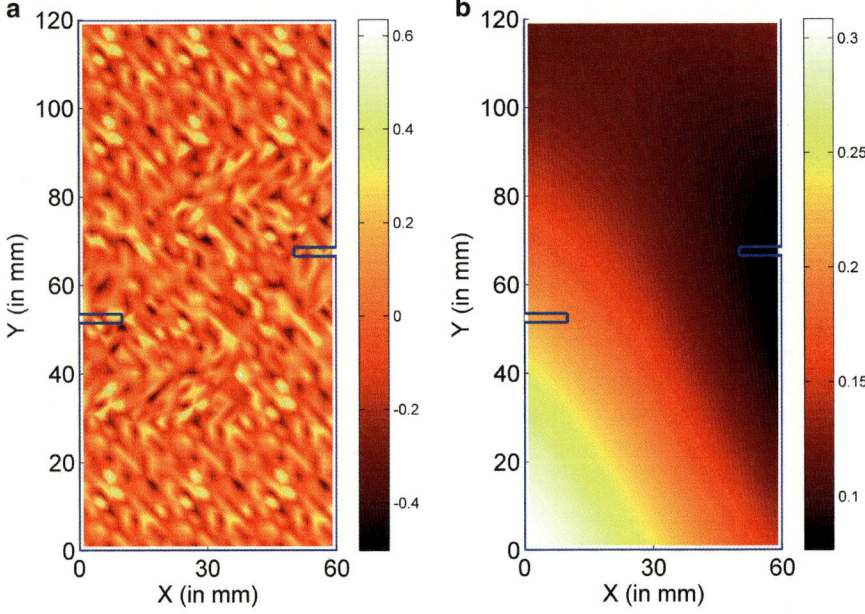

Fig. 3 Realizations of a lognormal field for $\sigma = 20\,\%$ and (**a**) $b = 1.2$, (**b**) $b = 120$

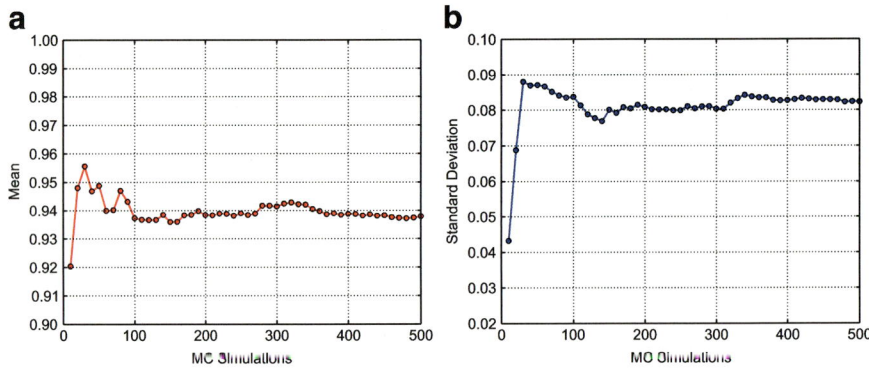

Fig. 4 Statistical convergence for (**a**) mean and (**b**) standard deviation of the peak load (E, f_t, G_f fully correlated with $\sigma = 10\,\%$ and $b = 120$)

a lognormal field for $\sigma = 20\,\%$ and $b = 1.2$, $b = 120$ are shown in Fig. 3a, b, respectively. The case of anisotropic correlation ($b_1 \neq b_2$) has also been examined to highlight its effect on the response variability, which is computed using direct MCS with a sample size equal to 500. The statistical convergence achieved within this number of samples is illustrated in Fig. 4 where the mean value and standard deviation of the peak load are plotted as a function of the number of simulations.

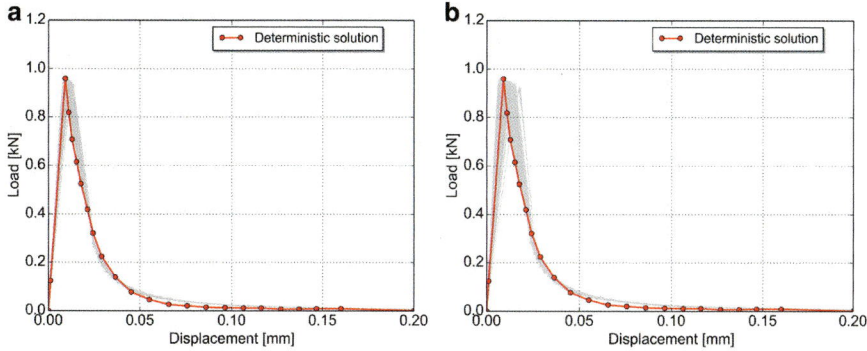

Fig. 5 Load-displacement curves for stochastic parameter E with (**a**) $\sigma = 10\%$ and (**b**) $\sigma = 20\%$ ($b = 120$)

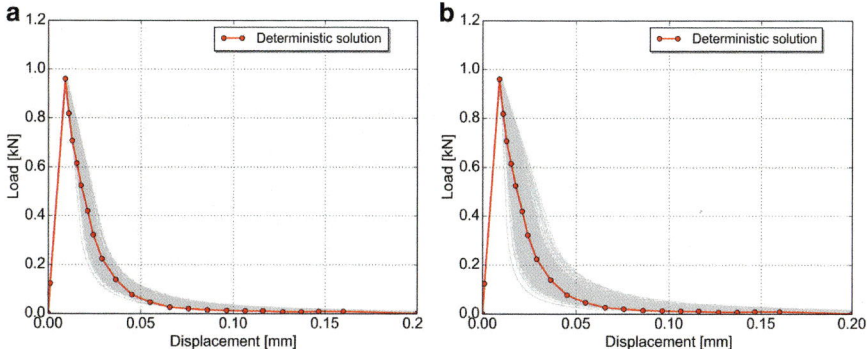

Fig. 6 Load-displacement curves for stochastic parameter G_f with (**a**) $\sigma = 10\%$ and (**b**) $\sigma = 20\%$ ($b = 120$)

Figures 5 and 6 show the load-displacement curves obtained from different stochastic simulations with variable E, G_f. Comparisons with the deterministic nonlinear solution of Nguyen (2008) are provided in these figures. As shown in Fig. 5, the variation of E affects the stiffness of the structure. The results obtained with the assumption of anisotropic correlation were very similar and therefore isotropic correlation ($b = b_1 = b_2$) is finally adopted. As a final step, two cases of combined variation of E, f_t, G_f are considered. In the first case, the lognormal stochastic fields representing the three parameters are fully correlated while in the second case there is no cross-correlation between them. The corresponding load-displacement curves shown in Fig. 7 are highly variable and thus lead to a large probability of failure p_f of the structure (defined as the probability of the peak load not exceeding that of the deterministic solution, which means that the structure fails at a smaller load). For $b = 1.2$, the peak load of all realizations is smaller than the deterministic one, while p_f is equal to 87 and 61 % for $b = 12$ and 120, respectively (case of fully correlated properties).

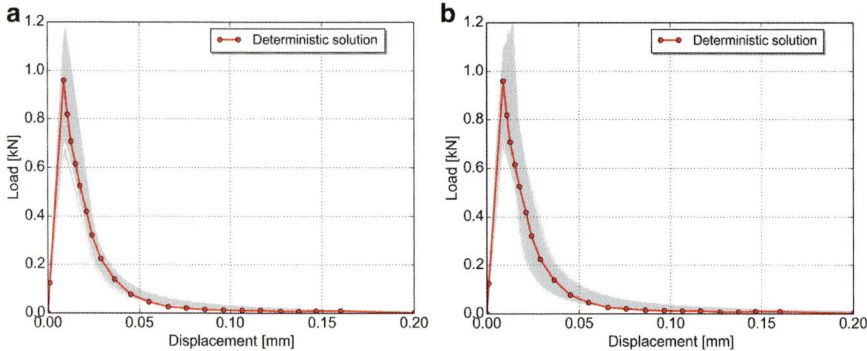

Fig. 7 Load-displacement curves for combined variation of the three parameters for $\sigma = 10\%$ ($b = 120$): (**a**) E, f_t, G_f fully correlated, (**b**) E, f_t, G_f uncorrelated

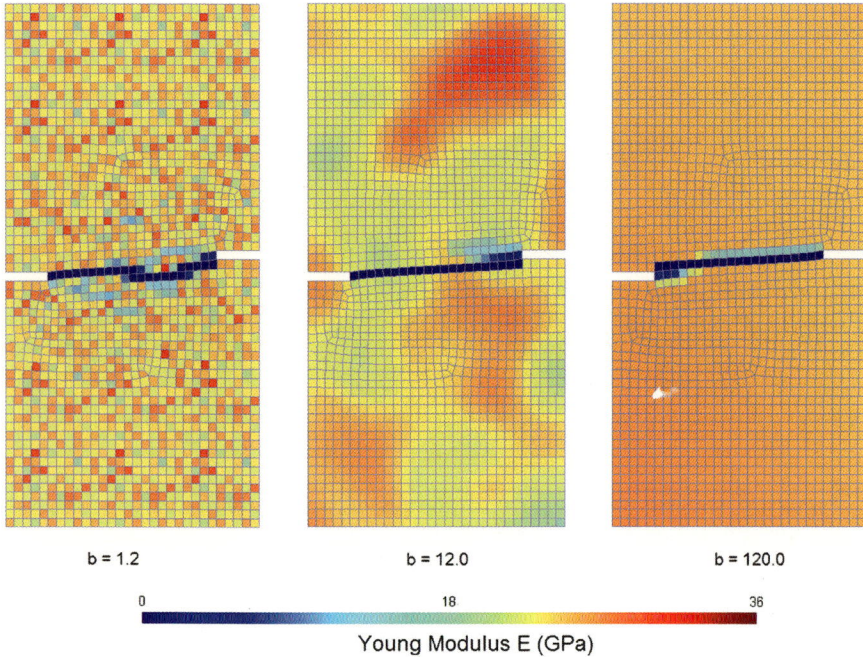

Fig. 8 Crack paths for a randomly selected realization and for different values of correlation length b (E, f_t, G_f fully correlated)

Finally, crack paths for a randomly selected realization and for different values of correlation length b are shown in Fig. 8 (the crack paths are formed by elements with zero stiffness at the end of SLA). The unrealistic crack pattern obtained for $b = 1.2$ is due to the high variability of the elastic modulus in this case which leads to neighboring elements with substantially different E.

6   Conclusions

In this work, the sequentially linear analysis is implemented in the framework of a stochastic setting to investigate the influence of uncertain spatially varying material properties on the fracture behavior of structures with softening materials. The proposed approach constitutes an efficient procedure avoiding the convergence problems encountered in regular nonlinear FE analysis. The uncertain properties are described by homogeneous stochastic fields using the spectral representation method in conjunction with translation field theory. The response variability is computed by means of direct MCS. The influence of the variation of each random parameter as well as of the coefficient of variation and correlation length of the stochastic fields has been quantified. The analysis of a benchmark structure has shown that the load-displacement curves, the crack paths and the probability of failure are affected by the statistical characteristics of the stochastic fields. The extension of SLA to the stochastic framework offers an efficient means to perform parametric investigations of the fracture behavior of structures with variable material properties. The possibility of using variability response functions as an alternative to MCS for computing the response variability of structures with softening materials in the framework of SLA is currently under investigation.

Acknowledgements This work is implemented within the framework of the research project "MICROLINK: Linking micromechanics-based properties with the stochastic finite element method: a challenge for multiscale modeling of heterogeneous materials and structures" – Action "Supporting Postdoctoral Researchers" of the Operational Program "Education and Lifelong Learning" (Action's Beneficiary: General Secretariat for Research and Technology), and is co-financed by the European Social Fund (ESF) and the Greek State. The provided financial support is gratefully acknowledged. M. Papadrakakis acknowledges the support from the European Research Council Advanced Grant "MASTER-Mastering the computational challenges in numerical modeling and optimum design of CNT reinforced composites" (ERC-2011-ADG 20110209).

References

Bažant ZP, Cedolin L (2010) Stability of structures: elastic, inelastic, fracture and damage theories. World Scientific, Hackensack

De Borst R, Crisfield MA, Remmers JJC, Verhoosel CV (2012) Non-linear finite element analysis of solids and structures, 2nd edn. Wiley, Chichester

DeJong MJ, Belletti B, Hendriks MA, Rots JG (2009) Shell elements for sequentially linear analysis: lateral failure of masonry structures. Eng Struct 31(7):1382–1392

Eliáš J, Vořechovský M, Le JL (2014) Fracture simulations of concrete using discrete meso-level model with random fluctuations of material parameters. In: Proceedings of the IUTAM symposium on multiscale modeling and uncertainty quantification of materials and structures. Springer, Santorini Island, Greece

Georgioudakis M, Stefanou G, Papadrakakis M (2014) Stochastic failure analysis of structures with softening materials. Eng Struct 61:13–21

Grigoriu M (1998) Simulation of stationary non-Gaussian translation processes. J Eng Mech 124(2):121–126

JCSS (2001) Probabilistic model code, part 3: resistance models. Joint Committee on Structural Safety. Published online at: http://www.jcss.byg.dtu.dk/

Jirásek M (1998) Nonlocal models for damage and fracture: comparison of approaches. Int J Solids Struct 35(31–32):4133–4145

Mariani S, Perego U (2003) Extended finite element method for quasi-brittle fracture. Int J Numer Methods Eng 58(1):103–126

Moës N, Dolbow J, Belytschko T (1999) A finite element method for crack growth without remeshing. Int J Numer Methods Eng 46(1):131–150

Nguyen GD (2008) A thermodynamic approach to non-local damage modelling of concrete. Int J Solids Struct 45(7–8):1918–1934

Oliver J (1996) Modelling strong discontinuities in solid mechanics via strain softening constitutive equations. Part 1: fundamentals. Int J Numer Methods Eng 39(21):3575–3600

Oliver J, Huespe A, Dias I (2012) Strain localization, strong discontinuities and material fracture: matches and mismatches. Comput Methods Appl Mech Eng 241–244:323–336

Papadopoulos V, Stefanou G, Papadrakakis M (2009) Buckling analysis of imperfect shells with stochastic non-Gaussian material and thickness properties. Int J Solids Struct 46(14–15):2800–2808

Rots J (2001) Sequentially linear continuum model for concrete fracture. In: de Borst R, Mazars J, Pijaudier-Cabot G, van Mier J (eds) Fracture mechanics of concrete structures. Balkema, Lisse, pp 831–839

Rots JG, Invernizzi S (2004) Regularized sequentially linear saw-tooth softening model. Int J Numer Anal Methods Geomech 28(7–8):821–856

Rots JG, Belletti B, Invernizzi S (2008) Robust modeling of RC structures with an "event-by-event" strategy. Eng Fract Mech 75(3–4):590–614

Shi C, van Dam A, van Mier J, Sluys L (2000) Crack interaction in concrete. In: Wittmann F (ed) Materials for buildings and structures. EUROMAT – vol 6. Wiley, Weinheim, pp 125–131

Shields M, Deodatis G, Bocchini P (2011) A simple and efficient methodology to approximate a general non-Gaussian stationary stochastic process by a translation process. Probab Eng Mech 26(4):511–519

Shinozuka M, Deodatis G (1996) Simulation of multi-dimensional Gaussian stochastic fields by spectral representation. Appl Mech Rev 49(1):29–53

Stefanou G (2009) The stochastic finite element method: past, present and future. Comput Methods Appl Mech Eng 198:1031–1051

Vořechovský M (2008) Simulation of simply cross correlated random fields by series expansion methods. Struct Saf 30(4):337–363

Yang Z, Xu XF (2008) A heterogeneous cohesive model for quasi-brittle materials considering spatially varying random fracture properties. Comput Methods Appl Mech Eng 197(45–48): 4027–4039

Part II
Homogenization

A Coupling Method for the Homogenization of Stochastic Structural Models

Régis Cottereau

Abstract We describe a numerical homogenization method that yields the beam parameters corresponding to a homogenized stochastic solid model in a slender domain. This method is based on a novel volume coupling technique for random solid models and deterministic beam models, in the Arlequin framework, that is also described in this paper. The homogenization technique allows to use beam models, that are more practical from a numerical point of view for many industrial applications, constrained by information obtained at the micro-structure level, where beam mechanics cannot be reasonably applied. Two approaches are presented, extending the classical Kinematical and Statical Uniform Boundary Conditions used in classical numerical homogenization.

Keywords Stochastic homogenization • Multiscale mechanics • Euler-Bernoulli beam model • Arlequin method

1 Introduction

Thin and elongated structures are widely encountered in the industry. Beams in the construction industry and shells and plates in the automotive and aeronautical industries are typical examples. Their modeling with the Finite Element (FE) Method is expensive when using a classical formulation, based on the straightforward volume momentum equation and 3D displacement fields. Indeed, the constraint on the aspect ratio of the finite elements implies that the small thickness of the physical domain controls the minimum size of the elements overall. The cost of solving the corresponding FE system may then become prohibitive. This difficulty

R. Cottereau (✉)
Laboratoire MSSMat UMR 8579, École Centrale Paris, CNRS,
Chatenay-Malabry Cedex, France
e-mail: regis.cottereau@ecp.fr

M. Papadrakakis and G. Stefanou (eds.), *Multiscale Modeling and Uncertainty Quantification of Materials and Structures*, DOI 10.1007/978-3-319-06331-7__3,
© Springer International Publishing Switzerland 2014

can be circumvented by starting from formulations specifically designed for thin elements. Based on appropriate kinematical hypotheses on the displacement fields, such as assuming rigid sections for a beam, lower-dimensional formulations can be obtained. Hence a 3D elasticity problem in an elongated domain becomes a 1D beam problem over the mid-fiber. Likewise, a 3D elasticity problem in a flat domain becomes a 2D shell problem over the mid-surface. In these theories, the structural parameters (e.g. mass per unit length, moment of inertia) are analytically derived from the solid parameters (e.g. Young's modulus, Lamé parameters) and the geometry. In the sense that the full-scale 3D elastic model is transformed into a lower-dimensionality structural model, this transformation can be seen as an upscaling process.

When the solid parameters are heterogeneous, the derivation of a homogenized structural model is not so obvious. Analytical solutions exist for specific structures, such as periodic (Caillerie and Nedelec 1984; Kohn and Vogelius 1984; Buannic and Cartraud 2001; Cecchi and Sab 2002; Cartraud and Messager 2006; Grédé et al. 2006; Mistler et al. 2007; Mercatoris et al. 2009) or laminated (Hohe and Becker 2001; Rabczuk et al. 2004; Liu et al. 2006). However, these techniques work well for specific structures and in a particular range of application (for instance, when the mechanical functions of the core and faces of a composite structure are well differentiated). When the fluctuations of the parameters do not present any such simple structure (as in concrete for example), computational homogenization can still be used (Coenen et al. 2010). It is a direct extension of the classical numerical techniques for approximating homogenized coefficients in solid mechanics. The main difference lies in the geometry of the samples, which imposes that the typical sample spans the entire structure along the small dimension(s) and the test boundary conditions (Dirichet, Neumann or periodic) are only applied along the large dimension(s). Finally, when the heterogeneous parameters are modeled as random fields, to the best of our knowledge, there has been no proposal in the literature as to how to treat homogenization of structural models. This case would be the equivalent for solid-to-beam homogenization of the classical solid-to-solid random homogenization problem (Papanicolaou and Varadhan 1981; Huet 1990; Sab 1992; Bourgeat and Piatnitski 2004; Tartar 2009).

This papers aims at proposing a numerical homogenization technique that yields the homogenized parameters of a structural model from the given stochastic fields of parameters of an underlying solid model. It should be pointed out that the method proposed works for solid-to-beam homogenization in all the cases discussed above (homogeneous and heterogeneous, deterministic and stochastic). The main idea is to start from a chosen (a priori erroneous) set of homogenized parameters for the parameters of a structural model, and couple this model to a solid model with the input (stochastic) set of parameters, in a simple geometrical and loading numerical setup. If the coupled system yields the same solution as a mono-model structural parameter with the same set of (chosen) parameters, it is assumed to mean that the chosen model does correspond to the homogenized model. Else, a new set of structural parameters is chosen and the same experiment is repeated until convergence. This method is an application, for structural models, of a numerical homogenization technique introduced recently (Cottereau 2013a,b) for more classical random homogenization of solid models.

The core of this homogenization technique is a coupling method for structural and solid models. Many of those have been developed in the past: enforcing directly the structural hypothesis of rigid sections at the interface between the two models through so-called transition elements (Surana 1980; Bathe and Bolourchi 1980; Cofer and Will 1991; Gmür and Schorderet 1993; Dávila 1994), possibly adding some elasticity to the interface (Osawa et al. 2007; Xue et al. 2009; Song and Hodges 2010), enforcing continuity of the mechanical work at the interface (McCune et al. 2000; Shim et al. 2002), or using the Arlequin method in which the coupling is localized in a volume rather than over a surface (Ben Dhia 1998; Ben Dhia and Rateau 2001; Rateau 2003; Ben Dhia and Rateau 2005; Ben Dhia 2008; Barthel and Gabbert 2010; Rousseau et al. 2010; Qiao et al. 2011; Ghanem et al. 2013). The "weakness" of the coupling (in the sense of the strength of the kinematical constraint imposed by the homogeneous structural model onto the heterogeneous solid model) is essential for the success of the homogenization experiment, so we will consider here the Arlequin coupling.

The next section (Sect. 2) describes the two models that will be used in our method: (i) the stochastic heterogeneous solid model that we are trying to homogenize over an elongated domain, and (ii) the deterministic homogeneous beam model that is the target model. We will limit ourselves in this paper to a beam model, but extension to shell and plate models is expected to be straightforward. The following section (Sect. 3) describes the Arlequin coupling method for these two models. It is somehow a union of previous papers on the Arlequin coupling of deterministic beam and solid models (Rateau 2003; Ben Dhia and Rateau 2005; Barthel and Gabbert 2010; Rousseau et al. 2010; Qiao et al. 2011; Ghanem et al. 2013) on the one hand, and deterministic and stochastic solid models (Cottereau et al. 2010, 2011; Zaccardi et al. 2013; Le Guennec et al. 2013; Cottereau 2013b) on the other hand. Although this coupling method is not the main objective of this paper, and because no similar method can be found in the literature, we believe it is interesting to describe it to some level of detail. Finally, Sect. 4 describes the core of the paper, which is the homogenization method.

2 Description of the Mono-models

In this section, we describe the two models that will be considered: a stochastic continuum mechanics (solid) model and a deterministic Timoshenko beam model. We also highlight the kinematical hypothesis that allows to go from the solid model (when it is assumed deterministic and homogeneous) to the beam model. Throughout, we will indicate quantities related to the solid model with an 's' index, and the quantities related to the beam model with a 'b' index.

2.1   Stochastic Solid Model

Let Ω_s be a domain of $\mathbb{R}^3$ with a smooth boundary $\partial\Omega_s$, separated into a partition $\partial\Omega_s = \Gamma_D \cup \Gamma_N$. The domain Ω_s is filled with an elastic and isotropic solid, loaded in the bulk by $\boldsymbol{f}_s$ and on the surface Γ_N by $\boldsymbol{g}_s$ (both assumed deterministic), and kinematically constrained along Γ_D. The Lamé parameters λ and μ of the solid are modeled as positive, second-order, mean-square continuous stochastic fields indexed on $\mathbb{R}^3$, and defined on probability spaces $(\varXi, \mathscr{A}, P)$, where $\varXi$ is a set of events, $\mathscr{A}$ is a σ-algebra of elements of $\varXi$ and P is a probability measure over $\mathscr{A}$. Under a small perturbations hypothesis, the weak formulation of the stochastic boundary value problem reads: find $\boldsymbol{u}_s \in \mathscr{V}_s$ such that:

$$\mathbb{E}\left[\int_{\Omega_s} \boldsymbol{\sigma}[\boldsymbol{u}_s] : \boldsymbol{\varepsilon}[\boldsymbol{v}_s] d\boldsymbol{x}\right] = \int_{\Omega_s} \boldsymbol{f}_s \cdot \mathbb{E}[\boldsymbol{v}_s] d\boldsymbol{x} + \int_{\Gamma_N} \boldsymbol{g}_s \cdot \mathbb{E}[\boldsymbol{v}_s] d\boldsymbol{x}, \quad \forall \boldsymbol{v}_s \in \mathscr{V}_s, \quad (1)$$

where $\boldsymbol{\varepsilon}[\boldsymbol{u}] = 1/2(\nabla\boldsymbol{u} + \nabla^T\boldsymbol{u})$ is the infinitesimal strain tensor, the superscript ^{T} denotes the transpose operator, $\boldsymbol{\sigma}[\boldsymbol{u}] = \lambda \operatorname{Tr}\boldsymbol{\varepsilon}[\boldsymbol{u}]\boldsymbol{I} + 2\mu\boldsymbol{\varepsilon}[\boldsymbol{u}]$ is the Cauchy stress tensor, $\boldsymbol{I}$ is the identity tensor, and Tr denotes the trace operator. Equivalently, one can use the Young's modulus E and the Poisson's ratio ν instead of the Lamé constants. These pairs of parameters are linked through:

$$\lambda = \frac{E\nu}{(1+\nu)(1-2\nu)}, \quad \mu = \frac{E}{2(1+\nu)}. \quad (2)$$

The functional space is $\mathscr{V}_s = \mathscr{L}^2(\varXi, \mathscr{H}_0^1)$, with $\mathscr{H}_0^1 = \{\boldsymbol{v} \in (\mathscr{H}^1(\Omega_s))^3, \boldsymbol{v}_{|\Gamma_D} = \boldsymbol{0}\}$. Endowed with the appropriate inner product and norm, $\mathscr{V}_s$ is a Hilbert space. Using Lax-Milgram theorem, it can be proved that the problem (1) has a unique solution $\boldsymbol{u}_s$ (see for instance Babuška et al. 2004). An approximation of that solution can then be obtained, for example, by using a Stochastic FE method (Ghanem and Spanos 1991; Stefanou 2009) or a Monte Carlo approach (Robert and Casella 2004).

2.2   Deterministic Beam Model

A beam is a structure whose axial extension is much larger than any dimension orthogonal to it. The cross-sections are defined by intersecting the beam with planes orthogonal to its axis. We define the neutral fiber of this beam as the line joining the centers of mass of all the sections. For simplicity, we consider here a beam whose neutral fiber $\mathscr{F}_b$ is straight and with constant sections, that we denote $\mathscr{S}$. The beam occupies the domain $\Omega_b = \mathscr{F}_b \times \mathscr{S} \in \mathbb{R}^3$. Under the hypotheses of homogeneous symmetric cross-sections, the geometrical centers are coincident to the centers of mass and inertia (see Fig. 1a for an illustrative example). We assume throughout that the neutral fiber is aligned along $\boldsymbol{e}_1$ at rest. A Cartesian reference system is adopted, with base vectors $\boldsymbol{e}_i$ $1 \leq i \leq 3$, and corresponding coordinates $\boldsymbol{x} \in \mathscr{F}_b$

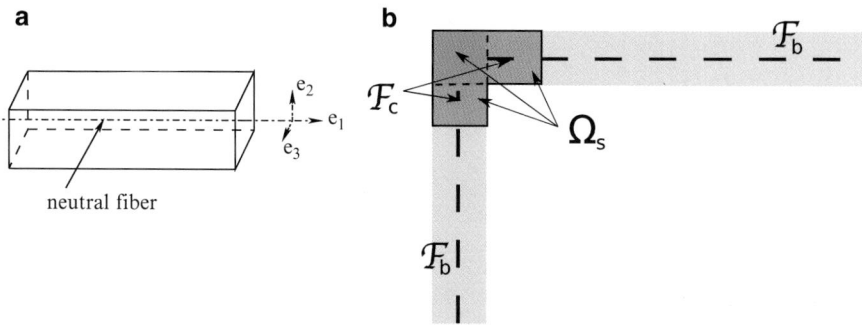

Fig. 1 An example of straight beam with a rectangular cross-section and of a coupled solid-beam model in 2D. (**a**) Beam model. (**b**) Beam-solid model

and $(y, z) \in \mathscr{S}$. Any vector v can be developed into its axial and section parts as $v = v_1 e_1 + v_\perp$. We limit ourselves all along to small transformations around the initial position.

The classical Timoshenko beam theory (Timoshenko 1922; Oñate 2013) assumes that the cross-sections behave as rigid bodies, although they do not necessarily remain perpendicular to the neutral fiber. This kinematical hypothesis leads to parameterize the 3D displacement field $u_b(x, y, z)$ of the beam as a function of two 1D functions: the displacement of the neutral fiber $u_0(x)$ and the rotation vector of the cross-sections $\theta(x)$:

$$u_b(x, y, z) = u_0(x) + \theta(x) \times x_\perp, \tag{3}$$

where $x_\perp = [0 \ y \ z]^T$ gives the location of a point in a cross-section.

The beam is subjected to a linear force f_l and moment c_l (that would be equal respectively to $\int_{\mathscr{S}} f_s dx + \int_{\partial \mathscr{S}} g_s dx$ and $\int_{\mathscr{S}} x_\perp \times f_s dx + \int_{\partial \mathscr{S}} x_\perp \times g_s dx$ if the beam were modeled as a solid). On the Neumann extremities of the mean fiber $\mathscr{F}_b^N$, force F_b and moment C_b loads are also enforced (that would correspond to $\int_{\mathscr{S}} g_{s|\Gamma_N} dx$ and $\int_{\mathscr{S}} g_{s|\Gamma_N} \times x_\perp dx$, respectively). Assuming an isotropic and elastic behavior, the balance of momentum for each section leads to the following weak formulation: Find $u_0 = u_1 e_1 + u_\perp \in \mathscr{V}_b$ and $\theta = \theta_1 e_1 + \theta_\perp \in \mathscr{V}_b$ such that for all $v_0 = v_1 e_1 + v_\perp \in \mathscr{V}_b$ and $\gamma = \gamma_1 e_1 + \gamma_\perp \in \mathscr{V}_b$:

- Axial momentum equation

$$\int_{\mathscr{F}_b} E_b S u_1' v_1' dx = \int_{\mathscr{F}_b} f_1 v_1 dx + (F_{b1} v_1)_{\partial \mathscr{F}_b}; \tag{4}$$

- Torsion momentum equation

$$\int_{\mathscr{F}_b} \mu_b J_1 \theta_1' \gamma_1' dx = \int_{\mathscr{F}_b} c_1 \gamma_1 dx + (C_{b1} \gamma_1)_{\partial \mathscr{F}_b}; \tag{5}$$

- Bending momentum equation

$$\int_{\mathscr{F}_b} E_b J \theta'_\perp \cdot \gamma'_\perp + G_b S(u'_\perp + e_1 \times \theta_\perp) \cdot (v'_\perp + e_1 \times \gamma_\perp) dx$$

$$= \int_{\mathscr{F}_b} f_l \cdot v^c_\perp + c_l \cdot \gamma_\perp dx + (F_{b\perp} \cdot v_\perp + C_{b\perp} \cdot \gamma_\perp)_{\partial \mathscr{F}_b}, \qquad (6)$$

where all the integrals are one-dimensional and the notation a' denotes the derivative of a quantity a with respect to the variable x. The geometrical parameters are

$$S = \int_{\mathscr{S}} dx, \quad J = \int_{\mathscr{S}} (||x_\perp||^2 I - x_\perp \otimes x_\perp) dx, \qquad (7)$$

and $J_1 = J : (e_1 \otimes e_1) = \int_{\mathscr{S}} ||x_\perp||^2 dx$, where I is the identity tensor in $\mathbb{R}^3$. If the beam formulation were derived from a solid model with homogeneous mechanical parameters, the beam mechanical parameters would be $E_b = E$, $\mu_b = \mu$ and $G_b = \tau \mu$, where τ is a shear reduction parameter accounting for the non-uniformity of the shear stress along the cross-section (Oñate 2013). For a non-homogeneous solid model, the derivation of the parameters of the corresponding beam model is not obvious. Assuming (for notational simplicity) homogeneous Dirichlet boundary conditions for both the displacement and rotation fields, the functional space is $\mathscr{V}_b = \{v \in (\mathscr{H}^1(\mathscr{F}_b))^3, v_{\partial \mathscr{F}_b} = 0\}$. We also define $\mathscr{W}_b = \{w = u + \theta \times x_\perp, u \in \mathscr{V}_b, \theta \in \mathscr{V}_b\}$. Endowed with the inner product of $\mathscr{H}^1(\Omega_b), (w_1, w_2)_b = \int_{\Omega_b} w_1 \cdot w_2 + \nabla w_1 : \nabla w_2 dx$, and the corresponding norm, $\mathscr{W}_b$ is a Hilbert space. Using Lax-Milgram theorem (Ern and Guermond 2004), the problem (4)–(6) can be shown to have a unique solution $(u_0, \theta) \in \mathscr{W}_b$. This unique solution can be approximated by the Finite Element method (Hughes 1987; Zienkiewicz and Taylor 2005).

3   Coupling Method in a Stochastic Framework

In this section, we consider a mechanical problem posed over a domain $\Omega \in \mathbb{R}^3$, and a quantity of interest that can be estimated using the stochastic heterogeneous solid model described in Sect. 2.1. Further, we assume that the complexity of this model is only required over a limited region in order for the quantity of interest to be well evaluated. Hence, we propose to use a coupled model: (i) fine-scale stochastic heterogeneous model over part of the domain, and (ii) coarser deterministic homogeneous beam model over the rest of the domain. The coupled model is developed in the Arlequin framework. This framework is based on three ingredients: (i) splitting of the domain into overlapping subdomains to which different models are attached, (ii) introduction of weight functions to dispatch the global energy among the models, (iii) imposition of a weak compatibility constraint between the solutions of the different models.

3.1 Arlequin Formulation

The domain Ω is divided into two overlapping subdomains Ω_s and $\Omega_b = \mathscr{F}_b \times \mathscr{S}$ such that $\Omega_s \cup \Omega_b = \Omega$ (see Fig. 1b). We select a coupling volume $\Omega_c \subset (\Omega_s \cap \Omega_b) \in \mathbb{R}^3$, over which the two models are assumed to exchange information, and introduce its mean fiber $\mathscr{F}_c$ such that $\Omega_c = \mathscr{F}_c \times \mathscr{S}$. For notational simplicity, we assume Dirichet boundary conditions only on the beam model $\Gamma_D \subset \partial \mathscr{F}_b$. Forces are imposed in the bulk f_s and on the boundary g_s (on $\Gamma_N \subset \partial \Omega_s$) for the solid model and along $\mathscr{F}_b$ for the beam model (f_l and c_l). The mixed Arlequin problem reads: find $(u_s, u_b, \Phi) \in \mathscr{V}_s \times \mathscr{W}_b \times \mathscr{W}_c$ such that

$$\begin{cases} a_s(u_s, v) + C(\Phi, \Pi(v)) = \ell_s(v), & \forall v \in \mathscr{V}_s \\ a_b(u_b, v_b) - C(\Phi, v_b) = \ell_b(v_b), & \forall v_b \in \mathscr{W}_b , \\ C(\Psi, \Pi(u_s) - u_b) = 0, & \forall \Psi \in \mathscr{W}_c \end{cases} \tag{8}$$

where the forms $a_s : \mathscr{V}_s \times \mathscr{V}_s \to \mathbb{R}$, $a_b : \mathscr{W}_b \times \mathscr{W}_b \to \mathbb{R}$, $C : \mathscr{W}_c \times \mathscr{W}_c \to \mathbb{R}$ are defined by:

$$a_s(u, v) = \mathbb{E}\left[\int_{\Omega_s} \alpha_s \, \sigma[u] : \varepsilon[v] dx \right], \tag{9}$$

$$a_b(u_b, v_b) = \int_{\mathscr{F}_b} \alpha_b \left\{ E_b S u_1' v_1' + \mu_b J_1 \theta_1' \gamma_1' + E_b J \theta_\perp' \cdot \gamma_\perp' \right.$$
$$\left. + G_b S(u_\perp' + e_1 \times \theta_\perp) \cdot (v_\perp' + e_1 \times \gamma_\perp) \right\} dx \tag{10}$$

with $u_b = (u_1 e_1 + u_\perp) + (\theta_1 e_1 + \theta_\perp) \times x_\perp$ and $v_b = (v_1 e_1 + v_\perp) + (\gamma_1 e_1 + \gamma_\perp) \times x_\perp$, and

$$C(u_b, v_b) = \mathbb{E}\left[\int_{\Omega_c} (u_b \cdot v_b + \kappa \varepsilon[u_b] : \varepsilon[v_b]) \, dx \right] \tag{11}$$

where κ is a constant essentially introduced for dimensionality purposes (Ben Dhia and Rateau 2005). Note that for functions u_b and v_b of $\mathscr{W}_c$, decomposed as above, we have

$$C(u_b, v_b) = \int_{\mathscr{F}_b} \mathbb{E}\left[S u_1 v_1 + S u_\perp \cdot v_\perp + J_1 \theta_1 \gamma_1 + J \theta_\perp \cdot \gamma_\perp + \kappa \left(S u_1' v_1' + J_1 \theta_1' \gamma_1' \right.\right.$$
$$\left.\left. + J \theta_\perp' \cdot \gamma_\perp' + S(u_\perp' + e_1 \times \theta_\perp) \cdot (v_\perp' + e_1 \times \gamma_\perp) \right) \right] dx. \tag{12}$$

The projector $\Pi : \mathscr{V}_s \to \mathscr{W}_c$ is defined by

$$\Pi(v) = \langle v \rangle + \langle \nabla \times v + \frac{1}{2}(\nabla \cdot (e_1 \times w))e_1 \rangle \times x_\perp \tag{13}$$

where $\langle v \rangle = \int_{\mathcal{S}} v(\boldsymbol{x})d\boldsymbol{x}/S$ for any scalar, vector, or tensor v. Note that any rigid body displacement field in the form $\boldsymbol{u}_0(\boldsymbol{x}) + \boldsymbol{\theta}(\boldsymbol{x}) \times \boldsymbol{x}_\perp$ is conserved by the projection. The linear forms $\ell_s : \mathcal{V}_s \to \mathbb{R}$ and $\ell_b : (\mathcal{V}_b)^2 \to \mathbb{R}$ are defined, respectively, by

$$\ell_s(\boldsymbol{v}) = \int_{\Omega_s} \boldsymbol{f}_s \cdot \mathbb{E}[\boldsymbol{v}_s]d\boldsymbol{x} + \int_{\Gamma_N} \boldsymbol{g}_s \cdot \mathbb{E}[\boldsymbol{v}_s]d\boldsymbol{x}, \tag{14}$$

and (see Sect. 2.2 for the definition of the linear forces and moments)

$$\ell_b(\boldsymbol{v}, \boldsymbol{\theta}) = \int_{\mathcal{F}_b} \left\{ f_{l,1}v_1 + c_{l,1}\gamma_1 + \boldsymbol{f}_{l,\perp} \cdot \boldsymbol{v}_\perp^c + c_{l,\perp} \cdot \boldsymbol{\gamma}_\perp \right\} d\boldsymbol{x} \tag{15}$$

The weight functions $\alpha_s(x, y, z)$ and $\alpha_b(x)$ in Eqs. (9) and (10) are chosen such that:

$$\begin{cases} \alpha_s = 1 & \text{in } \Omega_s \backslash \Omega_b \\ \alpha_s(x, y, z) = \tilde{\alpha}_s(x) & \text{in } \Omega_s \cap \Omega_b, \\ \tilde{\alpha}_s, \alpha_b > 0 & \text{in } \Omega_s \cap \Omega_b, \\ \tilde{\alpha}_s + \alpha_b = 1 & \text{in } \Omega_s \cap \Omega_b. \end{cases} \tag{16}$$

The functional spaces are $\mathcal{V}_s = \mathcal{L}^2(\Xi, (\mathcal{H}^1(\Omega_s))^3)$ and $\mathcal{V}_b = \{\boldsymbol{v}, \boldsymbol{\theta} \in (\mathcal{H}^1(\mathcal{F}_b))^3, \boldsymbol{v}|_{\Gamma_D} = \boldsymbol{\theta}|_{\Gamma_D} = \boldsymbol{0}\}$, and the so-called mediator space $\mathcal{V}_c$ is defined as:

$$\mathcal{V}_c = \{(\boldsymbol{v}(x) + \boldsymbol{\xi}_T) + (\boldsymbol{\gamma}(x) + \boldsymbol{\xi}_R) \times \boldsymbol{x}_\perp \mid \boldsymbol{v}, \boldsymbol{\gamma} \in (\mathcal{H}^1(\mathcal{F}_c))^3,$$
$$\boldsymbol{\xi}_T, \boldsymbol{\xi}_R \in (\mathcal{L}^2(\Xi, \mathbb{R}))^3\}. \tag{17}$$

This choice of mediator space ensures that the resulting mixed formulation (8) is well-posed. The restriction of $\mathcal{V}_s$ to the coupling zone would be another possible option. The resulting mixed formulation would equally be well-posed, but the condition $C(\boldsymbol{\Psi}, \boldsymbol{u}_s - (\boldsymbol{u}_0 + \boldsymbol{\theta} \times \boldsymbol{x}_\perp)) = 0$ would be imposed in a much stronger manner, since the dimensionality of the mediator space when discretizing would be much larger. In particular, this would force the average of the solid solution to follow the kinematics of the beam model, which is not desirable because the mechanical parameters of the solid model are heterogeneous, so that it is not reasonable to assume that the sections remains rigid.

One can consider that the system (8) consists of three equations: (i) one governing the behavior of the stochastic solid model, weighted by $\alpha_s(x)$ and with a loading arising in the coupling volume Ω_c embodied in the operator C; (ii) one governing the behavior of the beam model, weighted by $\alpha_b(x)$ and with a loading opposite to the previous in the coupling volume Ω_c; and (iii) one enforcing the weak compatibility between the two solutions $\boldsymbol{u}_s(x)$ and $\boldsymbol{u}_b = \boldsymbol{u}_0(x) + \boldsymbol{\theta} \times \boldsymbol{x}_\perp$.

3.2   Finite Element Discretization

Based on the previous continuous weak formulation (8), we now consider the discretization and the resulting matrix system. The domain Ω_s is split into elements $\mathscr{E}_s$ giving a mesh $\mathscr{T}_s$ with n_s degrees of freedom (DOFs). The beam fiber $\mathscr{F}_b$ is split into elements $\mathscr{E}_b$ to form a mesh $\mathscr{T}_b$ with n_b DOFs. Finally, $\mathscr{F}_c$ is split into elements to form a mesh $\mathscr{T}_c$ with n_c DOFs. All the fields in System (8) are approximated by fields that are globally continuous and polynomials by parts over the elements of the relevant meshes. In particular, we consider the following finite-dimensional functional spaces: $\mathscr{H}^{1,H}(\Omega_s) = \{v \in \mathscr{C}^0(\Omega_s), v \in \mathbb{P}^1(\mathscr{E}_s)\}$, $\mathscr{H}^{1,H}_s = (\mathscr{H}^{1,H}(\Omega_s))^3$, $\mathscr{V}^H_s = \mathscr{L}^2(\varXi, \mathscr{H}^{1,H})$, $\mathscr{H}^{1,H}(\mathscr{F}_b) = \{v \in \mathscr{C}^0(\mathscr{F}_b), v \in \mathbb{P}^2(\mathscr{E}_b)\}$, $\mathscr{V}^H_b = \{v, \theta \in (\mathscr{H}^{1,H}(\mathscr{F}_b))^3, v_{|\Gamma_D} = \theta_{|\Gamma_D} = 0\}$, $\mathscr{H}^{1,H}(\mathscr{F}_c) = \{v \in \mathscr{C}^0(\mathscr{F}_c), v \in \mathbb{P}^2(\mathscr{E}_c)\}$, and $\mathscr{V}^H_c = \{(v + \xi_T) + (\gamma + \xi_R) \times x_\perp, v, \gamma \in (\mathscr{H}^{1,H}(\mathscr{F}_c))^3, \xi_T, \xi_R \in (\mathscr{L}^2(\varXi, \mathbb{R}))^3\}$, where $\mathbb{P}^1(A)$ and $\mathbb{P}^2(A)$ represent, respectively, the sets of linear and quadratic polynomials over domain A. The consideration of quadratic polynomials over the beam elements simplifies the discretization of the projection operator $\varPi$. Note that we discretize here only along the space dimension, because we will use the Monte Carlo method (Robert and Casella 2004) for the random dimension. The (scalar) bases associated respectively with $\mathscr{H}^{1,H}(\Omega_s)$, $\mathscr{H}^{1,H}(\mathscr{F}_b)$ and $\mathscr{H}^{1,H}(\mathscr{F}_c)$, are denoted by: $\{v_i^s(x)\}_{1 \le i \le n_s}$, $\{v_i^b(x)\}_{1 \le i \le n_b}$, and $\{v_i^c(x)\}_{1 \le i \le n_c}$. After space discretization, the mixed system (8) may be written:

$$\mathbb{E}[A(\xi)U(\xi)] = F \tag{18}$$

where ξ indicates dependency on $\varXi$, and where

$$A = \begin{bmatrix} A_s(\xi) & 0 & PC & PC_\xi & 0 \\ 0 & A_b & -C & -C_\xi & 0 \\ (PC)^T & -C^T & 0 & 0 & S_c^T \\ (PC_\xi)^T & -C_\xi^T & 0 & 0 & 0 \\ 0 & 0 & S_c & 0 & 0 \end{bmatrix}. \tag{19}$$

In that matrix, we have, for the stiffness matrix of the solid model, for $1 \le i,k \le n_s$ and $1 \le j,\ell \le 3$:

$$A_{s,(ij,k\ell)}(\xi) = \int_{\Omega_s} \alpha_s \sigma[v_i^s e_j] : \varepsilon[v_k^s e_\ell] dx , \tag{20}$$

and, the stiffness matrix of the beam model:

$$A_b = \begin{bmatrix} E_b S A_b^{1s} & 0 & 0 & 0 \\ 0 & G_b S A_b^1 & 0 & G_b S A_b^{1v} \\ 0 & 0 & \mu_b J_1 A_b^{1s} & 0 \\ 0 & G_b S A_b^{1vT} & 0 & G_b S A_b^0 + E_b(JA)_b^1 \end{bmatrix}, \tag{21}$$

with, for $1 \leq i,k \leq n_b$, and $2 \leq j,\ell \leq 3$, $A_{b,ik}^{1s} = \int_{\mathscr{F}_b} \alpha_b (v_i^b)' (v_k^b)' dx$,
$A_{b,(ij,k\ell)}^0 = (e_\ell \cdot e_j) \int_{\mathscr{F}_b} \alpha_b v_i^b v_k^b dx$, $A_{b,(ij,k\ell)}^1 = (e_\ell \cdot e_j) A_{b,ik}^{1s}$, $A_{b,(ij,k\ell)}^{1v} = (e_\ell \times e_j) \cdot e_1 \int_{\mathscr{F}_b} \alpha_b (v_i^b)' v_k^b dx$ and $(JA)_{b,(ij,k\ell)}^1 = J : (e_\ell \otimes e_j) A_{b,ik}^{1s}$. Note
that the transpose sign in Eq. (21) is defined in the sense that $A_{b,(ij,k\ell)}^{1vT} = A_{b,(k\ell,ij)}^{1v}$.

The coupling matrix for the beam model is given by

$$
C = \begin{bmatrix}
SC_1^s & 0 & 0 & 0 \\
0 & SC_1 & 0 & SC_1^v \\
0 & 0 & J_1 C_1^s & 0 \\
0 & SC_1^{vT} & 0 & SC_0 + (JC)_1
\end{bmatrix}
\tag{22}
$$

with, for $1 \leq i,k \leq n_c$, and $2 \leq j,\ell \leq 3$, $C_{1,ik}^s = \int_{\mathscr{F}_b} v_i^b v_k^b + \kappa (v_i^b)' (v_k^b)' dx$,
$C_{1,(ij,k\ell)} = (e_\ell \cdot e_j) C_{1,ik}^s$, $C_{0,(ij,k\ell)} = (e_\ell \cdot e_j) \int_{\mathscr{F}_b} v_i^b v_k^b dx$, $C_{1,(ij,k\ell)}^v = (e_\ell \times e_j) \cdot e_1 \int_{\mathscr{F}_b} (v_i^b)' v_k^b dx$ and $(JC)_{1,(ij,k\ell)} = J : (e_\ell \otimes e_j) C_{1,ik}^s$. Observe that the structure
of the coupling matrix is very close to that of the beam stiffness matrix, with unit
material parameters and an additional block diagonal contribution. We additionally
introduce the projection matrix

$$
P = [P^{u_1} \ P^{u_\perp} \ P^{\theta_1} \ P^{\theta_\perp}]
\tag{23}
$$

where $P_{(ij,k)}^{u_1} = (e_j \cdot e_1) \langle v_i^s(x_k, x_\perp) \rangle$, $P_{(ij,k\ell)}^{u_\perp} = (e_j \cdot e_\ell) \langle v_i^s(x_k, x_\perp) \rangle$, $P_{(ij,k)}^{\theta_1} = \langle \nabla \cdot v_i^s(x_k, x_\perp)(e_1 \times e_j) \rangle / 2$, and $P_{(ij,k\ell)}^{\theta_\perp} = \langle \nabla \cdot v_i^s(x_k, x_\perp)(e_j \times e_\ell) \rangle$. We also get
C_ξ by summing the coordinates of C over all the DOFs in the columns (because ξ_T
and ξ_R are constant functions (in space) over the coupling domain).

The vector S_c in Eq. (19) is used to remove the over-parameterization of the
functional space $\mathscr{V}_c$. Indeed we note that the subspace of constant (in space)
deterministic functions of $\mathscr{V}_c$ can be described either only with v or ξ_T. We therefore
impose a condition that $\int_{\mathscr{F}_c} v dx = 0$. It is also possible to impose rather $\mathbb{E}[\xi_T] = 0$
but this would be less trivial in a Monte-Carlo-based simulation. Likewise, there
is redundancy between the elements $\theta \times x_\perp$ and $\xi_R \times x_\perp$, and we impose
$\int_{\mathscr{F}_c} \theta dx = 0$. We therefore obtain S_c by summing C along all the DOFs of the
lines.

The load vector is $F^T = [F_s \ F_b \ 0 \ 0 \ 0]$, where

$$
F_{s,(ij)} = \int_{\Omega_s} \alpha_s v_i^s f_s \cdot e_j dx + \int_{\Gamma_N} \alpha_s v_i^s g_s \cdot e_j dx,
\tag{24}
$$

and

$$
F_b^T = [F_b^{u_1} \ F_b^{u_\perp} \ F_b^{\theta_1} \ F_b^{\theta_\perp}],
\tag{25}
$$

with $F_{b,i}^{u_1} = \int_{\mathscr{F}_b} \alpha_b v_i^b f_1 dx$, $F_{b,(ij)}^{u_\perp} = \int_{\mathscr{F}_b} \alpha_b v_i^b f_1 \cdot e_j dx$, $F_{b,i}^{\theta_1} = \int_{\mathscr{F}_b} \alpha_b v_i^b c_1 dx$ and $F_{b,(ij)}^{\theta_\perp} = \int_{\mathscr{F}_b} \alpha_b v_i^b c_l \cdot e_j dx$ for $1 \leq i \leq n_b$ and $2 \leq j \leq 3$. Finally the unknown vector is decomposed as:

$$U(\xi)^T = [U_s(\xi)\ U_b\ U_c\ U_\xi(\xi)\ \Lambda] \tag{26}$$

where $U_b^T = [U_b^{u_1}\ U_b^{\theta_1}\ U_b^{u_\perp}\ U_b^{\theta_\perp}]$, $U_c^T = [U_c^{u_1}\ U_c^{\theta_1}\ U_c^{u_\perp}\ U_c^{\theta_\perp}]$ and $U_\xi^T = [U_{\xi_T}\ U_{\xi_R}]$.

The System (18) can then be solved by the Monte Carlo approach, or through a condensation technique of the deterministic part of A onto its random part. More details on these two approaches can be found in Cottereau et al. (2011) and Le Guennec et al. (2013). The approximate solution of the solid model is then:

$$u_s(x, \xi) = \sum_{i=1}^{n_s} \sum_{j=1}^{3} U_{s,ij}(\xi) v_i^s(x) e_j, \tag{27}$$

and the approximate solution of the beam model $u_b(x) = u_0(x) + \theta(x) \times x_\perp$ is:

$$u_b(x) = \sum_{i=1}^{n_s} v_i^s(x) \left\{ U_{b,i}^{u_1} e_1 + U_{b,i}^{\theta_1} e_1 \times x_\perp + \sum_{j=2}^{3} \left(U_{b,(ij)}^{u_\perp} e_j + U_{b,(ij)}^{\theta_\perp} e_j \times x_\perp \right) \right\}. \tag{28}$$

4 Homogenization of a Stochastic Solid Model into a Beam Model

In the previous section, we have described a way of coupling a solid model with stochastic fluctuating mechanical parameters to a beam model with deterministic mechanical parameters. Supposedly, the parameters of the beam model are an upscaled version of the solid parameters. However, as discussed in the introduction, it is not clear how to define the parameters of the deterministic beam given the parameters fields of the solid model. This section aims at proposing a method to do so.

The criterion that we will consider for selecting the upscaled model (the beam model) is based on the idea that its behavior would be the same if considered alone or coupled to its micro-scale version (the solid model). In some sense, if the beam model is well chosen, its mechanical behavior at the macro-scale (for quantities of interest relative to the beam model) should not feel any difference if coupled or not to the solid model it upscales. Note that we do not pretend that this criterion ensures uniqueness of the upscaled model, although it does seem reasonable. The technique we propose here is an extension of the technique proposed in Cottereau (2013a,b).

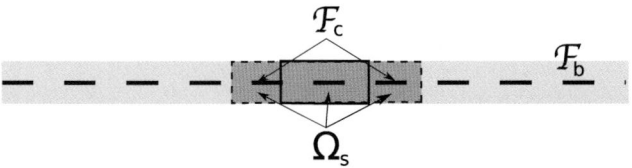

Fig. 2 Coupling configuration for the homogenization problem: the beam model is present everywhere $\mathscr{F}_b \times \mathscr{S} = \Omega$, and the sample microstructure Ω_s is placed in the middle

In terms of implementation, we propose a very simple iterative approach. Starting from an initial guess of parameter vector $\boldsymbol{p}_b = [E_b \; \mu_b \; G_b]^T$, we choose a set of boundary conditions, and we compute for each boundary conditions:

1. The solution of a coupled beam-solid model numerically using the Arlequin technique presented in the previous section (Fig. 2); and
2. The solution of a beam model alone, analytically.

The two solutions are then compared in terms of energies and the values of the parameter vector $\boldsymbol{p}_b$ are updated in order to decrease that difference, for instance using the Nelder-Mead technique (Lagarias et al. 1998).

As in the classical numerical homogenization technique, the set of boundary conditions that is chosen influences the homogenized vector that is obtained after convergence. At least two approaches can be proposed, using Dirichlet or Neumann boundary conditions, generalizing classical results of homogenization in elastic media. Finally, the choice of initial condition for the parameter vector might also influence the convergence value, or at least the rate of convergence. Two reasonable initial choices would be

$$\begin{cases} \boldsymbol{p}_b = & [\mathbb{E}[E] \; \mathbb{E}[\mu] \; \tau\mathbb{E}[\mu]] \\ \boldsymbol{p}_b = & [\mathbb{E}[E^{-1}]^{-1} \; \mathbb{E}[\mu^{-1}]^{-1} \; \tau\mathbb{E}[\mu^{-1}]^{-1}] \end{cases} \tag{29}$$

generalizing the classical Hashin and Shtrikman bounds in linear elasticity (Huet 1990). The general pattern of the numerical homogenization scheme is summarized in Algorithm 1, considering Dirichlet boundary conditions and arithmetic averages for the mechanical parameters.

5   Conclusion

We have proposed in this paper a new coupling technique and a new homogenization method. The coupling technique deals with a stochastic solid model and a deterministic beam model, while the homogenization method allows to upscale a stochastic solid beam into a deterministic beam model. Numerical simulations of the proposed homogenization technique will have to be performed in order to

Algorithm 1: Algorithmic description of the proposed iterative technique for the numerical homogenization of a random solid model into a beam model

---

Data: N realizations of random solid parameters $(E(x,\xi), \mu(x,\xi))$ and choice of τ
Result: Dirichlet Arlequin estimate of parameters of the beam model $p_b* = [E_b^a, \mu_b, G_b]^T$
Initialization: $p_b^0 \longleftarrow [\mathbb{E}[E] \; \mathbb{E}[\mu] \; \tau \mathbb{E}[\mu]]$;
While $\| p_b^i - p_b^{i-1} \| > criterion$ **do**
    set the beam parameters: $[E_b^a, \mu_b, G_b]^T \longleftarrow p_b^i$;
    solve the Arlequin coupled system (8) with Dirichlet boundary conditions on $\mathscr{F}_b$ to estimate u_b;
    solve (analytically) the (mono-model) beam problem to get u_b^{beam} ;
    update p_b^{i+1} to minimize $|\mathscr{E}_b^{\text{arl}}[u_b] - \mathscr{E}_b^{\text{beam}}[u_b^{\text{beam}}]|$ (using Nelder-Mead technique)
end
Store estimate: $p_b* = p_b^i$.

---

discuss the influence of the various numerical parameters involved (choice of initial parameters, boundary conditions). Extension to nonlinear beam models will also be considered, in the context of seismic engineering. Indeed, numerical studies at the micro-scale have shown that strong apparent damping appears in free vibrations of heterogeneous concrete beams (Jehel and Cottereau 2012). Understanding the upscaling of such a micro-scale nonlinear solid model into a beam nonlinear model would allow to reduce significantly the numerical costs associated with the simulation of full-scale buildings.

References

Babuška I, Tempone R, Zouraris GE (2004) Galerkin finite element aproximations of stochastic elliptic partial differential equations. SIAM J Numer Anal 42(2):800–825. doi:10.1137/S0036142902418680

Barthel C, Gabbert U (2010) Application of the Arlequin method in the virtual engineering design process. In: Wieners C (ed) Proceedings in applied mathematics and mechanics: 81st annual meeting of the international association of applied mathematics and mechanics (GAMM), Karlsruhe, vol 10, pp 141–142. doi:10.1002/pamm.201010063

Bathe KJ, Bolourchi S (1980) A geometric and material nonlinear plate and shell element. Comput Struct 11(1–2):23–48. doi:10.1016/0045-7949(80)90144-3

Ben Dhia H (1998) Multiscale mechanical problems: the Arlequin method. C R Acad Sci Ser IIB Mech Phys Astron 326(12):899–904. doi:10.1016/S1251-8069(99)80046-5

Ben Dhia H (2008) Further insights by theoretical investigations of the multiscale Arlequin method. Int J Multiscale Comput Eng 6(3):215–232. doi:10.1615/IntJMultCompEng.v6.i3.30

Ben Dhia H, Rateau G (2001) Mathematical analysis of the mixed Arlequin method. C R Acad Sci Ser I Math 332(7):649–654. doi:10.1016/S0764-4442(01)01900-0

Ben Dhia H, Rateau G (2005) The Arlequin method as a flexible engineering design tool. Int J Numer Methods Eng 62(11):1442–1462. doi:10.1002/nme.1229

Bourgeat A, Piatnitski A (2004) Approximations of effective coefficients in stochastic homogenization. Ann Inst Henri Poincaré 40:153–165. doi:10.1016/j.anihpb.2003.07.003

Buannic N, Cartraud P (2001) Higher-order effective modeling of periodic heterogeneous beams. I. Asymptotic expansion method. Int J Solids Struct 38(40–41):7139–7161. doi:10.1016/S0020-7683(00)00422-4

Caillerie D, Nedelec JC (1984) Thin elastic and periodic plates. Math Methods Appl Sci 6(1):159–191. doi:10.1002/mma.1670060112

Cartraud P, Messager T (2006) Computational homogenization of periodic beam-like structures. Int J Solids Struct 43(3–4):686–696. doi:10.1016/j.ijsolstr.2005.03.063

Cecchi A, Sab K (2002) Out of plane model for heterogeneous periodic materials: the case of masonry. Eur J Mech A/Solids 21(5):715–746. doi:10.1016/S0997-7538(02)01243-3

Coenen EWC, Kouznetsova VG, Geers MGD (2010) Computational homogenization for heterogeneous thin sheets. Int J Numer Methods Eng 83(8–9):1180–1205. doi:10.1002/nme.2833

Cofer WF, Will KM (1991) A three-dimensional, shell-solid transition element for general nonlinear analysis. Comput Struct 38(4):449–462. doi:10.1016/0045-7949(91)90041-J

Cottereau R (2013a) Numerical strategy for unbiased homogenization of random media. Int J Numer Methods Eng 95(1):71–90. doi:10.1002/nme.4502

Cottereau R (2013b) A stochastic-deterministic coupling method for multiscale problems. Application to numerical homogenization of random materials. Procedia IUTAM 6:35–43. doi:10.1016/j.piutam.2013.01.004

Cottereau R, Ben Dhia H, Clouteau D (2010) Localized modeling of uncertainty in the Arlequin framework. In: Langley R, Belyaev A (eds) Vibration analysis of structures with uncertainties. IUTAM bookseries. Springer, pp 477–488. doi:10.1007/978-94-007-0289-9_33

Cottereau R, Clouteau D, Ben Dhia H, Zaccardi C (2011) A stochastic-deterministic coupling method for continuum mechanics. Comput Methods Appl Mech Eng 200:3280–3288. doi:10.1016/j.cma.2011.07.010

Dávila CG (1994) Solid-to-shell transition elements for the computation of interlaminar stresses. Comput Syst Eng 5(2):193–202. doi:10.1016/0956-0521(94)90050-7

Ern A, Guermond JL (2004) Theory and practice of finite elements. Applied mathematical sciences, vol 159. Springer, New York

Ghanem RG, Spanos PD (1991) Stochastic finite elements: a spectral approach. Springer, New York

Ghanem A, Torkhani M, Mahjoubi N, Baranger TN, Combescure A (2013) Arlequin framework for multi-model, multi-time scale and heterogeneous time integrators for structural transient dynamics. Comput Methods Appl Mech Eng 254:292–308. doi:10.1016/j.cma.2012.08.019

Gmür TC, Schorderet AM (1993) A set of three-dimensional solid to shell transition elements for structural dynamics. Comput Struct 46(4):583–591. doi:10.1016/0045-7949(93)90387-S

Grédé A, Tie B, Aubry D (2006) Elastic wave propagation in hexagonal honeycomb sandwich panels: physical understanding and numerical modeling. J Phys IV 134:507–514. doi:10.1051/jp4:2006134078

Hohe J, Becker W (2001) Effective stress-strain relations for two-dimensional cellular sandwich cores: homogenization, material models, and properties. Appl Mech Rev 55(1):61 87. doi:10.1115/1.1425394

Huet C (1990) Application of variational concepts to size effects in elastic heterogeneous bodies. J Mech Phys Solids 38(6):813–841. doi:10.1016/0022-5096(90)90041-2

Hughes TJ (1987) The finite element method. Linear static and dynamic finite element analysis. Prentice-Hall, Englewood Cliffs

Jehel P, Cottereau R (2012) On damping created by the heterogeneity of the mechanical properties in RC frame seismic analysis. In: Proceedings of the 15th world conference on earthquake engineering, Lisbon

Kohn RV, Vogelius M (1984) A new model for thin plates with rapidly varying thickness. Int J Solids Struct 20(4):333–350. doi:10.1016/0020-7683(84)90044-1

Lagarias J, Reeds JA, Wright MH, Wright PE (1998) Convergence properties of the Nelder-Mead simplex method in low dimensions. SIAM J Optim 9(1):112–147

Le Guennec Y, Cottereau R, Clouteau D, Soize C (2013) A coupling method for stochastic continuum models at different scales. Probab Eng Mech. In print. doi:10.1016/j.probengmech.2013.10.005

Liu T, Deng ZC, Lu TJ (2006) Design optimization of truss-cored sandwiches with homogenization. Int J Solids Struct 43(25–26):7891–7918. doi:10.1016/j.ijsolstr.2006.04.010

McCune RW, Armstrong CG, Robinson DJ (2000) Mixed-dimensional coupling in finite element models. Int J Numer Methods Eng 49(6):725–750. doi:10.1002/1097-0207(20001030)49:6<725::AID-NME967>3.0.CO;2-W

Mercatoris BCN, Bouillard P, Massart TJ (2009) Multi-scale detection of failure in planar masonry thin shells using computational homogenisation. Eng Fract Mech 76(4):479–499. doi:10.1016/j.engfracmech.2008.10.003

Mistler M, Anthoine A, Butenweg C (2007) In-plane and out-of-plane homogenisation of masonry. Comput Struct 85(17–18):1321–1330. doi:10.1016/j.compstruc.2006.08.087

Oñate E (2013) Structural analysis with the finite element method: linear statics. Lecture notes on numerical methods in engineering and sciences, volume 2: beams, plates and shells. Springer, Dordrecht/London

Osawa N, Hashimoto K, Sawamura J, Nakai T, Suzuki S (2007) Study on shell-solid coupling FE analysis for fatigue assessment of ship structure. Mar Struct 20(3):143–163. doi:10.1016/j.marstruc.2007.04.002

Papanicolaou GC, Varadhan SR (1981) Boundary value problems with rapidly oscillating random coefficients. In: Fritz J, Lebowitz JL (eds) Proceedings of the conference on random fields, North Holland. Seria colloquia mathematica societatis Janos Bolyai, vol 2, pp 835–873

Qiao H, Yang QD, Chen WQ, Zhang CZ (2011) Implementation of the Arlequin method into ABAQUS: basic formulations and applications. Adv Eng Softw 42(4):197–207. doi:10.1016/j.advengsoft.2011.02.005

Rabczuk T, Kim JY, Samaniego E, Belytschko T (2004) Homogenization of sandwich structures. Int J Numer Methods Eng 61(7):1009–1027. doi:10.1002/nme.1100

Rateau G (2003) Méthode arlequin pour les problèmes mécaniques multi-échelles. PhD thesis, École Centrale Paris, Châtenay-Malabry

Robert CP, Casella G (2004) Monte Carlo statistical methods. Springer, New York

Rousseau J, Marin P, Daudeville L, Potapov S (2010) A discrete element/shell finite element coupling simulating impacts on reinforced concrete structures. Eur J Comput Mech 19(1–3):153–164. doi:10.3166/ejcm.19.153--164

Sab K (1992) On the homogenization and the simulation of random materials. Eur J Mech A/Solids 11(5):585–607

Shim KW, Monaghan DJ, Armstrong CG (2002) Mixed dimensional coupling in finite element stress analysis. Eng Comput 18(3):241–252. doi:10.1007/s003660200021

Song H, Hodges DH (2010) Rigorous joining of advanced reduced-dimensional beam models to 2D finite element models. In: 51st AIAA/ASME/ASCE/AHS/ASC structures, structural dynamics and materials conference, Orlando, pp 1–18. doi:10.2514/6.2010-2545

Stefanou G (2009) The stochastic finite element method: past, present and future. Comput Methods Appl Mech Eng 198(9–12):1031–1051. doi:10.1016/j.cma.2008.11.007

Surana KS (1980) Transition finite element for three-dimensional stress analysis. Int J Numer Methods Eng 15(7):991–1020. doi:10.1002/nme.1620150704

Tartar L (2009) The general theory of homogenization: a personalized introduction. Lecture notes of the Unione Matematica Italiana, vol 7. Springer, Heidelberg/New York

Timoshenko SP (1922) On the transverse vibrations of bars of uniform cross-section. Philos Mag Ser 6 43(253):125–131. doi:10.1080/14786442208633855

Xue K, Li YX, Shi DY, Maharjan S, Zhang L (2009) Study on shell-solid coupling method of trunk structure for efficient FE analysis. Key Eng Mater 419–420:217–220. doi:10.4028/www.scientific.net/KEM.419-420.217

Zaccardi C, Chamoin L, Cottereau R, Ben Dhia H (2013) Error estimation and model adaptation for stochastic-deterministic coupling in the Arlequin framework. Int J Numer Methods Eng 96(2):87–109. doi:10.1002/nme.4540

Zienkiewicz OC, Taylor RL (2005) The finite element method for solid and structural mechanics, 6th edn. Butterworth-Heinemann, Amsterdam/Boston

Adaptive Strategy for Stochastic Homogenization and Multiscale Stochastic Stress Analysis

Sei-ichiro Sakata

Abstract This paper discusses the probabilistic analysis of a multiscale problem of heterogeneous materials, such as composite materials, for estimating the probabilistic characteristics of their homogenized equivalent elastic properties and their macroscopic and microscopic stress fields. For this purpose, a function approximation-based stochastic homogenization method or a perturbation-based multiscale stochastic analysis method is employed. When using these methods for the probabilistic analyses, an appropriate set of samples must be selected for the approximation and the approximation order must be appropriately determined. For this problem, to improve the accuracy of a lower-order approximation-based analysis, some adaptive strategies for the multiscale stochastic analysis are introduced. One is based on the approximation of a response function with the adaptive weighted least-squares method and the other is a piecewise linear approximation with the adaptive expansion of a response function. As a numerical example, a stochastic homogenization and multiscale stochastic stress analysis of a glass particle-reinforced composite material is solved. On the basis of the results, the effectiveness of the proposed approaches is discussed.

Keywords Stochastic homogenization • Multiscale stochastic stress analysis • Adaptive strategy • Perturbation • Function approximation • Composite material

1 Introduction

Recently, uncertainty quantification for heterogeneous materials has become an important topic in the field of mechanical engineering. Because heterogeneous advanced materials such as composites have complex microstructures, the analysis

S. Sakata (✉)
Department of Mechanical Engineering, Kinki University, Osaka, Japan
e-mail: sakata@mech.kindai.ac.jp

M. Papadrakakis and G. Stefanou (eds.), *Multiscale Modeling and Uncertainty Quantification of Materials and Structures*, DOI 10.1007/978-3-319-06331-7_4,
© Springer International Publishing Switzerland 2014

of uncertainty propagation through different scales plays an important role in the reliability evaluation of a composite structure. Furthermore, this analysis is a key process in the verification and validation of numerical simulations based on computational mechanics. This type of problem is called a stochastic homogenization problem or a multiscale stochastic stress analysis problem, and several studies related to these problems have been published.

In the early stages of research in this field, the Monte Carlo simulation was sometimes used for analysis (Kaminski and Kleiber 1996; Sakata et al. 2010a). Because the Monte Carlo simulation does not require much assumption about the analysis problem, an acceptable result is obtained under most problem settings. However, it is known that the Monte Carlo simulation generally involves excessive computational resources, such as CPU time or storage, and that it includes a bias in the estimated result; therefore, a more efficient method needs to be developed.

For example, the perturbation-based stochastic homogenization method was reported (Kaminski and Kleiber 2000; Sakata et al. 2008a). This method has been extended to thermal conductivity problems (Kaminski 2001), thermoelastic problems (Sakata et al. 2010b, 2013), multiscale stress analysis problems (Sakata et al. 2011) and multiscale failure probability analyses (Sakata et al. 2012). Moreover, several other approaches were proposed (e.g. Xu and Brady 2006; Xu et al. 2009; Tootkaboni and Brady 2010). In other studies, an approximate stochastic homogenization method with kriging (Sakata et al 2008b) and a polynomial-based approximation (Kaminski 2009) were proposed to enable multiscale stochastic analysis by using existing commercial or in-house simulation software.

These methods for stochastic homogenization analysis or multiscale stochastic homogenization analysis have aimed to provide more accurate and efficient analyses, but a more accurate method can sometimes be more complex. For example, some methods are based on an approximation theory, and a higher order approximation or a complex series with many terms may provide a more accurate estimation. However, such a complex method or higher order approximation is sometimes instable or difficult to use; therefore, a simple approach should also be studied, e.g., a lower order perturbation-based or polynomial-based approximation. In general, a lower-order approximation is more robust and easier to use.

From this viewpoint, in this study, simple approaches were developed for improving the accuracy of analysis with a lower-order approximation. One approach is based on the approximation of a response function with the adaptive weighted least-squares method and the other is a piecewise linear approximation with the adaptive expansion of a response function.

In this paper, these adaptive methods are introduced, in addition to the perturbation- or polynomial approximation-based approaches. A numerical example is provided of the stochastic homogenization and multiscale stochastic stress analysis of a glass particle-reinforced composite material. In accordance with the results, the effectiveness of the proposed approach is discussed.

2 Methodology

2.1 Homogenization Method

For the multiscale stochastic analysis of the elastic problem of a composite structure, the homogenization theory is employed in this study. From a general formulation of the homogenization theory (Guedes and Kikuchi 1990), a homogenized macroscopic elastic tensor, E^H, can be computed as

$$E^H = \frac{1}{|Y|} \int_Y E \left(I - \frac{\partial \chi}{\partial y} \right) dY, \tag{1}$$

where E is an elastic tensor of a microstructure, the superscript H indicates a homogenized quantity, $|Y|$ is the volume of a unit cell, I is a unit tensor, and dY is a small volume element in the microstructure. χ is a characteristic displacement, which can be obtained as a solution of the following characteristic equation:

$$\frac{\partial}{\partial y} E \frac{\partial \chi}{\partial y} dY = \frac{\partial}{\partial y} E \, dY. \tag{2}$$

Additionally, the microscopic stress, σ, can be computed from

$$\sigma = E \left(I - \frac{\partial \chi}{\partial y} \right) \varepsilon^{macro}, \tag{3}$$

where the superscript *macro* indicates a macroscopic quantity. The macroscopic strain, ε^{macro}, is computed from a conventional single-scale elastic analysis by using the homogenized equivalent elastic properties of a composite material obtained from Eq. 1.

2.2 Monte Carlo Simulation for Multiscale Stochastic Analysis

In multiscale stochastic analyses, it is assumed that a microscopic quantity such as the elastic property or geometry of a microstructure has a certain random variation. In this case, realization of the microscopic random variable, X, can be expressed by the following equation:

$$X^* = (1 + \beta) \times X^0, \tag{4}$$

where the superscript 0 means the expected value, * indicates an observed value of X, and β is the normalized random variable, and it is assumed that β is distributed according to Gaussian distribution.

If a microscopic quantity such as an elastic property of a component material includes a random variation, the observed value of the homogenized elastic tensor can be expressed as a probabilistic response of the microscopic random variation as follows:

$$E^{H*} = \frac{1}{|Y|} \int_Y E^* \left(I - \frac{\partial \chi^*}{\partial y} \right) dY, \tag{5}$$

The realization of the characteristic displacement against the microscopic random variable can be obtained by solving the characteristic equation, which, considering the stochastic response, can be rewritten as

$$\frac{\partial}{\partial y} E^* \frac{\partial \chi^*}{\partial y} dY = \frac{\partial}{\partial y} E^* dY, \tag{6}$$

Similarly, the observed value of the microscopic stress can be computed from the following equation:

$$\sigma^* = E^* \left(I - \frac{\partial \chi^*}{\partial y} \right) \varepsilon^{macro*}, \tag{7}$$

The observed value of the macroscopic strain is computed from the single-scale stochastic stress analysis against the microscopic random variation; therefore, both the stochastic homogenization analysis and the conventional single-scale stochastic stress analysis will be needed for the microscopic stochastic stress analysis.

As examples of the probabilistic characteristics, the expectation or variance of the homogenized elastic tensor can be expressed as

$$E\left[E^{H*}\right] = \int_{-\infty}^{\infty} E^{H*} f\left(E^H\right) dE^H, \tag{8}$$

$$\mathrm{Var}\left[E^{H*}\right] = \int_{-\infty}^{\infty} \left(E^{H*} - E\left[E^{H*}\right]\right)^2 f\left(E^H\right) dE^H, \tag{9}$$

where $E[E^{H*}]$ and $\mathrm{Var}[E^{H*}]$ are the expectation and variance of the homogenized elastic tensor, and $f(E^H)$ is the probabilistic density function of the homogenized elastic tensor.

In the Monte Carlo simulation, the observed value of the homogenized elastic tensor becomes a function of the normalized random variable β, and the expected value, Exp[ ], and variance, Var[ ], of the homogenized elastic tensor can be approximately computed as

$$\mathrm{Exp}\left[E^{H*}\right] \approx \frac{1}{n} \sum_i E^{H*}(\beta_i), \tag{10}$$

$$\text{Var}\left[E^{H*}\right] \approx \frac{1}{n-1} \sum_i \left(E^{H*}(\beta_i) - \text{Exp}\left[E^{H*}\right]\right)^2, \tag{11}$$

where n is the number of trials.

From these quantities, the coefficient of variance CV[] can be computed as

$$\text{CV}\left[E^{H*}\right] = \sqrt{\text{Var}\left[E^{H*}\right]}/\text{Exp}\left[E^{H*}\right], \tag{12}$$

When assuming the Gaussian random variable, the random number is generated with the following formula:

$$\beta = \sqrt{-2s^2 \log U_1} \times \sin 2\pi U_2, \tag{13}$$

where $0 < U_1 \leq 1$ and $0 \leq U_2 \leq 1$ are observed values of a uniform random variable and s is the standard deviation of the random variable β.

2.3 Multiscale Stochastic Analysis with a Function Approximation Method

2.3.1 Perturbation-Based Multiscale Stochastic Analysis

Because the computation of Eqs. 10 and 11 involves a huge number of trials, an approximation-based stochastic homogenization and multiscale stress analysis method has been proposed. In this study, the perturbation-based stochastic homogenization method is employed as the analysis method with single-point expansion.

If random variation of a material property or geometrical parameter of a microstructure is taken into account, Eq. 5 can be approximated by applying the asymptotic expansion of the random response as follows:

$$\begin{aligned} E^{H*} &\approx E^{H0} + E^{H1}\beta + \cdots + E^{Hi}\beta^i + \cdots \\ &= \frac{1}{|Y|} \int_Y \left(E^0 + \cdots + E^i\beta^i + \cdots\right) \left\{I - \frac{\partial}{\partial y}\left(\chi^0 + \cdots + \chi^i\beta^i + \cdots\right)\right\} dY, \end{aligned} \tag{14}$$

where the superscript i indicates the ith order perturbation term. To obtain the perturbation terms of the characteristic displacement, the following simultaneous equation is solved:

$$\left. \begin{aligned} K^{Y0}\chi^0 &= F^{Y0} \\ (K^{Y1}\chi^0 + K^{Y0}\chi^1) &= F^{Y1} \\ &\vdots \end{aligned} \right\}, \tag{15}$$

where K^Y and F^Y are the microscopic stiffness matrix and load vector, respectively, for the characteristic equation obtained from Eq. 2.

Substituting the first-order approximation of the microscopic elastic tensor and the characteristic displacement into Eq. 14, and comparing the coefficients of the same order random variables with each other, the perturbation terms of the homogenized elastic tensor can be obtained. For example, the first-order perturbation term of the homogenized elastic tensor and the microscopic stress for material property variation can be computed as

$$E^{H1} = \frac{1}{|Y|} \int_Y E^1 \left(I - \frac{\partial \chi^0}{\partial y} \right) dY - \frac{1}{|Y|} \int_Y E^0 \frac{\partial \chi^1}{\partial y} dY, \qquad (16)$$

$$\sigma^1 = E^0 \left(I - \frac{\partial \chi^0}{\partial y} \right) \varepsilon^{macro1} - \left\{ E^1 \left(I - \frac{\partial \chi^0}{\partial y} \right) - E^0 \frac{\partial \chi^1}{\partial y} \right\} \varepsilon^{macro0}, \qquad (17)$$

With these perturbation terms, the expectation and variance of the homogenized elastic property or microscopic stress of a composite material considering a microscopic random variation can be approximately computed with the first-order second moment method. For example, the expectation and variance of the microscopic stress can be estimated as follows:

$$\left. \begin{array}{l} \text{Exp}\left[E^{H*} \right] = E^{H0} \\ \text{Var}\left[E^{H*} \right] = \sum_i \sum_j \left[E^{H1} \right]_i \left[E^{H1} \right]_j \text{cov}\left[\beta_i, \beta_j \right] \end{array} \right\}, \qquad (18)$$

2.3.2 Polynomial Approximation Approach

For a highly nonlinear response function, the perturbation-based approach may sometimes be inefficient. For this reason, a multipoint approximation-based approach was developed. To construct a surrogate model for the stochastic homogenization analysis, when using a polynomial-based model, a set of coefficients of the polynomial function can be conventionally determined by minimizing an approximation error. The optimization problem in the case of the homogenized Young's modulus approximation can be written as follows:

$$\left. \begin{array}{ll} \text{find} & c_i \quad (i = 0, 1, \ldots, m) \\ \text{to minimize } F_{obj} = \sum_{j=1}^k w_j \left(\widehat{E^{H*}}_j - E^H(\beta_j) \right)^2 \\ s.t. & \widehat{E_j^{H*}} = \sum_{i=0}^m c_i \beta_j^i \end{array} \right\}, \qquad (19)$$

where $E^H(\beta_j)$ shows an observed value of the stochastic response function at the jth sampling location of the random variable β_j, k is the number of samples, m is the order of approximation function, w_j is the weighting coefficient for each sample, and in a general least-squares method, $w_j = 1$.

After constructing the polynomial-based surrogate model, the Monte Carlo simulation or the first-order second moment method can be used for the multiscale stochastic analysis by using the surrogate at a lower computational cost.

2.3.3   Piecewise Linear Approximation Approach

Because a higher-order polynomial-based approximation sometimes fluctuates, a low-order function is preferable for stability. In this study, a piecewise linear approximation is employed for the multiscale stochastic analysis. For example, the observed microscopic stress can be simply represented with a piecewise linear form as

$$\sigma^* = \sigma\left(\beta\right) \approx \sigma^0{}_i + \sigma^1{}_i \beta \quad (\beta_i \leq \beta \leq \beta_{i+1}), \tag{20}$$

where $\sigma^0{}_i$ and $\sigma^1{}_i$ are the zero and first-order perturbation terms for the ith interval, namely,

$$\sigma^0{}_i = \sigma^0\big|_{\beta=\beta_i}, \quad \sigma^1{}_i = \sigma^1\big|_{\beta=\beta_i}, \tag{21}$$

After obtaining each term of the piecewise linear approximations of the microscopic stresses, their probabilistic properties can be estimated. The expectation and variance of the microscopic stress can be computed with the piecewise linear approximation as follows:

$$\mathrm{E}\left[\sigma^*\right] \approx \int_{p_0}^{p_1} \left(\sigma^0{}_0 + \sigma^1{}_0\beta\right) f\left(\beta\right) \mathrm{d}\beta + \int_{p_1}^{p_2} \left(\sigma^0{}_1 + \sigma^1{}_1\beta\right) f\left(\beta\right) \mathrm{d}\beta, \tag{22}$$

$$\mathrm{Var}\left[\sigma^*\right] \approx \int_{p_0}^{p_1} \left(\sigma^0{}_0 + \sigma^1{}_0\beta - \mathrm{Exp}\left[\sigma^*\right]\right)^2 f\left(\beta\right) \mathrm{d}\alpha$$

$$+ \int_{p_1}^{p_2} \left(\sigma^0{}_1 + \sigma^1{}_1\beta - \mathrm{Exp}\left[\sigma^{H*}\right]\right)^2 f\left(\beta\right) \mathrm{d}\beta + \cdots, \tag{23}$$

where $P_i = [p_i, p_{i+1}]$, $p_0 \ll \mu_\beta, p_{n+1} \gg \mu_\beta$, p_i is the ith expansion point, μ_β is the expected value of the normalized microscopic random variable of β.

3   Adaptive Strategies for Approximation-Based Multiscale Stochastic Analysis

Multipoint approximation-based multiscale stochastic analysis methods, like the approximation function approach or the piecewise linear approximation approach, have the potential to improve the accuracy and efficiency of multiscale stochastic analyses, but some problems still exist. One is the determination of the appropriate approximation order or the location of expansion points. In general, these are problem dependent, and a better computational condition should be chosen for different problem settings. For this problem, adaptive strategies are introduced for the approximate multiscale stochastic analysis.

3.1 Polynomial-Based Approximate Stochastic Homogenization with Adaptive Weight

When using the polynomial-based approximate stochastic homogenization method, Eq. 19 must be solved to determine a set of appropriate coefficients of the approximated response function. In general, this minimization yields a surrogate model having the minimum average approximation error at the sample point. However, the approximation error is not independently referred to each sample point; therefore, the approximation error distribution is not controlled. On the other hand, in the approximate stochastic analysis, the probabilistic density of a random variable is given, which means that the appearance frequency depends on location. Because of this, the estimation error should be smaller when the probabilistic density is higher. For this purpose, sampling location and density can be optimized.

However, the computation of sampling values requires additional computational costs, and resampling for a more accurate estimation or different computational conditions will be expensive. From this viewpoint, an adaptive strategy for constructing a polynomial based surrogate is developed.

As explained previously, the weight $w_j = 1$ is adopted in a conventional least-squares method. In this study, this weight is assumed to control the distribution of the approximation error and is determined to be adaptive to the probabilistic density of the microscopic random variable.

In this case, a Gaussian distribution-based weight according to the probabilistic density function of the random variable, which is called the adaptive weight, as expressed by Eq. 24 is applied to the approximate stochastic homogenization analysis.

$$
w_j = w\left(\beta_j\right) = \begin{cases} \dfrac{1}{\sqrt{2\pi}s} \exp\left(-\dfrac{\beta_j^2}{2s^2}\right)^2 & : \beta_j \neq 0 \\ \infty & : \beta_j = 0 \end{cases},
\tag{24}
$$

Figure 1 shows an example of the approximation error between the exact and approximated function values. A linear approximation is constructed with using the weight $w_j = 1$ and Eq. 24 for expressing a relationship between an equivalent Young's modulus and the Poisson's ratio of the resin of a particle-reinforced composite material. According to Fig. 1, an approximation error becomes smaller around the expected value of the microscopic random variable ($\beta = 0$) when using

Fig. 1 Approximation error distribution

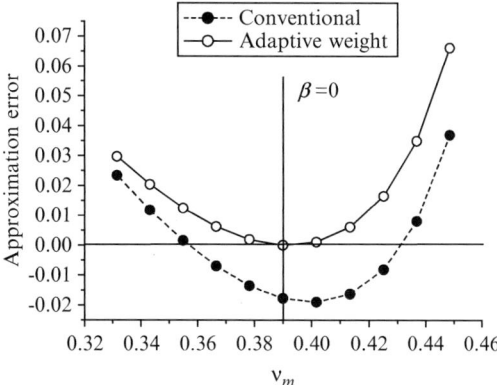

the adaptive weight, whereas the approximation error is larger near the expected value when using the conventional least-squares method. With this approach, the stochastic homogenization analysis is attempted in this study.

3.2 Adaptive Strategy for the Perturbation-Based Multiscale Stochastic Analysis

The response function of a multiscale stochastic problem becomes nonsmooth in some cases. In this case, an approximation function-based approach with a smooth function, such as the polynomial approximation or Taylor series expansion, is not appropriate as a surrogate model and the aforementioned piecewise linear approximation may be applicable.

The dense division of P_i in Eqs. 22 and 23 will provide a more accurate estimation of the variance, but the computational cost is higher. For this reason, fewer expansion points are preferred in terms of computational efficiency. In this case, the selection of an appropriate expansion point is important for improving accuracy. For this purpose, an adaptive strategy for the piecewise linear approximation approach is developed to estimate the coefficient of variance (CV) of the stress for a microscopic random variation.

This study considers a piecewise linear approximation for a nonsmooth continuous response function. In this case, estimating the nonsmooth point is important for the determination of the expansion points, so as to construct a better surrogate model. The algorithm of the proposed adaptive approach for this problem is summarized in the following.

1. Compute the perturbation term of the microscopic stresses for the expected microstructure.
2. Construct a surrogate model for estimating the stresses for an arbitrary value of the microscopic random variable.

Fig. 2 A schematic view of
polynomial and piecewise
linear approximation

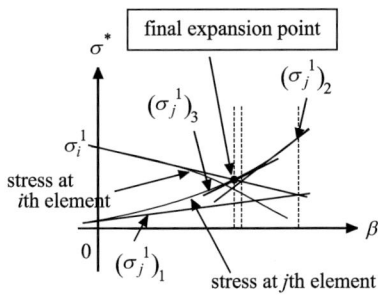

3. Search the maximum value of the stress for constructing the approximation at the first region. This approximated function is called the first approximation.
4. From the set of surrogate models, the element for which stress becomes larger than the first approximation is searched. The approximated function of the stress in the element is called the second approximation.
5. Investigate the intersection point of the first and second approximations, and compute the perturbation term for the second approximation at the intersection point.
6. Iterate steps 3 and 4, and the intersection point is updated until converged.
7. After convergence, construct the second approximation with the perturbation term at the converged intersection points.
8. Search a larger stress, estimated with the surrogate with the perturbation term of another element, and iterate steps 3–6 until an element having larger stress is not found.

Figure 2 shows a conceptual view of the presented adaptive piecewise linear approximation.

4 Numerical Examples

4.1 Problem Settings

In this study, the stochastic homogenization and multiscale stochastic stress analysis problems of a particle-reinforced composite material are considered. Probabilistic properties are estimated, such as expectation, variance, and CV of equivalent elastic properties and microscopic stresses in a particle-reinforced plastic. A schematic view of the composite and a finite-element model of the unit cell of the microstructure are shown in Fig. 3.

The properties of the employed particle and matrix correspond to E-glass and epoxy resin. The volume fraction of particle (Vp) is 0.5 in this example. The elastic properties of the component materials are listed in Table 1.

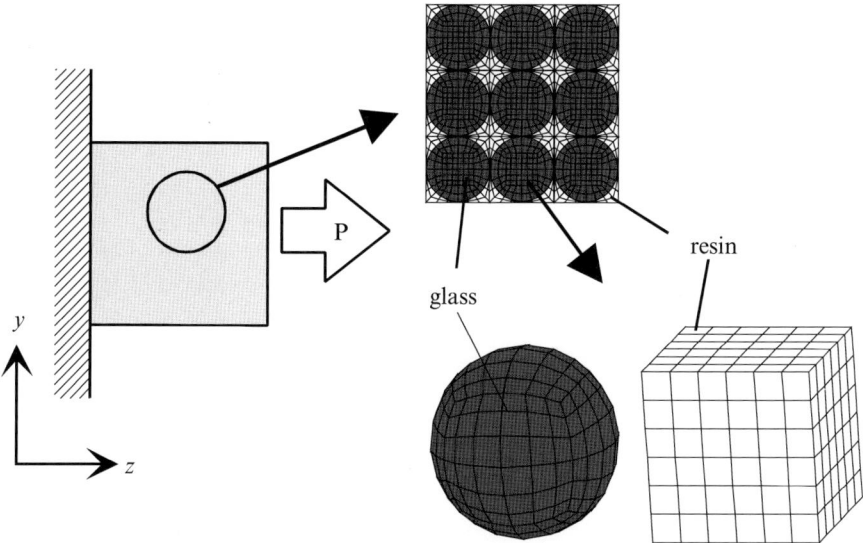

Fig. 3 Schematic view of the particle reinforced composite material

Table 1 Elastic properties of
each component material

	Young's modulus [GPa]	Poisson's ratio
Epoxy	4.5	0.39
E-glass	73.0	0.22

4.2 Stochastic Homogenization of Particle-Reinforced Composite Material with the Polynomial Approximation

For the stochastic homogenization analysis of a particle-reinforced composite material, the polynomial approximation-based stochastic homogenization method is employed. In this case, Young's modulus and Poisson's ratio of resin (E_m, ν_m) are considered as the microscopic random variables and the proposed approach is applied to the stochastic homogenization analysis of the particle-reinforced composite material.

Figure 4 shows the relative estimation errors between the approximation-based stochastic homogenization methods and the Monte Carlo simulation. In this case, the CV of the microscopic random variable is 0.05. Figure 4a shows the relative error in CV estimations of the equivalent elastic properties and estimated results for E_m variation; Fig. 4b shows those for ν_m variation. In the legend, "no weight," "single weight," "perturbation," and "adaptive" refer to the results obtained from the approximated functions, generated by a general least-squares method using $w_j = 1$, a single-point weight $w = \infty$ at $\beta = 0$, a single-point approximation with the perturbation method, and the proposed adaptive weight, respectively.

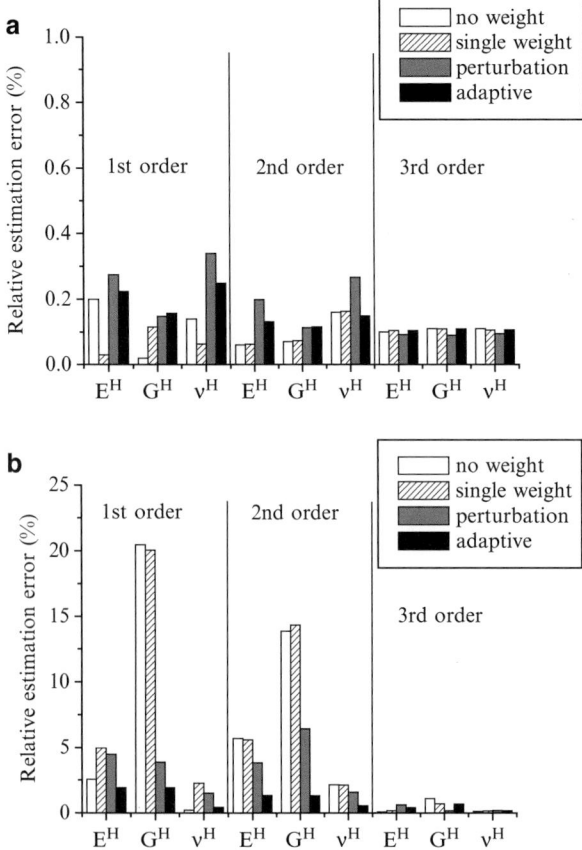

Fig. 4 Relative error in the estimated CVs of the equivalent elastic properties (**a**) For E_m variation, (**b**) For ν_m variation

According to Fig. 4a, all methods produce accurate results for E_m variation by using first, second, and third-order approximations, whereas the estimated CVs with the general least-squares method, single-point weighted least-squares method, and perturbation include large errors for ν_m variation, especially when using the first- or second-order approximations as shown in Fig. 4b. On the other hand, the proposed adaptive strategy–based approach provides more accurate results than the others when lower-order approximations are used. Although all methods provide an accurate estimation with the third-order approximation, the proposed adaptive approach is effective for lower-order approximation-based stochastic homogenization analysis.

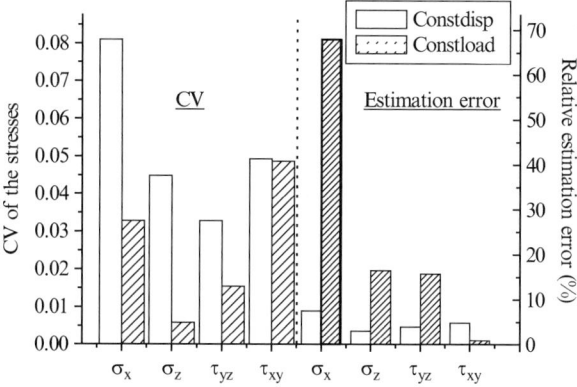

Fig. 5 Relative error in estimated CV of the maximum stresses

4.3 Multiscale Stochastic Stress Analysis of Particle-Reinforced Composite Material with the Piecewise Linear Approximation

The stochastic analysis of the maximum stress in the particle-reinforced composite material provides an example of multiscale stochastic stress analysis having a nonsmooth response function. In this case, v_m variation is considered to be a microscopic random variable. A unidirectional load or enforced displacement is applied to the rectangular composite structure in macroscale. The estimated CVs and the relative error in CV estimation of the maximum microscopic stress between the conventional first-order perturbation-based method and the Monte Carlo simulation are illustrated in Fig. 5. The legend "constload" and "constdisp" mean the cases that a constant load or enforced displacement is applied, respectively. As shown in this figure, the CV estimation of the maximum σ_x includes a large error.

To discuss one reason for this inaccuracy, the response function of the maximum σ_x for v_m variation is illustrated in Fig. 6. As shown in Fig. 6, the response function is nonsmooth; a reason for this nonsmoothness is that the finite element having the maximum stress alters according to the variation in v_m. In this case, a smooth function is not appropriate as a basic function of a surrogate and a single-point approximation cannot accurately estimate the response.

For this problem, the proposed adaptive piecewise linear approach is applied. Because the adaptive method requires some iteration steps for improving accuracy, the transition of the estimated CV of the maximum stress is illustrated in Fig. 7. The estimated CV with the first-order perturbation and the Monte Carlo simulation are also shown. As shown in Fig. 7, the proposed adaptive approach improves the estimation accuracy after four iterations and the final relative estimation error

Fig. 6 A response function
of the maximum stress σ_x for
ν_m variation

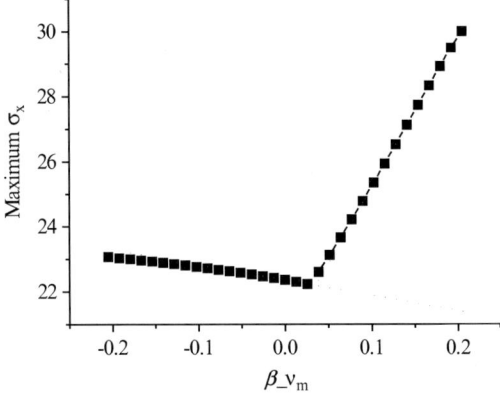

Fig. 7 Improvement of the
accuracy in CV estimation
with applying the proposed
approach

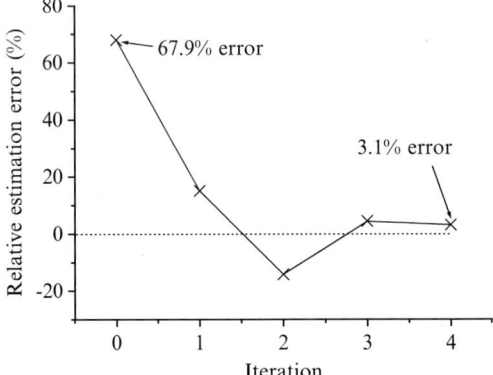

between the CVs obtained by the Monte Carlo simulation and the proposed
approach is 3 %. This result shows the effectiveness of the proposed adaptive
piecewise linear approximation approach for a multiscale stochastic problem having
a nonsmooth response function.

5   Conclusion

This paper discusses the stochastic homogenization and multiscale stochastic stress
analysis of composite materials. In particular, adaptive strategies are introduced
for improving the accuracy of the approximation-based stochastic homogenization
and multiscale stochastic stress analyses. The aim of the adaptive strategy for each
problem is the improvement of accuracy by using a lower-order approximation. One
is the weighted least-squares method, in which weights are determined adaptive
to the distribution of the microscopic random variable. The other is the adaptive

piecewise linear approximation, which determines an expansion point with iterative computation of the perturbation terms.

In the numerical examples, the stochastic homogenization and multiscale stochastic stress analyses are solved for a particle-reinforced composite material by considering a random variation of the elastic properties of a component material. On the basis of the numerical results, it is confirmed that the accuracy is efficiently improved with the proposed approaches. This kind of lower-order approximation with these adaptive methods will be helpful when performing a multiscale stochastic analysis.

Acknowledgments The author is pleased to acknowledge support in part by Grants-in-Aid for Young Scientists (B) (No.23760097) from the Ministry of Education, Culture, Sports Science and Technology, and MEXT-supported program for the Strategic Research Foundation at Private Universities, 2012–2014.

References

Guedes M, Kikuchi N (1990) Preprocessing and postprocessing for materials based on the homogenization method with adaptive finite element methods. Comput Meth Appl Mech Eng 83:143–198

Kaminski M (2001) Stochastic finite element method homogenization of heat conduction problem in fiber composites. Struct Eng Mech 11(4):373–392

Kaminski M (2009) Sensitivity and randomness in homogenization of periodic fiber-reinforced composites via the response function method. Int J Solids Struct 46(3–4):923–937

Kaminski M, Kleiber M (1996) Stochastic structural interface defects in fiber composites. Int J Solids Struct 33(20–22):3035–3056

Kaminski M, Kleiber M (2000) Perturbation based stochastic finite element method for homogenization of two-phase elastic composites. Comput Struct 78:811–826

Sakata S, Ashida F, Kojima T, Zako M (2008a) Three-dimensional stochastic analysis using a perturbation-based homogenization method for homogenized elastic property of inhomogeneous material considering microscopic uncertainty. Int J Solids Struct 45(3/4):894–907

Sakata S, Ashida F, Zako M (2008b) Kriging-based approximate stochastic homogenization analysis for composite material. Comput Meth Appl Mech Eng 197(21–24):1953–1964

Sakata S, Ashida F, Enya K (2010a) Stochastic analysis of microscopic stress in fiber reinforced composites considering uncertainty in a microscopic elastic property. J Solids Mech Mater Eng 4(5):568–577

Sakata S, Ashida F, Kojima T (2010b) Stochastic homogenization analysis for thermal expansion coefficient of fiber reinforced composites using the equivalent inclusion method with perturbation-based approach. Comput Struct 88(7–8):458–466

Sakata S, Ashida F, Enya K (2011) Perturbation-based stochastic stress analysis of a particle reinforced composite material via the stochastic homogenization analysis considering uncertainty in material properties. J Multi Comput Eng 9(4):395–408

Sakata S, Ashida F, Enya K (2012) A microscopic failure probability analysis of a unidirectional fiber reinforced composite material via a multiscale stochastic stress analysis for a microscopic random variation of an elastic property. Comput Mater Sci 62:35–46

Sakata S, Ashida F, Fujiwara K (2013) A stochastic homogenization analysis for a thermoelastic problem of a unidirectional fiber-reinforced composite material with the homogenization theory. J Therm Stress 36(5):405–425

Tootkaboni M, Brady LG (2010) A multi-scale spectral stochastic method for homogenization of multi-phase periodic composites with random material properties. Int J Num Meth Eng 83: 59–90

Xu XF, Brady LG (2006) Computational stochastic homogenization of random media elliptic problems using Fourier Galerkin method. Finite Elem Anal Des 42:613–622

Xu XF, Chen X, Shen L (2009) A green-function-based multiscale method for uncertainty quantification of finite body random heterogeneous materials. Comput Struct 87:1416–1426

Strength Properties of Porous Materials Influenced by Shape and Arrangement of Pores: A DLO Investigation Towards Material Design

Sebastian Bauer and Roman Lackner

Abstract The prediction of strength properties of porous materials, which in general are random in nature with varying spatial distribution and variation of pores and matrices, caused by the manufacturing processes, plays an important role with regard to the reliability of materials and structures. The recently developed discontinuity layout optimisation (DLO) and adaptive discontinuity layout optimisation (ADLO), which was used for determination of strength properties of materials and structures, are included in a stochastic limit analysis framework, using random variables. Therefore, different material properties influencing the overall strength of the porous material (e.g. matrix strength, shape, number, and distribution of pores) within a considered two-dimensional RVE are assumed to follow certain probability distributions. A sensitivity study for the identification of material parameters showing the largest influence on the strength of the considered porous materials is performed. The obtained results provide first insight into the nature of the reliability of strength properties of porous materials, paving the way to a better understanding and finally improvement of effective strength properties of porous materials.

Keywords Discontinuity layout optimization • Probability • Limit-analysis • Upscaling • Homogenisation of strength properties

1 Introduction

As engineering materials, such as composites and heterogeneous high strength – low weight materials, are continuously improving in engineering applications (e.g. aerospace, mechanical-engineering, civil-engineering, etc.),

S. Bauer (✉) • R. Lackner
Material Technology Innsbruck (MTI), University of Innsbruck, Technikerstrasse 19a,
6020 Innsbruck, Austria
e-mail: Sebastian.Bauer@uibk.ac.at; Roman.Lackner@uibk.ac.at

M. Papadrakakis and G. Stefanou (eds.), *Multiscale Modeling and Uncertainty Quantification of Materials and Structures*, DOI 10.1007/978-3-319-06331-7_5,
© Springer International Publishing Switzerland 2014

allowing their use in high-performance structures (e.g. airplanes, lightweight cars, high-rise buildings), the proper identification and understanding of the underlying effective strength properties are of crucial importance. Since the microstructure of these matrix-inclusion materials (e.g. ultra-high performance concrete, carbon fibre-reinforced materials, etc.) is in general random in nature with varying spatial distribution, size, and shape of pores, particles, matrices, and matrix-inclusion interfaces, a statistical framework becomes necessary for model-based analyses and finally prediction of material properties, see for instance (Koutsourelakis 2006, 2007; Al-Ostaz et al. 2007; Díaz et al. 2003; Dong et al. 2010; Du and Ostoja-Starzewski 2006; Frantziskonis 1998; Graham-Brady et al. 2006; Guilleminot et al. 2011; Guilleminot and Soize 2012; Liu et al. 2013; Mehrez et al. 2012). Hereby, the description of the random microstructure defines the quality of the employed model and, hence, strongly influences the obtained results. In addition to the volume fractions of considered inclusions (e.g. pores, particles, and fibres), their shape and more importantly their spatial arrangement are thought to strongly affect the strength properties of composite materials (Segurado and LLorca 2006).

The finite element method (FEM), employing nodes and finite elements for the spatial description of the material system under consideration, offers the possibility to represent the microstructure of composite materials in an appropriate manner. Accordingly, the FEM was successfully applied e.g. in Antretter and Fischer (2001) to determine the stress distribution inside the pore of RVEs containing a number of randomly oriented elliptical inclusions, and in Chakraborty and Rahman (2008) for the representation of microstructures with varying particle content and locations following a stochastic description in order to determine the fracture behaviour of functionally-graded materials. In Wriggers and Moftah (2006), the same approach was chosen for determination of the damage behaviour of concrete using a model with random aggregate arrangement at the mesoscopic level. Disadvantages of the FEM such as the dependency of the results on the underlying discretization and unsatisfactory modelling of the interface behaviour between inclusions and the matrix material were highlighted in Wriggers and Moftah (2006). In contrast to the numerical approach of the FEM, a continuum micromechanics-based approach, employing the concept of limit stress states, is proposed in Maghous et al. (2009), taking into account a random two-phase heterogeneous microstructure, giving access to the effective yield function and, hence, to strength properties of composite materials. However, neither the spatial arrangement of the inclusions within the RVE, nor the variation of shape of the inclusions are considered in this approach.

As a remedy, the application of the recently developed discontinuity layout optimisation (DLO) (Smith and Gilbert 2007) and adaptive discontinuity layout optimisation (ADLO) (Bauer and Lackner 2011) is proposed in this paper, taking the detailed microstructure of random heterogeneous materials into account. As a probabilistic extension of the ADLO towards material design, a framework using stochastic limit analysis for porous materials with varying material properties and morphologies (e.g. matrix strength, number, shape, and distribution of pores) following certain probability distributions is employed. The so-obtained results provide first insight into the nature of strength properties of porous materials with different volume fractions of circular and elliptical pores and varying properties of the pores, paving the way to a better understanding and finally improvement of effective strength properties of porous materials.

In the following section, the recently developed ADLO (Bauer and Lackner 2013, 2011) used within this work to predict the strength properties of porous materials and its extension towards a probabilistic framework is described. The so-obtained results are presented and discussed in Sect. 3. Finally, the work closes with conclusions and a brief outlook towards future work on material design.

2 Methodology

Within the DLO, the material system is represented by a random set of n nodes, where m discontinuities are generated as lines between these nodes by delaunay triangulation (de Berg et al. 2008). In addition to the discontinuity layout obtained from triangulation as shown in Fig. 1a (single layout), the double layout (Bauer and Lackner 2013) introducing additional diagonals as discontinuities is considered in this work (see Fig. 1b). The so-obtained discontinuity layout exhibits a larger variety of angles and, therefore, a larger variety of possible failure modes. The Mohr-Coulomb material model with the cohesion c and the friction angle φ is assigned to every discontinuity, taking different material properties for the matrix and pores into account. Every discontinuity may be a potential failure discontinuity and, thus, contribute to the failure mode. Considering the energy balance of the problem restricting it to live loads only, the objective is to minimize the internal energy E^{int} (dissipated energy along the failed discontinuities), which results in

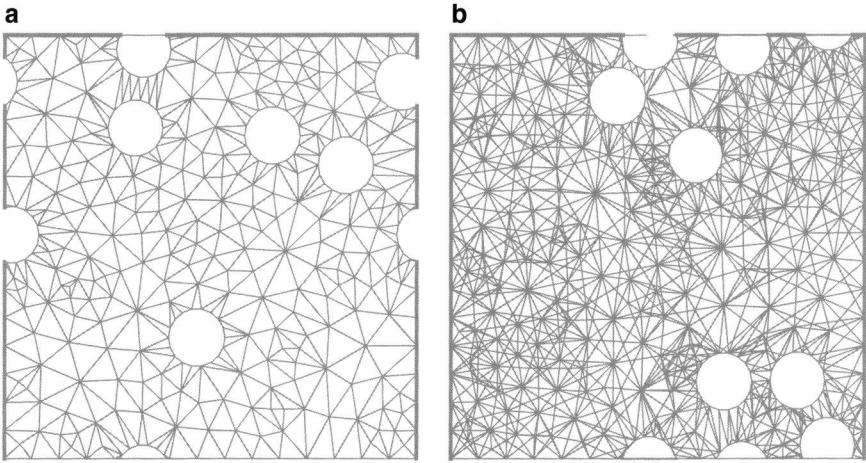

a **b**

Fig. 1 Random DLO layouts for RVEs with seven pores with equal node density: (**a**) single layout and (**b**) double layout

the minimization of the external work E^{ext} associated with the live load, yielding (Smith and Gilbert 2007; Bauer and Lackner 2013):

$$min\ E^{ext} = min\ E^{int}\ \rightarrow\ min\ \lambda \mathbf{f}_L^T \mathbf{d} = \mathbf{g}^T \mathbf{p}. \tag{1}$$

By employing linear programming, the set of failed discontinuities is calculated which yields the minimized internal energy of the material system, resulting in an upper-bound (UB) formulation with the following linear programming (LP) problem (for details, see Smith and Gilbert 2007; Bauer and Lackner 2011, 2013):

$$min\ \lambda \mathbf{f}_L^T \mathbf{d} = \mathbf{g}^T \mathbf{p},$$

subject to

$$\mathbf{Bd} = \mathbf{0}, \tag{2}$$

$$\mathbf{f}_L^T \mathbf{d} = 1,$$

$$\mathbf{Np} - \mathbf{d} = \mathbf{0},$$

$$\mathbf{p} \geq \mathbf{0},$$

where $\mathbf{f}_L$ [N] ($2m$) contains the shear and normal component for live load, λ is the failure load-factor, $\mathbf{g}$ [N] ($2m$) contains the product of length ℓ [m] and cohesive shear strength c [N/m] of the discontinuities, $\mathbf{B}$ [−] is a ($2n \times 2m$) compatibility matrix, and $\mathbf{N}$ [−] is a ($2m \times 2m$) plastic-flow matrix. In Eq. (2), $\mathbf{d}$ and $\mathbf{p}$ represent the unknowns of the LP problem, where $\mathbf{d}$ [m] is a ($2m$) vector of discontinuity displacements, and $\mathbf{p}$ [m] is a ($2m$) vector of plastic multipliers.

2.1 Probabilistic Formulation

In this work, the parameters of the matrix material such as cohesion c and the friction angle ψ, the shape and the volume fraction f_a of the pores are chosen to be deterministic. The orientation ϕ in case of elliptical pores and the size and spatial arrangement of the pores, on the other hand, are random variables. The pores are generated randomly within a two-dimensional RVE of size $L \times L$, with $L \approx 9r_{0.3}$, with $r_{0.3}$ as the radius of circular pores corresponding to a volume fraction of $f_a = 0.3$. Pores are placed one by one within the RVE, avoiding overlap with existing pores. Parts of pores intersecting the boundary of the RVE are considered at the opposite boundary of the material system, giving a periodic geometry.

For the following studies reported in this work, 500 different randomly-generated RVEs with 7 pores and a matrix material with $\varphi = 0°$ and $c = 0.5$ subjected to uniaxial tension were considered for each set of parameters:

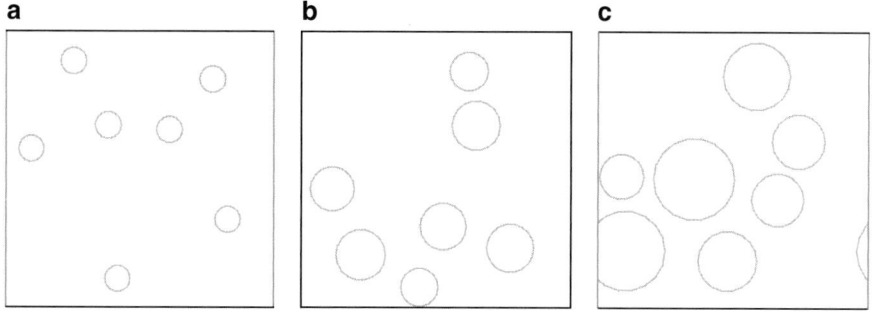

Fig. 2 RVEs with seven circular pores: (**a**) $f_a = 0.05$, $s_r/\bar{r} = 0.01$, (**b**) $f_a = 0.15$, $s_r/\bar{r} = 0.1$, and (**c**) $f_a = 0.3$, $s_r/\bar{r} = 0.2$

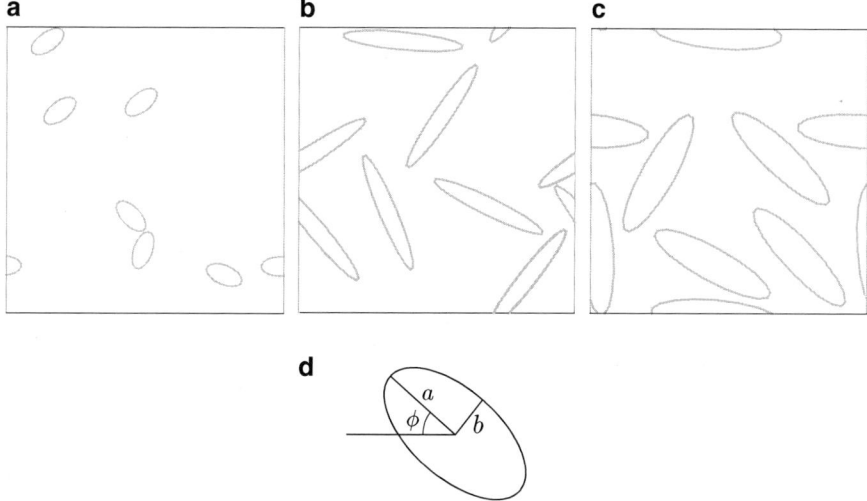

Fig. 3 RVEs with seven elliptical pores: (**a**) $f_a = 0.05$, $E = 2$, (**b**) $f_a = 0.15$, $E = 7$, (**c**) $f_a = 0.3$, $E = 4$, and (**d**) definition of axes and orientation of elliptical pore

- In the first study, considering varying volume fraction of the pores f_a and pore-size distribution $s_r/\bar{r}$, the tensile strength of f_t/f_t^M of RVEs with circular, randomly-distributed pores is predicted. Hereby, f_a is varied from 0.05 to 0.3 and $s_r/\bar{r}$ from 0.01 to 0.6, with s_r as the standard deviation and $\bar{r}$ as the mean value of the radii. Figure 2 shows three different layouts for the RVE.
- In a second study, considering variations in volume fraction of the pores f_a and the ellipse eccentricity E, effective strength properties of an RVE with elliptical, randomly-distributed, and randomly-oriented pores are determined (see Fig. 3a–c); f_a is varied from 0.05 to 0.3, and the ellipse eccentricity E from 1 to 7 (see Piat et al. 2006), where $E = a/b$, with a for the major semi-axis and b for the minor semi-axis (see Fig. 3d).

- Finally, considering varying volume fraction of the pores f_a and pore-size distribution $s_a/\bar{a}$, the influence on the effective strength of RVEs with elliptical, randomly-distributed, and randomly-oriented pores is investigated. Hereby, f_a is varied from 0.05 to 0.3 and $s_a/\bar{a}$ from 0.01 to 0.6, where s_a is the standard deviation and $\bar{a}$ the mean value of the semi-major axes.

3 Results and Discussion

In this section, the proposed stochastic framework for strength homogenisation is applied to three different aspects of porous materials: First, the effective strength properties of materials with circular pores of varying size are investigated, followed by the identification of the effective strength properties of materials with elliptical pores with different volume fractions. Finally, materials with elliptical pores with varying sizes are considered. The effective strength properties are illustrated by the mean value and the 5 % lower quantile of the calculated strength distribution, in analogy to specifications of existing codes (e.g. DIN EN 1990:2010-12 2010) as a characteristic/nominal design value for material properties. The variation of the strength properties is presented in terms of the standard deviation and a dimensionless standard deviation, with the latter being obtained from relating the standard deviation to the mean value.

3.1 Circular Pores: Variation of Size

Similar to the investigation in Bauer and Lackner (2013), Fig. 4a shows the mean value of f_t/f_t^M of the effective tensile strength distribution as a function of f_a for

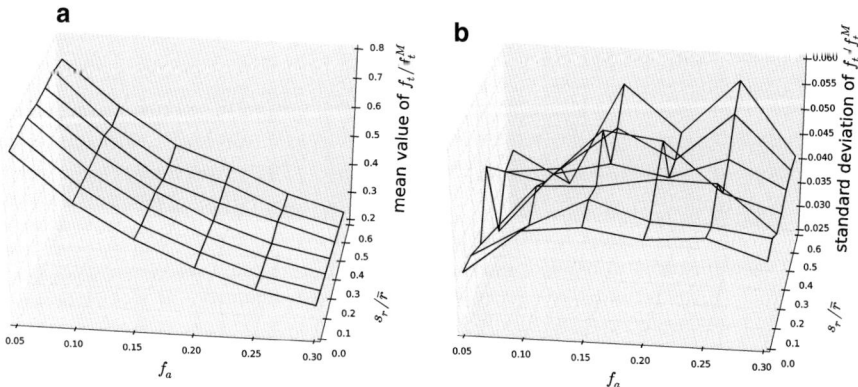

Fig. 4 Circular pores: (**a**) mean value and (**b**) standard deviation of f_t/f_t^M as a function of $s_r/\bar{r}$ and f_a (f_t^M: tensile strength of the matrix material)

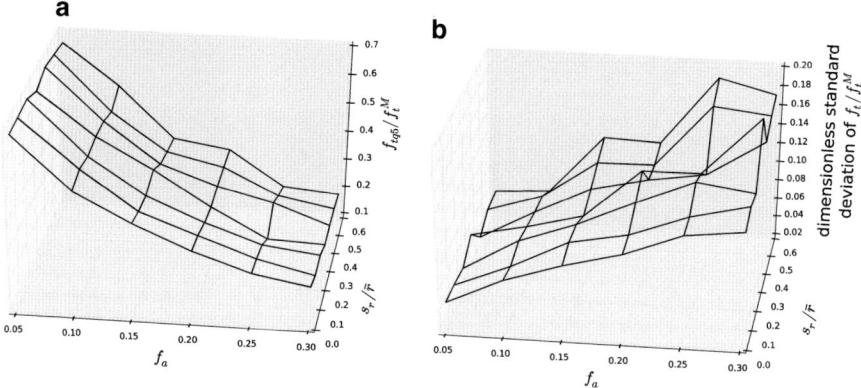

Fig. 5 Circular pores: (**a**) 5 % lower quantile value and (**b**) dimensionless standard deviation of f_t/f_t^M as a function of $s_r/\bar{r}$ and f_a (f_t^M: tensile strength of the matrix material)

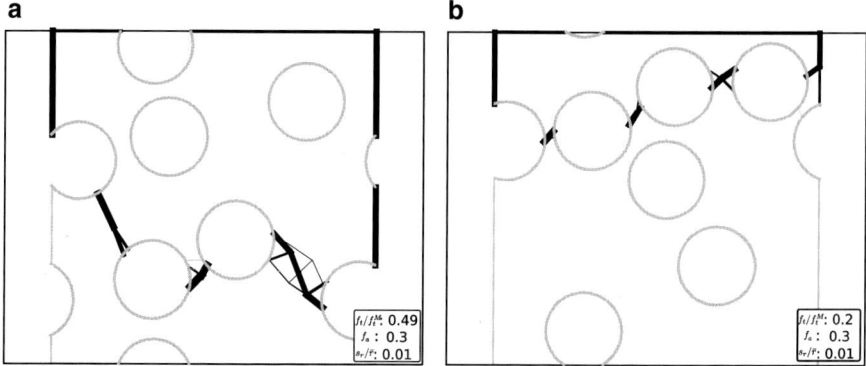

Fig. 6 Circular pores: failure mechanisms of RVEs with $f_a = 0.3$ and $s_r/\bar{r} = 0.01$, (**a**) high-strength and (**b**) low-strength configuration

different values of $s_r/\bar{r}$. As expected, strength properties decrease with increasing volume fraction of the pores f_a. The small values of the standard deviation of f_t/f_t^M (Fig. 4b), being one order of magnitude smaller than the mean values, results in an f_{tq5}/f_t^M distribution similar to the mean value (see Fig. 5a).

The dimensionless standard deviation increases almost linearly with increasing f_a, as illustrated in Fig. 5b, reflecting the increased variability in failure mechanisms introduced by the increasing pore space. The variation of the pore radii, on the other hand, has no influence of the dimensionless standard deviation of the strength properties. The failure mechanism corresponding to the maximum and minimum value of the effective strength of RVEs with $f_a = 0.3$, circular pores, and varying $s_r/\bar{r}$ are shown in Figs. 6 and 7.

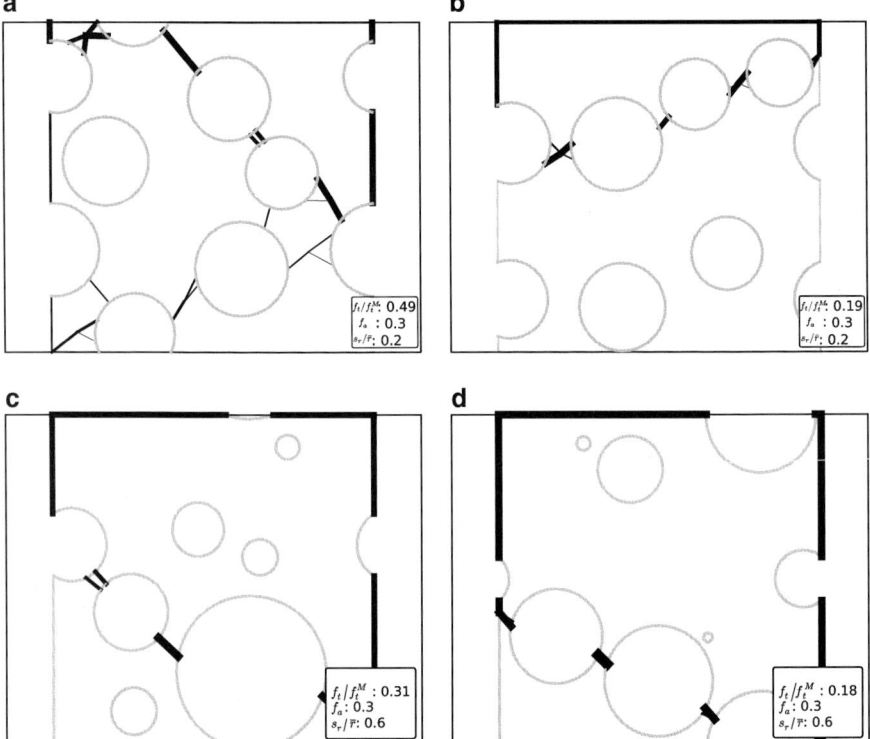

Fig. 7 Circular pores: failure mechanisms of RVEs with $f_a = 0.3$ and $s_r/\bar{r} = 0.2$, (**a**) high-strength and (**b**) low-strength configuration, and $s_r/\bar{r} = 0.6$, (**c**) high-strength and (**d**) low-strength configuration

3.2 Elliptical Pores: Variation of Volume Fraction

Figure 8a shows the mean value of the effective tensile strength distribution for the investigated porous RVEs with varying f_a and E. Hereby, the increase of both E and f_a result in a decrease of the effective strength properties. Both observations can be explained by the reduction of the effective cross-section area within the RVE, giving lower values for the effective strength. Similar to the previous study, the small values of the standard deviation of f_t/f_t^M (Fig. 8b), being one order of magnitude smaller than the mean values, result in an f_{tq5}/f_t^M distribution similar to the mean value (see Fig. 9a).

In Fig. 9b, the dimensionless standard deviation is given, showing an almost linear increase of the dimensionless standard deviation with increasing f_a and E. This behaviour reflects the increased variability in failure mechanism introduced by the increasing pore volume, on the one hand, and the additional degree of freedom introduced by the eccentricity, on the other hand. Exemplarily, the failure

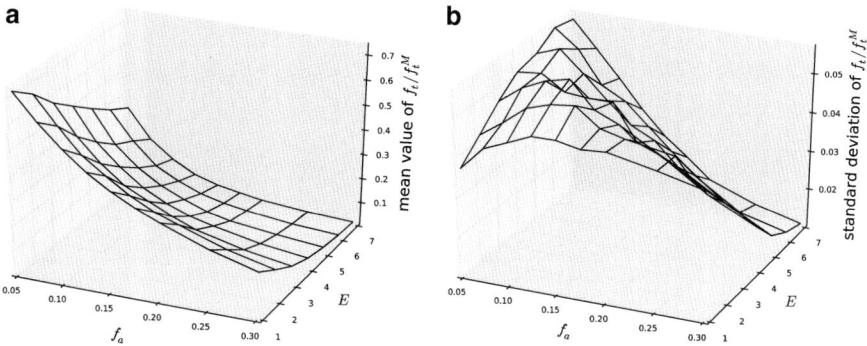

Fig. 8 Elliptical pores: (**a**) mean value and (**b**) standard deviation of f_t/f_t^M as a function of E and f_a (f_t^M: tensile strength of the matrix material)

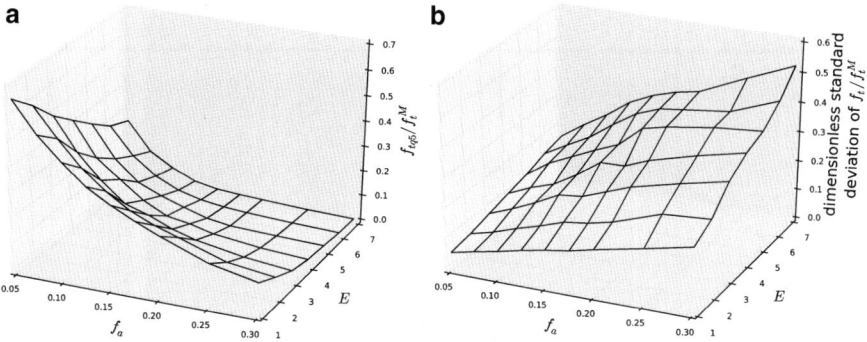

Fig. 9 Elliptical pores: (**a**) 5 % lower quantile value and (**b**) dimensionless standard deviation of f_t/f_t^M as a function of E and f_a (f_t^M: tensile strength of the matrix material)

mechanisms corresponding to the maximum (in case of low porosity) and minimum value (in case of high porosity) of the effective strength of RVEs with elliptical pores characterized by an eccentricity of $E = 7$ are shown in Fig. 10. Whereas the weakening effect on the cross-section in case for low porosities depends on the orientation and arrangement of the pores, a significant reduction of cross-section is obtained for almost all pore configurations in case of high porosities. The latter behaviour is reflected by the standard deviation given in Fig. 8b, showing its minimum value for high values of E and f_a.

Figures 11–13 show the failure mechanisms for high- and low-strength configurations for a volume fraction $f_a = 0.05$ with increasing E, illustrating the weakening effect of pores on the cross-section. For high values of E, this weakening effect strongly depends on the orientation and the arrangement of the pores, resulting in a high variation of strength properties (see Fig. 8b); for low values of E, the weakening effect depends mainly on the arrangement of pores, resulting in a lower variation of strength properties (see Fig. 9b).

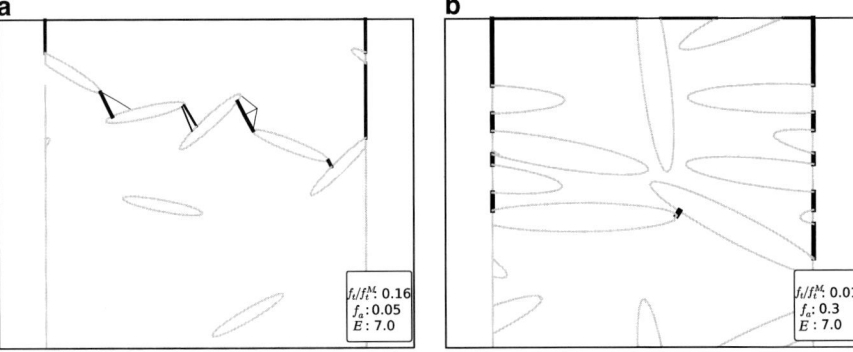

Fig. 10 Elliptical pores: failure mechanisms of RVEs with $E = 7$ with (**a**) $f_a = 0.05$ (maximum value for f_t/f_t^M) and (**b**) $f_a = 0.3$ (minimum value for f_t/f_t^M)

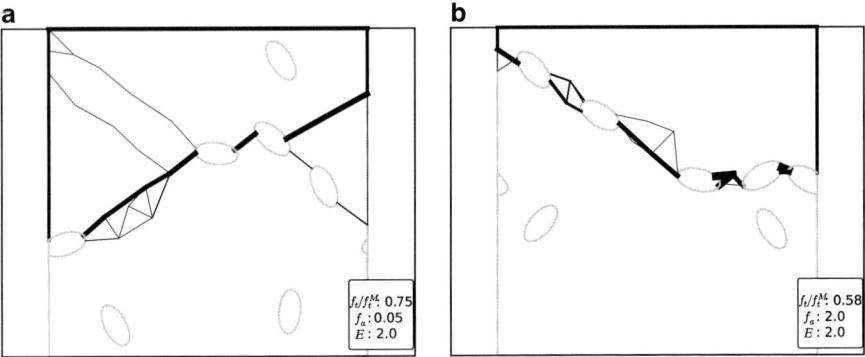

Fig. 11 Elliptical pores: failure mechanisms of RVEs with $E = 2$ and $f_a = 0.05$, (**a**) high-strength and (**b**) low-strength configuration

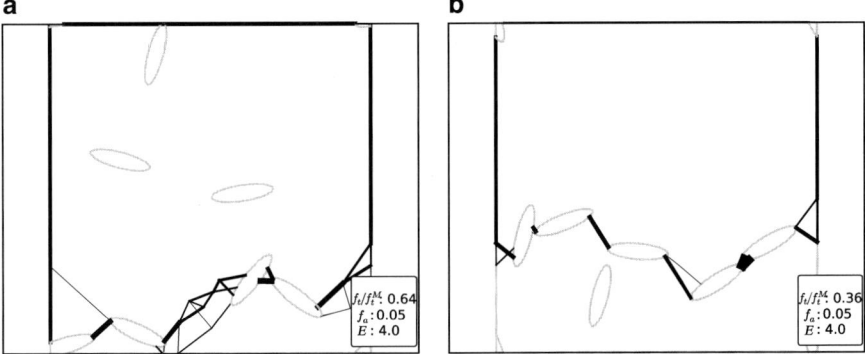

Fig. 12 Elliptical pores: failure mechanisms of RVEs with $E = 4$ and $f_a = 0.05$, (**a**) high-strength and (**b**) low-strength configuration

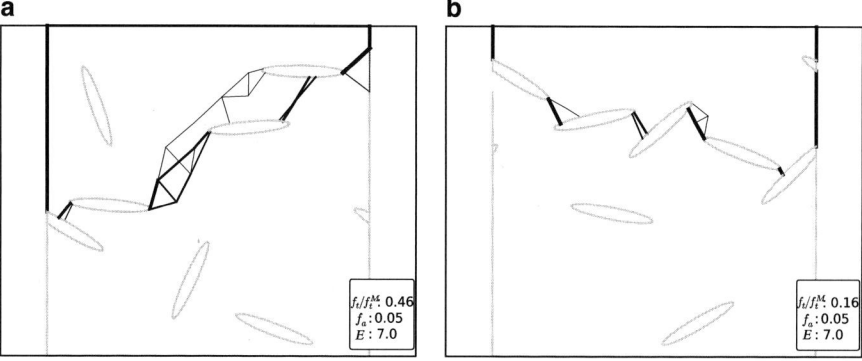

Fig. 13 Elliptical pores: failure mechanisms of RVEs with $E = 7$ and $f_a = 0.05$, (**a**) high-strength and (**b**) low-strength configuration

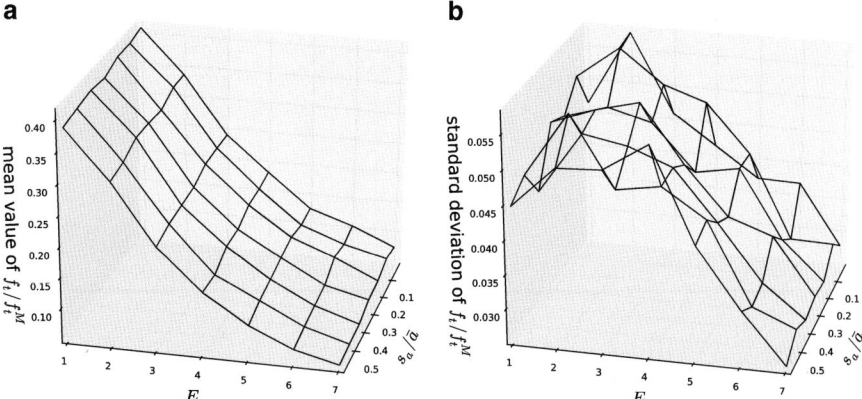

Fig. 14 Elliptical pores with $f_a = 0.15$: (**a**) mean value and (**b**) standard deviation of f_t/f_t^M as a function of $s_a/\bar{a}$ and E

3.3 Elliptical Pores: Variation of Size

Figures 14a and 16a show the mean value of f_t/f_t^M and the 5 % lower quantile value of the effective tensile strength distribution, f_{tq5}/f_t^M, as a function of E for different values of $s_a/\bar{a}$ considering a constant volume fraction of $f_a = 0.15$. As shown in the previous study, the strength properties decrease with increasing eccentricity of the pores E, which is explained by the decrease in the effective cross-section of the matrix material caused by longer pores (see Figs. 16 and 17). The standard deviation of f_t/f_t^M as a function of E and $s_a/\bar{a}$ is shown in Fig. 14b, following the same trend as the mean value for $E > 3$, decreasing with increasing E, while the variation of $s_a/\bar{a}$ has only a minor influence on the standard deviation

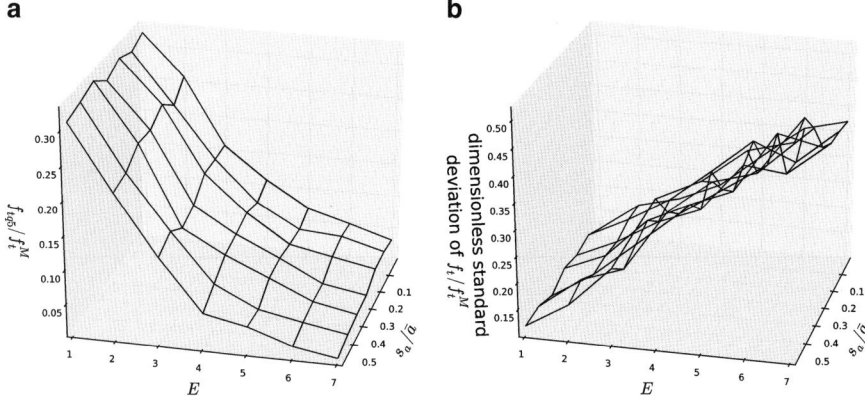

Fig. 15 Elliptical pores with $f_a = 0.15$: (**a**) 5 % lower quantile value and (**b**) dimensionless standard deviation of f_t/f_t^M as a function of $s_a/\bar{a}$ and E

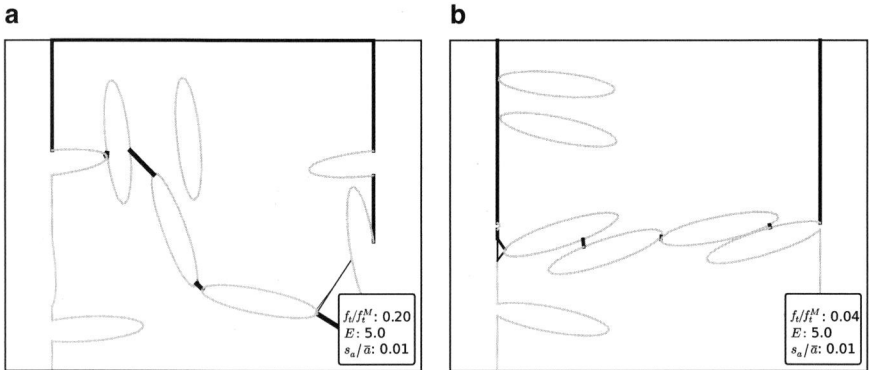

Fig. 16 Elliptical pores: failure mechanisms of RVEs with $f_a = 0.15$ and $s_a/\bar{a} = 0.01$, (**a**) high-strength and (**b**) low-strength configuration

of the tensile strength. The dimensionless standard deviation in Fig. 15b exhibits an almost linear increase with increasing E, where $s_a/\bar{a}$ shows again only a minor influence on the dimensionless standard deviation of f_t/f_t^M.

For low values of $s_a/\bar{a}$, the arrangement of the pores among one another influences the weakening of the cross-section (see Fig. 16); for high values of $s_a/\bar{a}$, on the other hand, the weakening of the cross-section is caused by larger pores (see Fig. 17) and depends on their arrangement. For both cases, the cross-sections and the length of the failure modes are similar, resulting in a similar standard deviation.

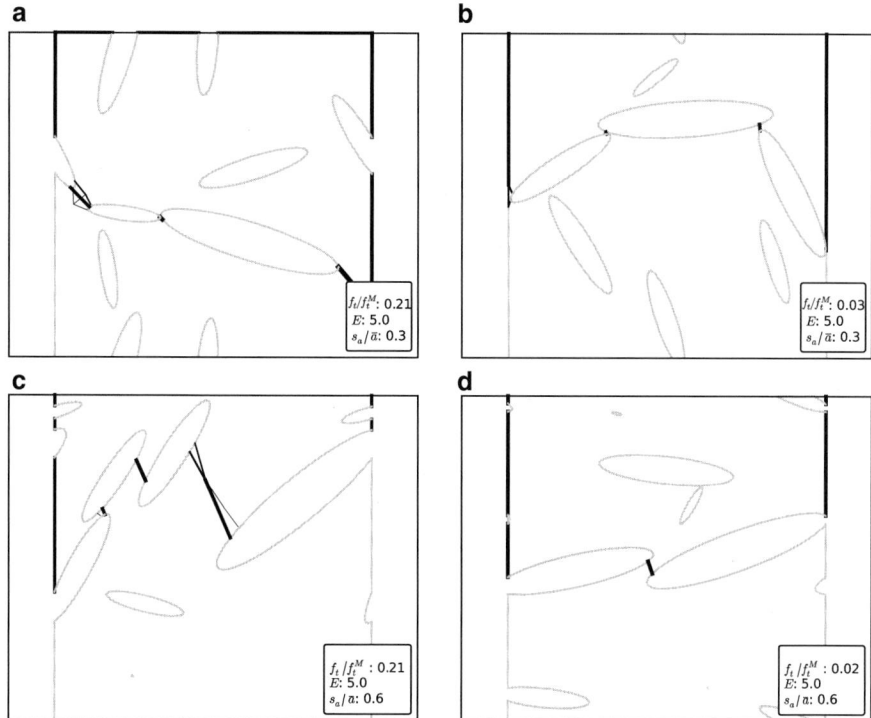

Fig. 17 Elliptical pores: failure mechanisms of RVEs with $f_a = 0.15$ and $s_a/\bar{a} = 0.3$, (**a**) high-strength and (**b**) low-strength configuration and $s_a/\bar{a} = 063$, (**c**) high-strength and (**d**) low-strength configuration

4 Conclusions and Outlook

In this work, a statistical framework using stochastic limit analysis was proposed for investigating the effect of randomly distributed/oriented circular and elliptical pores, on the effective strength properties, discontinuity layout optimisation was employed. Based on the results of the performed studies, the following conclusions can be drawn:

- The mean value of the effective strength of the considered heterogeneous porous material mainly depends on the length of the failure mode. An increase in length may be accomplished by the decrease of pore content and/or avoidance of elliptical pores with high eccentricity ratio.
- The 5 % lower quantile value showed similar dependence as the mean value, which was explained by the small values of the standard deviation of f_t/f_t^M, being one order of magnitude smaller than the respective mean values.

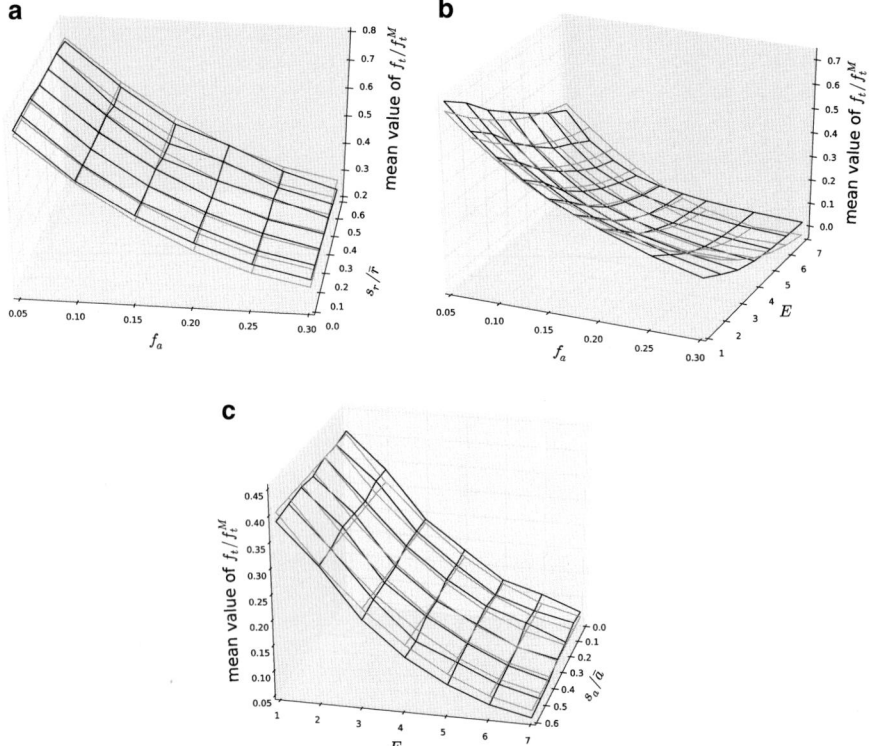

Fig. 18 Comparison of numerical solution and approximation: (**a**) circular pores: variation of size (Study 1), (**b**) elliptical pores: variation of volume fraction (Study 2), and (**c**) elliptical pores: variation of size (Study 3)

- A significant influence on the material properties caused by the variation of the size of the pores, while keeping the porosity and eccentricity constant, could not be observed.

Having identified the porosity and the eccentricity constant as the main two parameters controlling the effective strength, the influence of f_a and E on f_t/f_t^M is approximated by the following relations (see Fig. 18) using the least square regression:

$$\text{Study 1} : f_t/f_t^M = 5.5f_a^2 - 3.82f_a + 0.93 \tag{3}$$

$$\text{Study 2} : f_t/f_t^M = 8.19f_a^2 + 0.005E^2 - 4.5f_a - 0.09E + 0.99 \tag{4}$$

$$\text{Study 3} : f_t/f_t^M = 0.007E^2 - 0.11E + 0.51 \tag{5}$$

Fig. 19 Optimization strategies: (**a**) Case 1: reduction of f_a, $E =$ const., (**b**) Case 2: $f_a =$ const., reduction of E, (**c**) Case 3: reduction of both f_a and E

Table 1 Strength increase (in [%]) using different optimization strategies characterized by variation of f_a and E

f_a	E	f_a	E	Reduction of $f_a \downarrow 7.07\%$ and $E \downarrow 7.07\%$	Optimum
0.1	2	6.37	3.07	6.63	7.05 ($f_a \downarrow 8.91\%$ $E \downarrow 4.54\%$)
0.1	5	9.92	7.16	11.93	12.08 ($f_a \downarrow 7.99\%$ $E \downarrow 6.02\%$)
0.2	2	10.77	5.51	11.24	11.97 ($f_a \downarrow 8.99\%$ $E \downarrow 4.38\%$)
0.2	5	29.97	22.95	36.41	36.86 ($f_a \downarrow 7.99\%$ $E \downarrow 6.02\%$)

Alternatively, the relation representing the numerical results of Study 3 may be obtained from Eq. (4) by setting $f_a = 0.15$, giving

$$f_{t2}/f_t^M = 0.005E^2 - 0.09E + 0.50,$$ (6)

agreeing very well to the relation given in Eq. (5).

Figure 19 shows three optimization strategies for the improvement of the strength properties of the considered porous materials by varying pore-space properties: (i) reduction of pore content by Δf_a (Fig. 19a), (ii) reduction of eccentricity by ΔE (Fig. 19b), and (iii) a combination of the two cases (Fig. 19c). As a result of the optimization considering an overall change of pore-space properties Δf_a and ΔE of 10 %, reading

$$R = 10\% = 100\sqrt{\left(\frac{\Delta f_a}{f_a}\right)^2 + \left(\frac{\Delta E}{E}\right)^2},$$ (7)

the strength increase related to the original strength $\Delta f_t/f_t$ is given in Table 1. In the last column of Table 1, the maximum strength increase is given with the respective change of f_a and E given in parentheses.

Figure 20 shows the strength increase/decrease for a variation of f_a and E using the initial value of $f_a = 0.2$ and $E = 5$. Hereby, the light grey region indicates a strength increase, while the darkgrey region is characterized by a decrease of strength properties; additionally, two optimization paths are given as illustration.

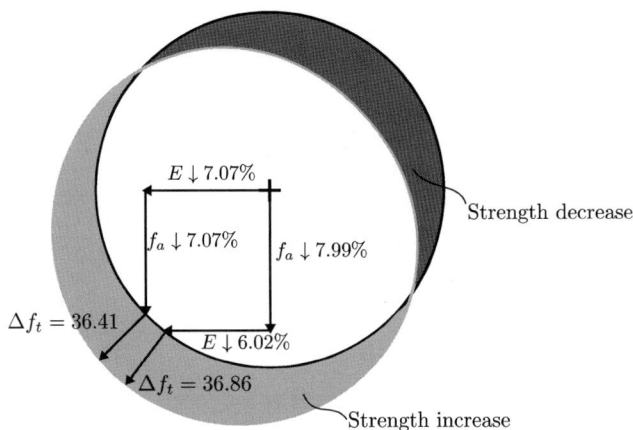

Fig. 20 Optimization for $f_a = 0.2$ and $E = 5$ for optimum ($\Delta f_t = 36.86$) and equal reduction ($\Delta f_t = 36.86$)

Acknowledgements The presented results were obtained within the research project 822 671 "Numerical model for predicting the strength evolution in cemented soil". Financial support by the Austrian Research Promotion Agency (FFG, Vienna, Austria) is gratefully acknowledged! Moreover, the authors thank Klaus Meinhard (Porr Technobau und Umwelt, Vienna, Austria) and Markus Astner (Geosystems Spezialbaustoffe GmbH, Gmunden, Austria) for fruitful discussions and helpful comments throughout the research work.

References

Al-Ostaz A, Diwakar A, Alzebdeh KI (2007) Statistical model for characterizing random microstructure of inclusion-matrix composites. J Mater Sci 42(16):7016–7030

Antretter T, Fischer FD (2001) The susceptibility to failure of the constituents of particulate two-phase composites. Int J Damage Mech 10(1):56–72

Bauer S, Lackner R (2011) Discontinuity layout optimization in upscaling of effective strength properties in matrix-inclusion materials. In: Proceedings of the 11th international conference on computational plasticity (COMPLAS XI), Barcelona, pp 1509–1516

Bauer S, Lackner R (submitted 2012, under review) Gradient-based adaptive discontinuity layout optimization for the prediction of strength properties in matrix-inclusion materials. Int J Solids Struct

Chakraborty A, Rahman S (2008) Stochastic multiscale models for fracture analysis of functionally graded materials. Eng Fract Mech 75(8):2062–2086

de Berg M, Cheong O, van Kreveld M, Overmars M (2008) Computational geometry: algorithms and applications, 3rd edn. Springer, Berlin

Díaz G, Kittl P, Rosales M (2003) Probabilistic design and quality control in probabilistic strength of materials. Int J Solids Struct 40(19):5001–5015

DIN EN 1990:2010-12 (2010) Eurocode: Grundlagen der Tragwerksplanung

Dong XN, Luo Q, Sparkman DM, Millwater HR, Wang X (2010) Random field assessment of nanoscopic inhomogeneity of bone. Bone 47(6):1080–1084

Du X, Ostoja-Starzewski M (2006) On the scaling from statistical to representative volume element in thermoelasticity of random materials. Netw Heterog Media 1(2):259–274

Frantziskonis GN (1998) Stochastic modeling of heterogeneous materials – a process for the analysis and evaluation of alternative formulations. Mech Mater 27:165–175

Graham-Brady L, Arwade S, Corr D, Gutiérrez M, Breysse D, Grigoriu M, Zabaras N (2006) Probability and materials: from nano- to macro-scale: a summary. Probab Eng Mech 21(3):193–199

Guilleminot J, Soize C (2012) Stochastic modeling of anisotropy in multiscale analysis of heterogeneous materials: a comprehensive overview on random matrix approaches. Mech Mater 44:35–46

Guilleminot J, Noshadravan A, Soize C, Ghanem R (2011) A probabilistic model for bounded elasticity tensor random fields with application to polycrystalline microstructures. Comput Methods Appl Mech Eng 200(17–20):1637–1648

Koutsourelakis P (2006) Probabilistic characterization and simulation of multi-phase random media. Probab Eng Mech 21(3):227–234

Koutsourelakis P (2007) Stochastic upscaling in solid mechanics: an excercise in machine learning. J Comput Phys 226(1):301–325

Liu Y, Steven Greene M, Chen W, Dikin DA, Liu WK (2013) Computational microstructure characterization and reconstruction for stochastic multiscale material design. Comput Aided Des 45(1):65–76

Maghous S, Dormieux L, Barthélémy J (2009) Micromechanical approach to the strength properties of frictional geomaterials. Eur J Mech A/Solids 28:179–188

Mehrez L, Doostan A, Moens D, Vandepitte D (2012) Stochastic identification of composite material properties from limited experimental databases, Part II: uncertainty modelling. Mech Syst Signal Process 27:484–498

Piat R, Tsukrov I, Mladenov N, Guellali M, Ermel R, Beck T, Schnack E, Hoffmann M (2006) Material modeling of the CVI-infiltrated carbon felt II. Statistical study of the microstructure, numerical analysis and experimental validation. Compos Sci Technol 66(15):2769–2775

Segurado J, LLorca J (2006) Computational micromechanics of composites: the effect of particle spatial distribution. Mech Mater 38(8–10):873–883

Smith C, Gilbert M (2007) Application of discontinuity layout optimization to plane plasticity problems. Proc R Soc A 463:2461–2484

Wriggers P, Moftah S (2006) Mesoscale models for concrete: homogenisation and damage behaviour. Finite Elem Anal Des 42(7):623–636

Homogenization of Random Heterogeneous Media with Inclusions of Arbitrary Shape

George Stefanou, Dimitris Savvas, Manolis Papadrakakis, and George Deodatis

Abstract In this paper, the effective properties of random heterogeneous (two-phase) media with arbitrarily shaped inclusions are computed in the framework of the extended finite element method (XFEM) coupled with Monte Carlo simulation (MCS). The implementation of XFEM is particularly suitable for this type of problems since there is no need to generate a new finite element mesh at each MCS. The inclusions are randomly distributed and oriented within the medium while their shape is implicitly modeled by the iso-zero of an analytically defined random level set function, which also serves as the enrichment function in the framework of XFEM. Homogenization is performed based on Hill's energy condition and MCS. The homogenization involves the generation of a large number of random realizations of the microstructure geometry based on a given volume fraction of the inclusions and other parameters (shape, spatial distribution and orientation). The influence of the inclusion shape on the effective properties of the random media is

G. Stefanou (✉)
Institute of Structural Analysis & Antiseismic Research, National Technical University of Athens, 9 Iroon Polytechneiou, Zografou Campus, Athens 15780, Greece

Institute of Structural Analysis & Dynamics of Structures, Aristotle University of Thessaloniki, 54124 Thessaloniki, Greece

Department of Civil Engineering & Engineering Mechanics, Columbia University, New York, NY 10027, USA
e-mail: stegesa@mail.ntua.gr

D. Savvas • M. Papadrakakis
Institute of Structural Analysis & Antiseismic Research, National Technical University of Athens, 9 Iroon Polytechneiou, Zografou Campus, Athens 15780, Greece

G. Deodatis
Department of Civil Engineering & Engineering Mechanics, Columbia University, New York, NY 10027, USA

M. Papadrakakis and G. Stefanou (eds.), *Multiscale Modeling and Uncertainty Quantification of Materials and Structures*, DOI 10.1007/978-3-319-06331-7_6,
© Springer International Publishing Switzerland 2014

highlighted. It is shown that the statistical characteristics of the effective properties can be significantly affected by the shape of the inclusions especially in the case of large volume fraction and stiffness ratio.

Keywords Random media • Homogenization • Level set • XFEM • Monte Carlo simulation

1 Introduction

The mechanical behavior of heterogeneous and in particular of composite materials is governed by the mechanical properties of their individual components, their volume fractions and other parameters defining their spatial and size distribution. Although only the macroscopic mechanical behavior is of interest in many cases, the microstructure attributes of this type of materials are extremely important for a better understanding of their intrinsic properties. This is the reason for which the linking of micromechanical characteristics with the random variation of material properties at the macro-scale has gained particular attention during the last years (e.g. Torquato 2002; Ostoja-Starzewski and Wang 1999; Kamiński and Kleiber 2000; Xu and Graham-Brady 2005; Yvonnet et al. 2008; Hiriyur et al. 2011; Ma et al. 2011; Greene et al. 2013; Clément et al. 2013). In this framework, it is possible to explore in detail the impact of each assumption made at the microlevel on the structural behavior of the macroscopic continuum.

A complete deterministic analysis of a heterogeneous medium taking its microstructure into account would involve excessive computational effort and may not be feasible even with today's available computing power. It is therefore necessary to approximate the complex microstructures with equivalent effective homogeneous material properties. To obtain the effective homogeneous properties of composite materials, various methods have been proposed, either analytical (e.g. Torquato 2002) or numerical (e.g. Miehe and Koch 2002; Yuan and Fish 2008; Charalambakis 2010; Geers et al. 2010; Ma et al. 2011). A stochastic approach to homogenization accounts for the fact that a particular microstructure is just a single sample realization from a specified spatial random process. Monte Carlo based stochastic homogenization involves the computational analysis of a large number of randomly generated realizations of the composite medium. The results from these analyses are used to derive the effective properties of an equivalent homogeneous medium and quantify their inherent uncertainties.

Classical finite element (FE) methods are commonly used to analyze complex microstructures. In this case, the mesh conforms to the internal material interface boundaries that cause the strong or weak discontinuities in the displacement solution field. While fast meshing algorithms are available to discretize a domain with such internal features, this step still involves a significant computational effort. This is especially true when large number of simulations are to be performed to quantify the probability distributions involved, with reasonable confidence. The development of the extended finite element method (XFEM) (Moës et al. 1999) offers the possibility to use a regular mesh which does not have to be adapted to the internal details (cracks or material interfaces) of each random realization of the microstructure. XFEM is therefore particularly suitable to model the local heterogeneous material structure in a representative volume element (RVE) for the application of homogenization techniques (Belytschko et al. 2009; Lian et al. 2013). It is worth noting that some issues have been recently identified with regard to the accuracy of the estimated local fluxes in the vicinity of the interface when XFEM is applied to diffusion problems in a multi-phase setup (Diez et al. 2013).

This paper deals with the homogenization of random heterogeneous (two-phase) media with arbitrarily shaped inclusions in the framework of XFEM coupled with Monte Carlo simulation (MCS). In particular, the influence of the inclusion shape on the effective properties of the random media is highlighted. The inclusions are randomly distributed and oriented within the medium and their shape is implicitly modeled by the iso-zero of an analytically defined random level set function, which also serves as the enrichment function in the framework of XFEM. The analytical function used is a random "rough" circle defined by a set of independent identically distributed (i.i.d.) random variables and deterministic constants governing the roughness of the shape (Stefanou et al. 2009). Homogenization is performed based on Hill's energy condition and MCS (Miehe and Koch 2002). The homogenization involves the generation of a large number of random realizations of the microstructure geometry based on a given volume fraction of the inclusions and other parameters (shape, spatial distribution and orientation). It is shown that the statistical characteristics of the effective properties can be significantly affected by the shape of the inclusions especially in the case of large volume fraction and stiffness ratio.

2   Modeling Inclusions of Arbitrary Shape with XFEM

2.1   Problem Formulation

Consider a medium which occupies a domain $\Omega \subset \mathbb{R}^2$ whose boundary is represented by Γ. Let prescribed traction $\bar{t}$ applied on surface $\Gamma_t \subset \Gamma$ (natural boundary conditions) and prescribed displacements $\bar{u}$ applied on $\Gamma_u \subset \Gamma$ (essential boundary conditions). The medium contains an inclusion which occupies the domain Ω^+ and is surrounded by the internal surface $\Gamma_{incl} \subset \Gamma$ such that $\Omega = \Omega^+ \cup \Omega^-$ and $\Gamma = \Gamma_t \cup \Gamma_u \cup \Gamma_{incl}$ (Fig. 1). The governing equilibrium and kinematic equations for the elastostatic problem of the medium ignoring the body forces is:

$$\text{div}\boldsymbol{\sigma} = 0 \quad \text{in } \Omega \tag{1a}$$

$$u = \bar{u} \quad \text{in } \Gamma_u \tag{1b}$$

$$\boldsymbol{\sigma} \cdot n = \bar{t} \quad \text{in } \Gamma_t \tag{1c}$$

$$[\![\boldsymbol{\sigma} \cdot n_{incl}]\!] = 0 \quad \text{in } \Gamma_{incl} \tag{1d}$$

Fig. 1 Schematic of a
medium which occupies a
domain $\Omega = \Omega^+ \cup \Omega^-$,
contains an inclusion (Ω^+)
and is subjected to essential
and natural boundary
conditions on surfaces Γ_u and
Γ_t respectively

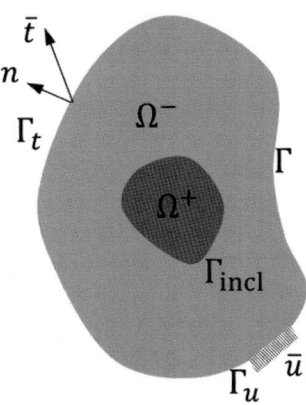

where n and n_{incl} are the unit normals to Γ_t and Γ_{incl}, respectively. Note that Eq. (1d) implies traction continuity along the material interface Γ_{incl}.

2.2   XFEM Discrete System

The weak form of the discrete problem, obtained from the differential equation (1a) through its transformation into a suitable variational form, can be stated as:

$$\text{find } u^h \in U^h \subset U \text{ such that } \forall v^h \in V^h \subset V,$$

$$a\left(u^h, v^h\right) = l\left(v^h\right) \tag{2}$$

where the trial function u and the test function v are represented as a linear combination of the same interpolation functions, and h stands for the characteristic size of the elements in the mesh. Note that to accurately capture a non-smooth solution resulting from material interfaces, the traditional FE method requires a mesh that conforms to the inclusion geometry. On the contrary, the XFEM eliminates the requirement of a conforming mesh by enriching the traditional FE approximation with a suitably constructed enrichment function. The XFEM displacement approximation for the trial and test functions can be decomposed into the standard FE part and the enriched part as follows:

$$u^h(\mathbf{x}) = u^h_{fem}(\mathbf{x}) + u^h_{enr}(\mathbf{x}) = \sum_{i \in I} N_i(\mathbf{x})\, u_i + \sum_{j \in J} N_j(\mathbf{x}) \{\psi(\mathbf{x}) - \psi(\mathbf{x}_j)\}\alpha_j$$

$$v^h(\mathbf{x}) = v^h_{fem}(\mathbf{x}) + v^h_{enr}(\mathbf{x}) = \sum_{i \in I} N_i(\mathbf{x})\, v_i + \sum_{j \in J} N_j(\mathbf{x}) \{\psi(\mathbf{x}) - \psi(\mathbf{x}_j)\}\beta_j$$

$$\tag{3}$$

where I is the set of all nodes in the mesh and J is the set of nodes that are enriched with the enrichment function ψ that satisfies the local character of the displacement field. A detailed description of the stochastic enrichment function that has been developed for arbitrarily shaped inclusions is provided in Sect. 2.3. To satisfy partition of unity, the enrichment function is enveloped by the original shape functions N_j and additional to the standard nodal variables u_i or v_i, enriched nodal variables α_j or β_j are introduced in the approximation equations for u^h or v^h, respectively. In order for the extended FE approximations to retain the Kronecker-δ property of the standard FE approximations so that at node j, $u^h(\mathbf{x}_j) = u_j$, a shifted enrichment function $S(\mathbf{x}) = \psi(\mathbf{x}) - \psi(\mathbf{x}_j)$ is used which was first suggested in Belytschko et al. (2001). By this shifting operator, the enrichment terms vanish at all nodes $j \in J$. Substituting Eq. (3) into the weak form of Eq. (2), we get a discrete system of algebraic equations:

$$\begin{bmatrix} K_{uu} & K_{u\alpha} \\ K_{\alpha u} & K_{\alpha\alpha} \end{bmatrix} \begin{bmatrix} u \\ \alpha \end{bmatrix} = \begin{bmatrix} F_u \\ F_\alpha \end{bmatrix} \tag{4}$$

where $[K_{uu}]_{ij} = a(N_i, N_j)$, $[K_{\alpha\alpha}]_{ij} = a(SN_i, SN_j)$ and $[K_{u\alpha}]_{ij} = [K_{\alpha u}]_{ji} = a(N_i, SN_j)$ are the stiffness matrices associated with the standard FE approximation, the enriched approximation and the coupling between them, respectively. The forces are expressed as $[F_u]_i = l(N_i)$ and $[F_\alpha]_j = l(SN_j)$. From the solution of the system, we finally obtain the nodal displacements u and enriched variables α.

2.3 Enrichment Function

Inclusions into a medium introduce a weak discontinuity in the displacement field (due to change in material properties) which shows a kink at the interface and a discontinuous first derivative. For modeling such fields in the framework of XFEM, usually a ramp function in the form of absolute distance function is used to enrich the approximation field (Krongauz and Belytschko 1998). XFEM is typically combined with the level set approach where a level set function ϕ is used to implicitly describe random inclusion geometry (Sukumar et al. 2001; Lang et al. 2013). While the level set method is often used to track moving interfaces on a fixed mesh (Sethian 1999), it is used herein to define the location of the inclusion interface and its stochastic variation. The location of the interface $\Gamma_{incl}(\theta)$ is implicitly defined by the iso-zero of the following random level set function representing a "rough" circle, which is taken as the signed radial distance function to the curve:

$$\phi(\mathbf{x}, \theta) = \|\mathbf{x} - \mathbf{c}\| - R(\alpha(\mathbf{x}), \theta) \tag{5}$$

where $\mathbf{x}$ is the spatial location of a point in the meshed domain, $\mathbf{c}$ is the center of the rough circle, $R(\alpha(\mathbf{x}), \theta)$ is a random field representing the radius of the rough circle, $\alpha(\mathbf{x}) \in [0, 2\pi]$ is the polar angle at position $\mathbf{x}$ and θ denotes the randomness of a quantity (Fig. 2).

Fig. 2 Schematic
representation of a rough
circle

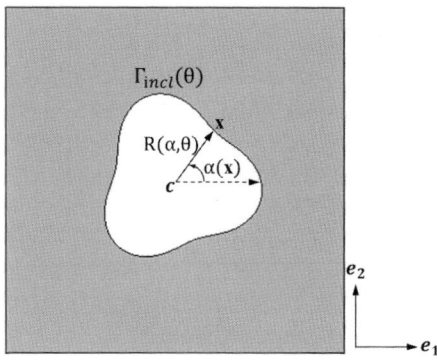

In this study, the following equation is used for the random radius (Stefanou et al. 2009):

$$R(\alpha, \theta) = 0.2 + 0.03Y_1(\theta) + 0.015\{Y_2(\theta)cos(k_1\alpha)$$
$$+ Y_3(\theta)sin(k_1\alpha) + Y_4(\theta)cos(k_2\alpha) + Y_5(\theta)sin(k_2\alpha)\} \quad (6)$$

where the i.i.d. uniform random variables $Y_i(\theta) \in U\left(-\sqrt{3}, \sqrt{3}\right)$, $i = 1, \ldots, 5$. k_1, k_2 are deterministic constants which define the period of oscillations of the random rough circle around the shape of the reference (perfect) circle. Finally the enrichment function which is used in Eq. (3) is formulated as the modified enrichment function proposed in Sukumar et al. (2001). This is the absolute value of the random level set function discretized according to the FE mesh of the spatial domain:

$$\psi(\mathbf{x}) = \left| \sum_{i \in I} N_i(\mathbf{x}) \phi_i \right| \quad (7)$$

where ϕ_i is the value of the level set function of Eq. (5) at node i and $N_i(\mathbf{x})$ are the FE nodal basis functions. Another choice for the enrichment function was introduced by Moës et al. (2003). This enrichment function has been used in Savvas et al. (2014) and leads to an improvement of the accuracy and convergence of the XFEM solution.

2.4 Convergence Study of XFEM Solution for Single Inclusion

In this section, three RVE models containing a single centered inclusion with different geometry are simulated both with the XFEM and standard FEM. Equations (5) and (6) are used for the construction of the inclusions where parameters (k_1, k_2) are chosen as $(0, 0)$, $(0, 3)$ and $(0, 6)$. All RVEs have a unit cell geometry with

Fig. 3 Schematic of
displacement type boundary
conditions on RVE model

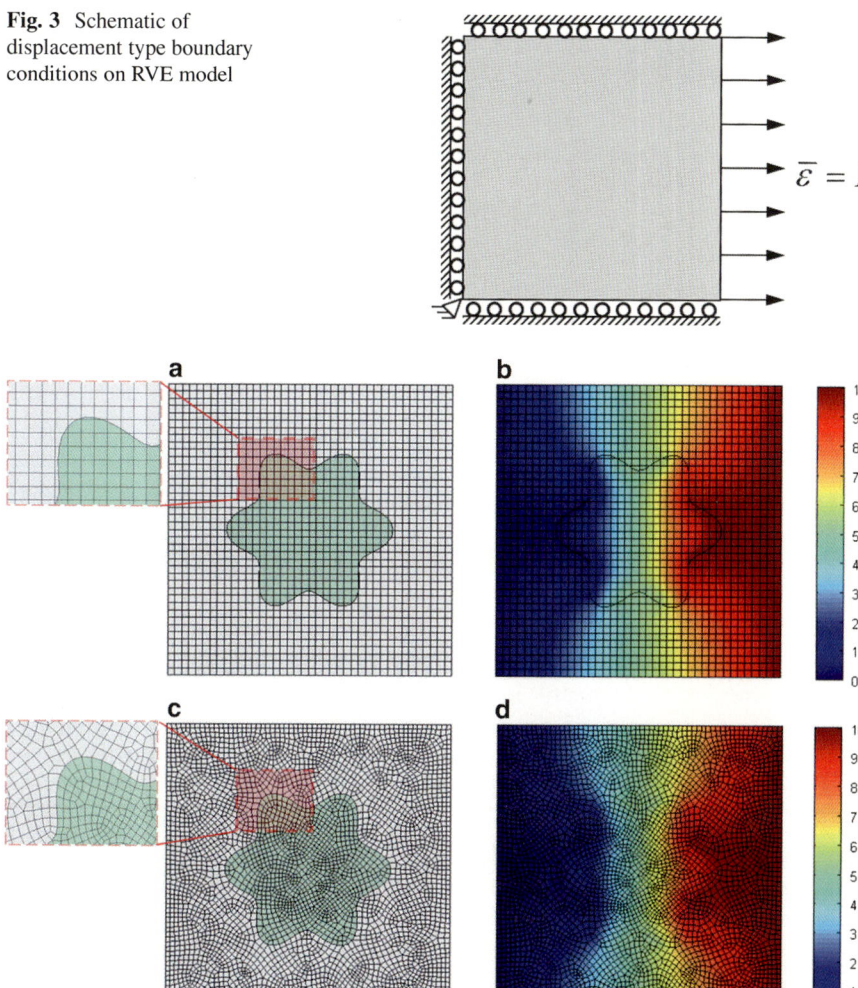

$\bar{\varepsilon} = 1$

Fig. 4 Comparison of displacement fields obtained from XFEM and FEM for RVE with inclusion
($k_1 = 0, k_2 = 6$): (**a**) XFEM mesh, (**b**) XFEM displacements, (**c**) FEM mesh and (**d**) FEM
displacements

dimensions 10×10 mm and a volume fraction (vf) of inclusions 30 %. In the
case of XFEM, a structured mesh of bilinear quadrilateral elements is used where
the inclusions are implicitly described through the enrichment function. The same
type of elements is used for FEM but in this case the mesh must conform to the
boundaries of the inclusions (see Fig. 4). The RVEs are subjected to displacement
type boundary conditions as shown in Fig. 3. A uniform strain $\bar{\varepsilon}$ is imposed on the
right edge of the models.

Fig. 5 Effect of elastic
moduli ratio on the accuracy
of the XFEM solution

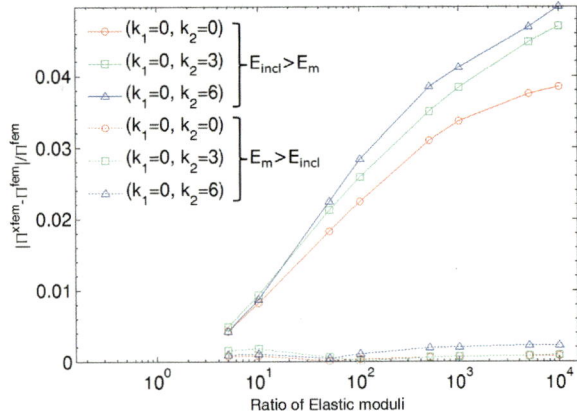

The matrix and the inclusions are modeled using linear elastic isotropic materials with Young's moduli $E_m = 1$ and $E_{incl} = 1{,}000$ GPa, respectively. The Poisson ratio for both materials is set equal to 0.3. Figure 4 displays the mesh and the displacement field obtained from XFEM and FEM for the RVE model with $k_1 = 0, k_2 = 6$. As it can be observed from the comparison of XFEM and FEM displacement fields, the results agree very well to each other. A convergence study with respect to the matrix-inclusion stiffness ratio is also conducted for the three RVE models described previously. Both stiff ($E_{incl} > E_m$) and compliant ($E_m > E_{incl}$) inclusions are considered for stiffness ratio values (E_{incl}/E_m or E_m/E_{incl}) ranging from 5 to 10,000. In Fig. 5, the relative difference between the XFEM and FEM solution in terms of strain energy Π is plotted against the corresponding ratio of elastic moduli. The relative error of the two solutions seems to increase as the stiffness ratio increases and is more significant for the case of stiff inclusions. For the elastic moduli ratio 10^3 used in the numerical examples (Sect. 4), the differences between XFEM and reference FEM solutions are at most 4.12 %.

3 Homogenization

3.1 Generation of Random Microstructures

In order to proceed to the stochastic homogenization procedure in the framework of MCS (Sect. 3.2), the first step is to generate a large number of random realizations of the microstructure geometry of the RVEs. For this purpose an efficient algorithm was used in Hiriyur et al. (2011), which has been appropriately modified here to account for arbitrarily shaped inclusions. A specific volume fraction (vf) and number of inclusions n_{incl} is assigned to each RVE with dimensions $X_1 \times X_2$. For the generation of arbitrarily shaped inclusions, Eqs. (5) and (6) are used with specific deterministic constants k_1, k_2 and random variables $Y_i(\theta)$ produced according to a

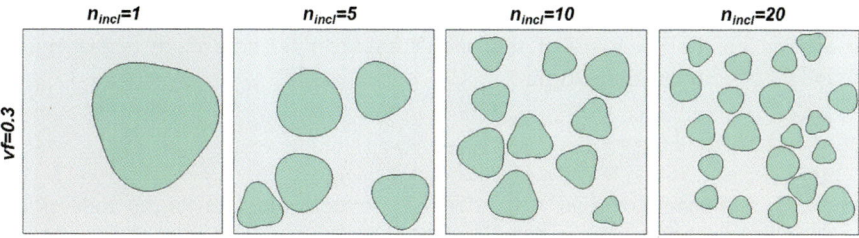

Fig. 6 Sample realizations of generated random microstructures with vf = 30 % and parameters $k_1 = 0$ and $k_2 = 3$

prescribed uniform PDF f_{Y_i}. The random boundary curve $\Gamma_{incl}(\theta)$ of an inclusion is constructed using N discrete points as follows:

$$\Gamma_{incl}(\theta) : \{\mathbf{x} = \mathbf{c} + R(\alpha, \theta)(cos\alpha \cdot \mathbf{e_1} + sin\alpha \cdot \mathbf{e_2})\} \qquad (8)$$

where $\mathbf{e_1}, \mathbf{e_2}$ are the unit vectors of the Cartesian coordinate system (Fig. 2). For each inclusion, a set of random coordinates representing the center of the rough circle and its random orientation angle are also generated according to prescribed uniform distributions f_c and f_β. In Fig. 6, a set of RVE realizations are shown, generated using the algorithm mentioned above. These RVEs have a volume fraction vf = 30 % and contain different number of inclusions with parameters $k_1 = 0$ and $k_2 = 3$.

3.2 Homogenization in the Framework of MCS

The homogenization scheme adopted in this paper is based on the fundamental assumption of statistical homogeneity of the heterogeneous medium (Hashin 1983) which means that all statistical properties of the state variables are the same at any material point and thus a representative volume element (RVE) can be identified. Effective homogeneous material properties, corresponding to the random microstructures generated by the algorithm in Sect. 3.1, are obtained by MCS. For this purpose, a sufficiently large number of elastic analyses are conducted where the RVEs are subjected to the displacement boundary conditions shown in Fig. 3. Although there is a constant homogenized material property within the RVE, this property changes from realization to realization making it a random variable. Assuming that the resulting homogeneous material will remain linear and isotropic, the effective Young's modulus E_{eff} and Poisson ratio ν_{eff} are the only parameters to be defined through the stochastic homogenization procedure.

Miehe and Koch (2002) proposed a computational procedure to exclusively define the overall macroscopic stresses and tangent moduli of a typical microstructure from the discrete forces and stiffness properties on the boundary nodes of the meshed RVE model. Following this procedure, a prescribed strain tensor $\bar{\varepsilon}$ is applied on the boundary of the microstructure models through displacement boundary conditions in the form:

$$u_q = \mathbb{D}_q^T \bar{\varepsilon} \tag{9}$$

where $\mathbb{D}_q$ is a geometric matrix that depends on the coordinates of the nodal point q which lies on the boundary of the model, defined by

$$\mathbb{D}_q = \frac{1}{2} \begin{bmatrix} 2x_1 & 0 \\ 0 & 2x_2 \\ x_2 & x_1 \end{bmatrix} \tag{10}$$

where $(x_1, x_2) \in Y$. The overall macroscopic stress $\bar{\sigma}$ is then calculated in an average manner from the nodal reaction forces f_q obtained by XFEM analysis as

$$\bar{\sigma} = \frac{1}{|V|} \sum_{q=1}^{M} \mathbb{D}_q f_q \tag{11}$$

where V is the volume of the RVE and M is the number of boundary nodes q. The macroscopic stress is related to the imposed macroscopic strain by a linear isotropic elastic constitutive matrix in the form

$$\begin{bmatrix} \bar{\sigma}_{11} \\ \bar{\sigma}_{22} \\ \bar{\sigma}_{12} \end{bmatrix} = \begin{bmatrix} C_{eff} & D_{eff} & 0 \\ D_{eff} & C_{eff} & 0 \\ 0 & 0 & G_{eff} \end{bmatrix} \begin{bmatrix} \bar{\varepsilon}_{11} \\ \bar{\varepsilon}_{22} \\ \bar{\varepsilon}_{12} \end{bmatrix} \tag{12}$$

where

$$C_{eff} = \begin{cases} \dfrac{E_{eff}}{1 - v_{eff}^2} & \text{plane stress} \\[3mm] \dfrac{(1 - v_{eff}) E_{eff}}{(1 + v_{eff})(1 - 2v_{eff})} & \text{plane strain} \end{cases} \tag{13}$$

$$D_{eff} = \begin{cases} \dfrac{v_{eff} E_{eff}}{1 - v_{eff}^2} & \text{plane stress} \\[3mm] \dfrac{v_{eff} E_{eff}}{(1 + v_{eff})(1 - 2v_{eff})} & \text{plane strain} \end{cases} \tag{14}$$

$$\text{and } G_{eff} = \frac{E_{eff}}{2(1 + v_{eff})} \tag{15}$$

The computation of the effective Young's modulus and Poisson ratio is accomplished by imposing the macrostrain vector $\bar{\varepsilon} = [1 \ 0 \ 0]^T$ in form of displacements (see Fig. 3). Thus $C_{eff} = \bar{\sigma}_{11}/\bar{\varepsilon}_{11}$ and $D_{eff} = \bar{\sigma}_{22}/\bar{\varepsilon}_{11}$ can be calculated from which E_{eff} and ν_{eff} are derived for each Monte Carlo sample.

4   Numerical Examples

The probability distribution of the effective elastic modulus and Poisson ratio for a plane-stress medium containing inclusions of arbitrary shape is obtained using the approach described in previous sections. As already stated, a linear isotropic material model is considered for both matrix and inclusions with Poisson ratio $\nu_m = \nu_{incl} = 0.3$. E_{eff}, ν_{eff} are computed through the coupled XFEM-MCS homogenization approach of Sect. 3.

A unit cell of size 10×10 mm subjected to the displacement boundary conditions shown in Fig. 3, is used in the analyses. To achieve statistical convergence, a total of 1,000 Monte Carlo simulations are performed for each volume fraction of inclusions considered ranging from 0.2 to 0.4. The number of inclusions in each MC sample is fixed to 15. Parametric investigations with respect to the stiffness ratio E_{incl}/E_m are conducted to highlight its effect on the results. It is noted that the computed E_{eff} is in all cases within the upper and lower bounds defined by the Voigt and Reuss models, respectively.

For stiffness ratio $E_{incl}/E_m = 10$ (stiff inclusions) or $E_m/E_{incl} = 10$ (compliant inclusions), the effect of the inclusion shape is negligible in all cases of shape roughness as the difference in the mean value of the effective elastic modulus between $(k_1 = 0, k_2 = 0)$ (perfect circle) and $(k_1 = 0, k_2 = 6)$ (arbitrary shape) is less than 5 % for both the stiff and compliant inclusions.

The influence of the inclusion shape on the effective properties becomes more pronounced in the case of large stiffness ratios. Figure 7 displays the histograms of E_{eff} for stiffness ratio $E_{incl}/E_m = 1,000$, three values of vf and different cases of inclusion shape roughness. The differences in the mean value of the effective elastic coefficient C_{eff} for various inclusion shapes and volume fractions are given in Table 1. An increase of about 30 % in C_{eff} can be observed in the case of vf=0.4 between $(k_1 = 0, k_2 = 0)$ and $(k_1 = 0, k_2 = 6)$. The shape of the histograms is significantly affected by the volume fraction and shape of the inclusions.

As shown in Fig. 8, a reduction of the effective elastic modulus occurs with the increase of shape roughness in the case of compliant inclusions ($E_m/E_{incl} = 1,000$). The effect of volume fraction and inclusion shape on the histograms of E_{eff} is less pronounced than in the case of stiff inclusions. The differences in the mean value of C_{eff} are still significant (Table 2). A decrease of about 16 % in C_{eff} can be noticed in the case of vf=0.4 between $(k_1 = 0, k_2 = 0)$ and $(k_1 = 0, k_2 = 6)$.

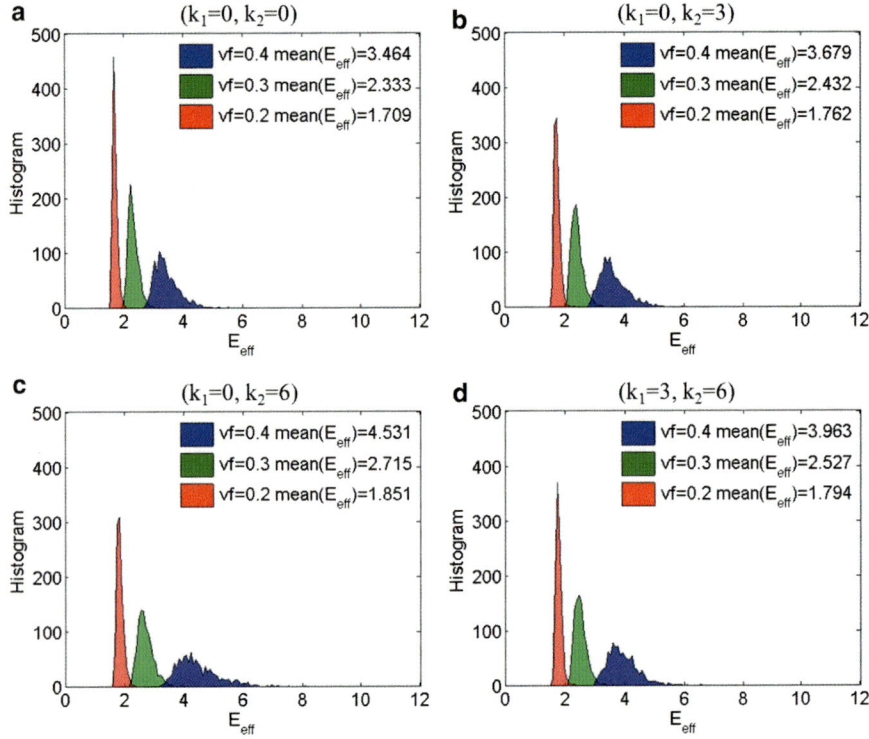

Fig. 7 Stiff inclusions ($E_{incl}/E_m = 1{,}000$) histograms of E_{eff}: (**a**) ($k_1 = 0, k_2 = 0$), (**b**) ($k_1 = 0, k_2 = 3$), (**c**) ($k_1 = 0, k_2 = 6$) and (**d**) ($k_1 = 3, k_2 = 6$) for vf = [0.2, 0.3, 0.4]

Table 1 Effect (% increase) of shape roughness (k_1, k_2) on mean(C_{eff}) for stiff inclusions ($E_{incl}/E_m = 1{,}000$)

vf	$(0,0) - (0,3)$	$(0,0) - (0,6)$	$(0,3) - (0,6)$	$(0,0) - (3,6)$
0.2	2.71	7.64	4.80	4.44
0.3	3.86	15.15	10.87	7.39
0.4	5.82	29.33	22.22	13.38

5   Conclusions

In this paper, the homogenization of random heterogeneous media with arbitrarily shaped inclusions was performed in the framework of XFEM coupled with Monte Carlo simulation. In particular, the influence of the inclusion shape on the effective properties of the random media was studied. The inclusions were randomly distributed and oriented within the medium and their shape was implicitly modeled by the iso-zero of an analytically defined random level set function ("rough" circle), which also served as the enrichment function in the framework of

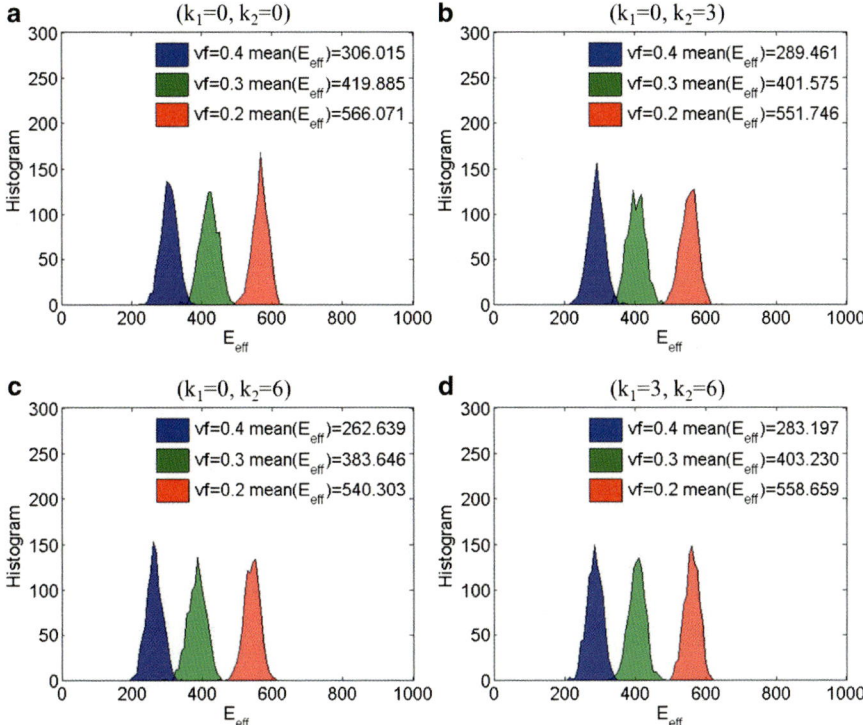

Fig. 8 Compliant inclusions ($E_m/E_{incl} = 1{,}000$) histograms of E_{eff}: (**a**) ($k_1 = 0, k_2 = 0$), (**b**) ($k_1 = 0, k_2 = 3$), (**c**) ($k_1 = 0, k_2 = 6$) and (**d**) ($k_1 = 3, k_2 = 6$) for vf = [0.2, 0.3, 0.4]

Table 2 Effect (% decrease) of shape roughness (k_1, k_2) on mean(C_{eff}) for compliant inclusions ($E_m/E_{incl} = 1{,}000$)

vf	$(0,0) - (0,3)$	$(0,0) - (0,6)$	$(0,3) - (0,6)$	$(0,0) - (3,6)$
0.2	3.26	5.03	1.83	1.75
0.3	5.39	9.57	4.42	4.78
0.4	7.00	15.95	9.62	8.91

XFEM. The formulation exploits the characteristic features of XFEM avoiding the regeneration of a new finite element mesh at each Monte Carlo simulation leading to accelerated computations. Parametric investigations with respect to the inclusion-matrix stiffness ratio and the inclusion volume fraction have been conducted. The numerical results have shown that the statistical characteristics of the effective properties can be significantly affected by the shape of the inclusions especially in the case of large volume fraction and stiffness ratio.

Acknowledgements This work is implemented within the framework of the research project "MICROLINK: Linking micromechanics-based properties with the stochastic finite element method: a challenge for multiscale modeling of heterogeneous materials and structures" – Action "Supporting Postdoctoral Researchers" of the Operational Program "Education and Lifelong Learning" (Action's Beneficiary: General Secretariat for Research and Technology), and is co-financed by the European Social Fund (ESF) and the Greek State. The provided financial support is gratefully acknowledged. M. Papadrakakis acknowledges the support from the European Research Council Advanced Grant "MASTER–Mastering the computational challenges in numerical modeling and optimum design of CNT reinforced composites" (ERC-2011-ADG 20110209). Special thanks are also due to Professor Haim Waisman for providing the computer code of the XFEM model for inclusions of elliptical shape.

References

Belytschko T, Moës N, Usui S, Parimi C (2001) Arbitrary discontinuities in finite elements. Int J Numer Methods Eng 50(4):993–1013

Belytschko T, Gracie R, Ventura G (2009) A review of extended/generalized finite element methods for material modeling. Model Simul Mater Sci Eng 17(4):1–24

Charalambakis N (2010) Homogenization techniques and micromechanics: a survey and perspectives. Appl Mech Rev 63(3):803

Clément A, Soize C, Yvonnet J (2013) Uncertainty quantification in computational stochastic multiscale analysis of nonlinear elastic materials. Comput Methods Appl Mech Eng 254:61–82

Diez P, Cottereau R, Zlotnik S (2013) A stable extended fem formulation for multi-phase problems enforcing the accuracy of the fluxes through lagrange multipliers. Int J Numer Methods Eng 96:303–322

Geers M, Kouznetsova V, Brekelmans W (2010) Multi-scale computational homogenization: trends and challenges. J Comput Appl Math 234(7):2175–2182

Greene MS, Xu H, Tang S, Chen W, Liu WK (2013) A generalized uncertainty propagation criterion from benchmark studies of microstructured material systems. Comput Methods Appl Mech Eng 254:271–291

Hashin Z (1983) Analysis of composite materials: a survey. J Appl Mech 50(2):481–505

Hiriyur B, Waisman H, Deodatis G (2011) Uncertainty quantification in homogenization of heterogeneous microstructures modeled by XFEM. Int J Numer Methods Eng 88(3):257–278

Kamiński M, Kleiber M (2000) Perturbation based stochastic finite element method for homogenization of two-phase elastic composites. Comput Struct 78(6):811–826

Krongauz Y, Belytschko T (1998) EFG approximation with discontinuous derivatives. Int J Numer Methods Eng 41(7):1215–1233

Lang C, Doostan A, Maute K (2013) Extended stochastic fem for diffusion problems with uncertain material interfaces. Comput Mech 51:1031–1049

Lian W, Legrain G, Cartraud P (2013) Image-based computational homogenization and localization: comparison between x-fem/levelset and voxel-based approaches. Comput Mech 51:279–293

Ma J, Temizer I, Wriggers P (2011) Random homogenization analysis in linear elasticity based on analytical bounds and estimates. Int J Solids Struct 48(2):280–291

Miehe C, Koch A (2002) Computational micro-to-macro transitions of discretized microstructures undergoing small strains. Arch Appl Mech 72(4–5):300–317

Moës N, Cloirec M, Cartraud P, Remacle JF (2003) A computational approach to handle complex microstructure geometries. Comput Methods Appl Mech Eng 192(28):3163–3177

Moës N, Dolbow J, Belytschko T (1999) A finite element method for crack growth without remeshing. Int J Numer Methods Eng 46:131–150

Ostoja-Starzewski M, Wang X (1999) Stochastic finite elements as a bridge between random material microstructure and global response. Comput Methods Appl Mech Eng 168(1):35–49

Savvas D, Stefanou G, Papadrakakis M, Deodatis G (2014) Homogenization of random heteroge-neous media with inclusions of arbitrary shape modelled by XFEM. Comput Mech (submitted)

Sethian JA (1999) Level set methods and fast marching methods, evolving interfaces in com-putational geometry, fluid mechanics, computer vision, and materials science. Cambridge University Press, Cambridge/New York

Stefanou G, Nouy A, Clement A (2009) Identification of random shapes from images through polynomial chaos expansion of random level set functions. Int J Numer Methods Eng 79(2):127–155

Sukumar N, Chopp DL, Moës N, Belytschko T (2001) Modeling holes and inclusions by level sets in the extended finite-element method. Comput Methods Appl Mech Eng 190(46):6183–6200

Torquato S (2002) Random heterogeneous materials: microstructure and macroscopic properties. Springer, New York

Xu XF, Graham-Brady L (2005) A stochastic computational method for evaluation of global and local behavior of random elastic media. Comput Methods Appl Mech Eng 194(42):4362–4385

Yuan Z, Fish J (2008) Toward realization of computational homogenization in practice. Int J Numer Methods Eng 73(3):361–380

Yvonnet J, Quang HL, He QC (2008) An XFEM/level set approach to modelling surface/interface effects and to computing the size-dependent effective properties of nanocomposites. Comput Mech 42(1):119–131

Part III
Inverse Problems–Identification

Using Experimentally Determined Resonant Behaviour to Estimate the Design Parameter Variability of Thermoplastic Honeycomb Sandwich Structures

Stijn Debruyne and Dirk Vandepitte

Abstract Honeycomb panels combine a high specific strength and stiffness with a low areal mass. Consequently, these structures are ideally suited for ground transportation vehicle purposes. They have a complex but regular geometry.

This paper describes the full process of estimating the variability of some of the panel design parameters of thermoplastic honeycomb structures. The uncertainty of the various stiffness parameters of the core and skin is estimated from the experimentally determined modal behaviour of a set of honeycomb beam and panel samples. This work thus deals with uncertainty quantification by considering an inverse problem. Variability analysis are carried out at different scales in order to obtain a full scope of the impact and origin (from the manufacturing process) of honeycomb design parameter variability.

Keywords Thermoplastic honeycomb sandwich panels • Finite element model update • Random field modelling • Epistemic uncertainty

1 Introduction

The use of composite materials has increased enormously during the past decades. Rising energy costs, finite material resources and a demanding market drive designers to shift from traditional materials to 'tailor made' structural materials

S. Debruyne
Department of Mechanical Engineering, KU Leuven Kulab, Zeedijk 101, 8400 Ostend, Belgium
e-mail: stijn.debruyne@kuleuven.be

D. Vandepitte (✉)
Department of Mechanical Engineering, KU Leuven, Celestijnenlaan, 300b,
3001 Heverlee, Belgium
e-mail: dirk.vandepitte@mech.kuleuven.be

M. Papadrakakis and G. Stefanou (eds.), *Multiscale Modeling and Uncertainty Quantification of Materials and Structures*, DOI 10.1007/978-3-319-06331-7__7,
© Springer International Publishing Switzerland 2014

Fig. 1 Honeycomb panel
structure

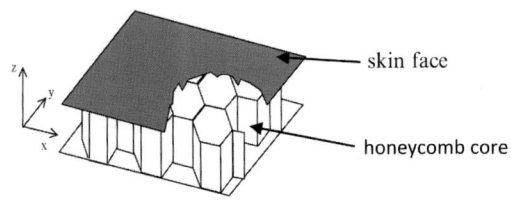

like composites, laminates and sandwiches (Zenkert 1997). Within this group of materials, honeycomb sandwich structures are widely used and appreciated.

Honeycombs are geometrically complex but regular structures. They consist of a honeycomb core that is bonded to skin face sheets. The structure of a typical honeycomb panel is shown in Fig. 1.

The honeycomb core is most often very light and only has significant strength and stiffness in the out-of-plane direction of a considered panel (Gibson and Ashby 1988). The skin faces are usually thin and have high membrane strength and stiffness. The combination of these types of core and skin yields a panel structure with a low areal mass but yet a very high specific bending stiffness.

Thermoplastic honeycomb panels consist of a thermoplastic core and skin. In this research, Monopan® (http://www.monopan.nl/) sandwich panels are used as test and modelling objects. These panels are entirely made from thermoplastic polypropylene (pp). The honeycomb core has cylindrical tubular cells and is attached to the skin faces by fusion bonding, using a welding foil. To increase the strength and stiffness of the skin faces, these are glass fibre weave reinforced using balanced 2/2 twintex® twill. The outer surfaces of these panels are smoothened by means of a pp finishing sheet.

In the last decade much research is carried out on the analytical prediction of natural frequencies of honeycomb beams and panels and in the finite element numerical modelling of the dynamic deformation of honeycomb structures. The elastic mechanical properties of a typical honeycomb core are described and analytically calculated by Gibson and Ashby (1988). They propose formulae for the calculation of the in-plane and out-of-plane elastic moduli and Poisson ratios of the core. The main work on the dynamics of sandwich panels is related to conventional foam-core structures. Nilsson and Nilsson (2002) tried to analytically predict natural frequencies of a honeycomb sandwich plate with free boundary conditions using Blevins (1984) formula in which areal mass and equivalent bending stiffness are frequency dependent. Another, more practical way to predict natural frequencies and mode shapes of a honeycomb panel is by means of finite element analysis (FEA). In recent years, different new approaches have been developed which incorporate high order shear deformation of the core. Work in this area has been carried out by Topdar (2003) and Liu (2001, 2002, 2007). The latter stated that the shear moduli of the core are important factors in the determination of the natural frequencies and the sequence of mode shapes, especially at high frequencies. At low frequencies natural frequencies are mostly determined by the bending stiffness of the panel.

The use of vibration measurement data for the identification of elastic material properties is studied by Lauwagie (2005). This work discusses in detail how Young's moduli, shear moduli and Poisson ratios of laminated materials can be obtained from modal data as resonance frequencies and mode shapes.

However, when modelling complex structures with many design variables there is always a discrepancy between the outcome of the numerical predictions and the experimental observations. In real life design parameters of a structure exert some variability evolving from the manufacturing process. Once the product is manufactured the parameter variability is untraceable in many cases. This of course reflects in some uncertainty on the elastic behaviour of the considered structure, static or dynamic. Since there are many uncertain design parameters involved with honeycomb panels, their dynamic behaviour can be regarded as an uncertain or stochastic process (Schuëller and Pradlwarter 2009). In recent years there has been a growing interest in the relation between the outcome of stochastic processes and their underlying governing parameters (Schenk and Schuëller 2005). Different analyzing methods have been developed to estimate the statistics of stochastic processes, systems and structural behaviour. In this area, little research is done on the uncertainty of the dynamic behaviour of honeycomb sandwich structures and the relation with the scatter on the various design variables involved.

The first part of this paper highlights the basic terminology on uncertainty quantification. It outlines the different types of uncertainty studied in this research and motivates the use of the random field method.

The second part briefly describes the test samples that are used throughout the research. It indicates that some of the beam and panel parameters show a significant scatter which influences the dynamic behaviour of the structures.

Part three deals with the process of experimental modal analysis (EMA). Considerations on boundary conditions and excitation methods are made and the process of modal parameter estimation is outlined.

Part four deals with the numerical modelling of the considered dynamic behaviour. The choice of the shell-volume-shell finite element model (SVS) as a homogenization method for the real honeycomb structure is motivated. Results of the numerically calculated resonance frequencies and mode shapes for the considered honeycomb beams and panels are discussed and compared with their experimental counterparts. Furthermore, this part discusses the updating process of the finite element models. This part also discusses the uncertainty involved and the need for stochastic model updating.

The sixth part discusses the results of applying the random field method to the various databases obtained through finite element model updating. For both skin and core of the considered beams and panels the relation between their real geometrical parameters and the parameters of the homogenized models is discussed in detail.

The last part summarizes the work and general conclusions are made. It discusses the shortcomings of the work done and gives some research prospects.

2 Uncertainty Quantification

When analyzing or simulating a process or system, one should always consider
the impact of uncertain parameters. Uncertainty quantification (UQ) (Loeve 1997;
Moens and Vandepitte 2005a; Pradlwarter and Schuëller 1997; Schuëller and
Pradlwarter 2009) deals with the propagation of process parameter variability
through the process mechanism. Recently, there is a growing interest in studying the
uncertainty of process parameters through observations of the process variability.
Such an inverse UQ is addressed in this research.

Related to this research, uncertainty can generally be divided in two different
types. The first kind of uncertainty is uncertainty that can be reduced as more
knowledge about the process or its governing parameters becomes available. This is
referred to as the epistemic uncertainty. In this research, the epistemic uncertainty
is directly related to the limited amount of experimental data that is available.
This is both due to the low number of test specimens and the limited number of
measurement points on a specific sample. The second type of uncertainty is the
physical or intrinsic uncertainty. This is also referred to as aleatory uncertainty. This
uncertainty is inherent to the physics of any process or system. Neither an exact
process model nor the exact governing parameters of the process exists. Unlike the
epistemic uncertainty, aleatory uncertainty cannot be reduced by gathering more
information on the process of its parameters.

Uncertainty quantification (UQ) studies the impact of uncertain data and errors
to end up with more reliable predictions of physical problems.

Experimental results and process observations always show some random characteristics. Statistical and probabilistic methods are developed for a rational treatment and analysis of these uncertainties. According to Schenk and Schuëller (2005;
Pradlwarter and Schuëller 1997), different types of uncertainty exist. Modelling
uncertainties can be reduced by gathering more knowledge on the physics of the
process or system. The physical uncertainties, inherent to the system or process,
cannot be reduced through this however. Uncertainty on material properties and
boundary conditions, geometry imperfections and load fluctuations are most often
not considered or not known. Each uncertain parameter is marked by a probability
of occurrence, quantified by a probability density function (PDF). Its characteristics,
such as mean value and variance, can be estimated by statistical procedures.
Many methods have been developed to deal with uncertainty. The most prominent
approach is to treat the deterministic system or process to be a stochastic one
(Schuëller and Pradlwarter 2009; Robert and Casella 2004; Yang 2009; Cooker
1990).

The generalized polynomial chaos method (GPC) is a recently developed generalization of the classical polynomial chaos (Ghanem 1991). This method expresses
stochastic process or system characteristics as orthonormal polynomials. It comes
down to a spectral projection (Schenk and Schuëller 2005) of the process or system
in a random space. Generalized polynomial chaos is a widely used method and will
be used in this research as a tool for stochastic modelling.

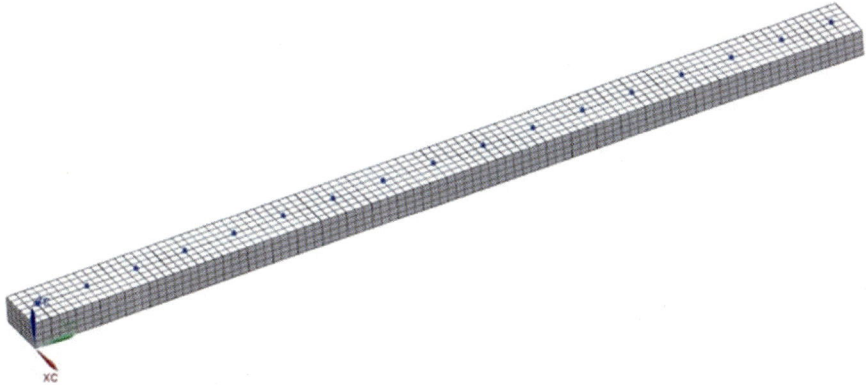

Fig. 2 Indication of a zone in a honeycomb beam sample

A random field (RF) can be regarded as an extension of a stochastic process. The index parameter t of expression (1) is related to a one-dimensional space (e.g. time). In case of random fields this index is expanded to multi-dimensional spaces.

$$
F(t), t \in D \subseteq \mathfrak{R}^n, t = \begin{Bmatrix} t_1 \\ t_2 \\ \vdots \\ t_n \end{Bmatrix} \tag{1}
$$

In this research, the quantities of interest are the elastic properties of the homogenized skin and core of honeycomb sandwich beams and plates. The dynamic behaviour of the considered honeycomb structures can be regarded as a stochastic process, governed by a set of random variables which are the various elastic constants. Recently, advanced methods using the random field methodology have been developed. Ghanem (1991, 2006), Desceliers and Soize (Desceliers et al. 2007; Arnst et al. 2010; Soize 2010) have contributed much to the development of these methods.

The low number of available test structures induces a high epistemic uncertainty. Mehrez and Doostan (2012a, b) applied methods to accurately estimate this uncertainty. The purpose of applying the random field method here is twofold. The first concern is to estimate the true probability density distributions of the considered databases and to exclude all variability that is not directly related to the real physically related variability. The second purpose is to estimate the epistemic uncertainty caused by the limited amount of experimental data available. The various parameters of interest are modelled as independent random fields. For example, consider the results of vibration tests on a set of honeycomb beams. The set consists of 22 specimens of the type, shown in Fig. 2.

Each specimen is considered to be a one-dimensional structure divided in 17 equal zones. For each considered design parameter the updated finite element model yields a database, with dimensions $M \times N_{\exp}$, with M being the sample size and $N_{\exp}$ the number of intervals in a specimen. Each database represents the discrete values of a stochastic random field.

The covariance matrix $[C_F]$ of such a database is spectrally decomposed using the Karhunen-Loève (KL) series expansion (Schenk and Schüller 2005). Therefore the eigenvalues and corresponding eigenvectors of $[C_F]$ need to be calculated, leading to the corresponding eigenvalue system described by (2).

$$[C_{\tilde{F}}]\varphi_k = \lambda_k \varphi_k, \forall k : 1 \leq k \leq N_{\exp} \tag{2}$$

The KL-series expansion (3) expresses the variability $\overline{F} - \underline{\tilde{F}}$ of the discretized random field $\overline{F}$ against its mean $\underline{\tilde{F}}$ using the eigenvalues and eigenvectors of the covariance matrix and using a finite set of random variables $\eta^{(k)}$ with zero mean and generally non Gaussian distribution.

$$\overline{F} - \underline{\tilde{F}} = \sum_{k=1}^{N_{\exp}} \eta^{(k)} \sqrt{\lambda_k} \varphi_k \tag{3}$$

For reasons of computational efficiency this series expansion is truncated after a number of terms that is smaller than $N_{\exp}$. Based on a convergence study of the normalized sum of the set of eigenvalues, the series can be truncated after μ terms. In this study this number is chosen so that the first μ eigenvalues of the covariance matrix cover at least 95 % of the variance.

Modelling the random field comes down to determining the statistics of the random vector $\eta^\mu = \{\eta^{(k)}\}$ which has a zero mean and the $\mu \times \mu$ identity matrix as covariance matrix. The joint density of η^μ however is not known and has to be estimated from the available random field realisations (experimental data). This estimation process uses a Hermite polynomial chaos (PC) expansion with Bayesian Inference (Schenk and Schüller 2005; Schuëller and Pradlwarter 2009; Ghanem 1991, 2006; Desceliers et al. 2007; Arnst et al. 2010). Expression (4) estimates the samples $\widehat{\eta}^{(k)}$ of the random variables $\eta^{(k)}$ as a one-dimensional Hermite polynomial chaos.

$$\widehat{\eta}^{(k)} = \sum_{\alpha=1}^{\infty} \gamma_\alpha^{(k)} H_\alpha(\xi_k) \approx \sum_{\alpha=1}^{q} \gamma_\alpha^{(k)} H_\alpha(\xi_k), k = 1, \ldots, \mu \tag{4}$$

The polynomial chaos expansion (PCE) of (4) is truncated after q terms, corresponding to the order of the PCE. The coefficients of the polynomial chaos are estimated using a Bayesian Inference scheme (BI) with a Metropolis-Hastings Markov Chain Monte Carlo algorithm (Mehrez and Doostan 2012a, b).

3   Thermoplastic Honeycomb Sandwich Samples

In this research different sets of MonoPan® (http://www.monopan.nl/) honeycomb beams and panels are used as test objects. Table 1 gives an overview of the nominal dimensional properties of the different samples. A set of seven panels is available. For the beam samples, 22 are available.

The MonoPan® structure is a sandwich structure with its different components originating from the production process. Geometrically seen, the skin consists of three layers: the outer finishing foil, the inner welding foil and the glass fibre weave reinforced layer in between. Two different materials are used in the skin: glass fibre and polypropylene (PP). Since there is no specific data available on the different materials, it is assumed that all PP used in the skin has identical properties. In particular, all parameters that exhibit a significant spatial variability at macro level (order of a whole panel) rather than at meso level (order of a unit cell) are of interest in this research.

As a first example, the skin thickness of honeycomb beam samples is experimentally determined. Using a vernier calliper a set of measurements is carried out, covering 650 measurements. Figure 3 shows the histogram of this experiment.

A mean skin thickness of 1.13 mm and a coefficient of variation (COV) of 7.23 % is obtained.

A second characteristic parameter is the orientation of the weft and warp fibre yarns of the skin's weave.

Table 1 Dimensional properties of the test panels

Sample thickness (mm)	Sample length (mm)	Sample width (mm)	Honeycomb core	Skin faces	Mass (kg/m²)
25	830	50	Cylindrical, PP80	1 mm, PP, Twintex twill 2/2 0.7 mm	5.3
25	2,500	1,200	Cylindrical, PP80	1 mm, PP, Twintex twill 2/2 0.7 mm	5.3

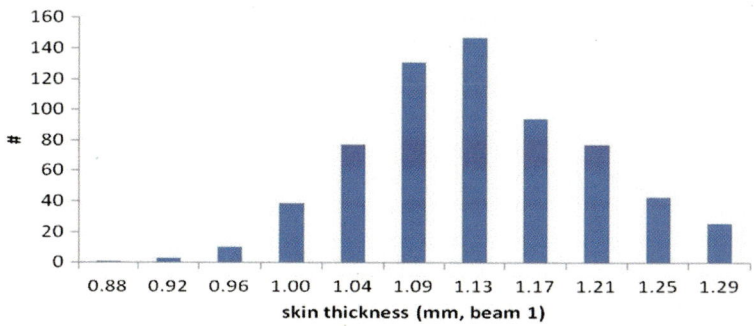

Fig. 3 Histogram of measured skin thickness

Fig. 4 Orientation of weft yarns: panel *upper left corner*

Altering the warp or weft yarn orientation of the weave affects the resulting stiffness properties of the skin face. By means of a sector the mean orientation angle of the weft yarns was determined at 21 evenly spaced locations on rectangular test panels. Across the test panels, COV's of 30 % are obtained for both warp and weft angles. Different zones of the panels show a large warp angle deviation. This is illustrated in Fig. 4.

Furthermore, this research studies panel parameters like sample thickness and honeycomb core cell geometry.

4 Experimental Modal Analysis

The relation between the design parameters of such a panel and its dynamic behaviour is calculated through finite element modelling (FEM). In order to validate the various numerical models, reliable experimental data is needed for comparison. The resonance frequencies and mode shapes of beams and panels under free-free boundary conditions describe the dynamic behaviour of interest.

Hammer excitation is used for the beam samples while an electro dynamic shaker excites the large panel samples. Structural responses are captured with lightweight accelerometers and a laser vibrometer respectively. Table 2 gives an overview of the measurement grids for both types of test samples.

Table 3 gives an overview of the obtained natural frequencies of the beam and panel samples.

Since experimental modal analysis makes use of both experimental and numerical techniques, there are two main sources that may cause errors on the finally obtained (or estimated) modal parameters of interest. First of all there are the errors (Moens and Vandepitte 2004, 2005b; De Gersem et al. 2005) that arise from the measurement itself. These can be traced indirectly by monitoring the coherence

Table 2 Overview of measurement grids

Test structure type	Structure dimensions	# measurement points	# grid rows	# grid columns	# excitation points	# response points
Beam	1D	17	1	17	17	2
Panel	2D	273	13	21	1	273

Table 3 Mean and standard deviation for the first eight resonance frequencies of honeycomb sandwich beam and panel samples

Beam mode	1	2	3	4	5	6	7	8
Mean (Hz)	117.8	296.8	520.3	759.3	1006.5	1258.8	1516.1	1758
Stdev	1.56	3.21	5.67	9.31	13.15	14.58	17.67	20.68
COV (%)	1.33	1.08	1.09	1.23	1.31	1.16	1.17	1.18
Panel mode	1	2	3	4	5	6	7	8
Mean (Hz)	10.45	11.43	23.68	31.96	45.50	57.93	61.89	72.39
Stdev	0.17	0.61	0.53	1.36	1.33	0.31	0.24	0.76
COV (%)	1.66	5.32	2.26	4.27	2.92	0.54	0.39	1.05

function (Ewins 1986; Heylen et al. 2003). The second type of errors is related to the process of modal parameter estimation (Cauberghe 2004; Randall 1987; LMS International) and can also be estimated. The mean resulting COV on experimentally obtained mode shapes is 0.21 % in case of the beam samples and 0.16 % in case of the panel samples. This variability has to be taken into account during further uncertainty quantification.

5   Numerical Modal Analysis and Model Updating

The finite element (FE) method is a well-known numerical method for static and dynamic numerical analysis of mechanical structures. In this research, FE models calculate resonance frequencies and mode shapes of honeycomb sandwich structures under free-free boundary conditions.

There are several methods of modelling honeycomb sandwich structures. The main difference between them is the degree of homogenization of the models. The method which is found to be suitable for this research homogenizes both skin faces and honeycomb core as uniform, orthotropic materials, having equivalent elastic (along the orthotropy axes) and mass properties as the original core and skin. According to Daniel and Ishai (2006), Mares et al. (2006), Berthelot (1996) and Govers and Link (2010), a fibre weave reinforced material can be modelled as a three layered sandwich.

The skins are modelled as shells while the honeycomb core is modelled as a volume. This modelling method is known as the shell-volume-shell method (SVS). In order to obtain realistic estimates for the homogenized core and skin elastic properties, a mixed experimental numerical approach is used.

Table 4 Comparison between EMA and FEM resonance frequencies and mode for beam and panel samples (mean values for all test samples)

Beam mode	Frequency (FEM, Hz)	Frequency diff. (%)	MAC (%)	Panel mode	Frequency (FEM, Hz)	Frequency diff. (%)	MAC (%)
1	110.2	2.71	99.13	1	10.25	5.46	96.43
2	282.3	−1.23	97.94	2	11	3.28	94.67
3	703.3	1.43	97.33	3	23.5	18.54	99.48
4	748.4	2.15	94.20	4	31	0.32	96.11
5	1004	0.73	91.16	5	44.5	9.50	99.53
6	1261	4.57	90.25	6	57.75	−13.36	91.07
7	1518	3.27	86.53	7	62	−9.09	94.54
8	2026	2.89	81.76	8	72.5	1.79	91.55

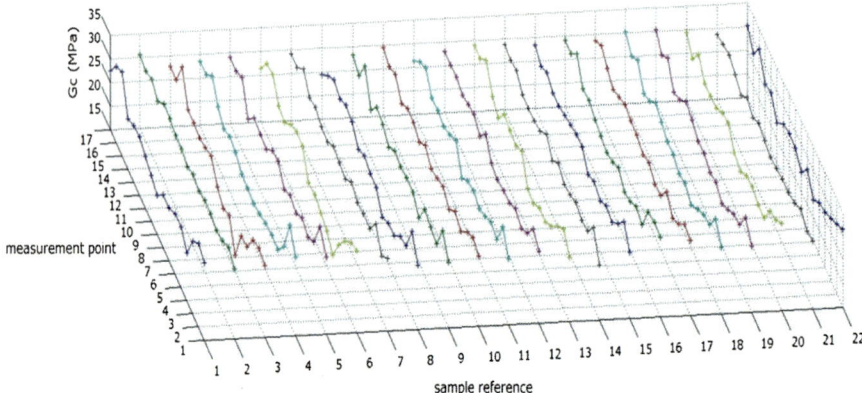

Fig. 5 Obtained values for G_c after FE model update (beam samples)

Table 4 compares the numerically and experimentally obtained modal parameters for both beams and panels. Natural frequencies are compared; the experimental-numerical mode shape pairs are validated using the MAC (Ewins 1986; Heylen et al. 2003; Cauberghe 2004; Randall 1987; LMS International).

For the considered honeycomb sandwich structures the in-plane stiffness moduli of the skin faces and the out-of-plane shear stiffness of the (homogenized) honeycomb core are important elastic parameters that determine the modal parameters of interest. The mean Finite Element models of the beam and panel samples are therefore updated with respect to these skin and core properties. This is done for all 22 beam samples and 7 panel samples using the experimentally determined natural frequencies and mode shapes. Referring to Table 2, the numerical models are divided into a number of zones, centered around the measurement points. In each zone the considered elastic parameters have a constant value. For each model zone, the model updating process yields values for the skin stiffness and core shear moduli. These are the updating parameters. Figure 5 illustrates the obtained parameter values for the core shear stiffness. Taking into account all test beams and

Table 5 Mean values for obtained updating variability of considered update variables, compared with their database variability (panel samples)

Update parameter	Lower skin E_{1s}	Lower skin E_{2s}	Upper skin E_{1s}	Upper skin E_{2s}	Core G_{c13}	Core G_{c23}
Model updating COV (%)	1.52	1.62	1.40	1.58	1.23	1.30
Model database COV (%)	7.79	5.24	5.16	7.93	3.16	3.84

all model zones, databases containing 22 by 17 stiffness values are obtained. Note that in this case no distinction is made between the stiffnesses of the two opposed skin faces.

For the panel samples, the two skin faces of a panel are considered as two separate entities. In this case the updating process of the Finite Element models yields databases for four orthotropic in-plane skin stiffness and two out-of-plane core shear stiffness (Berthelot 1996) properties. Taking into account all seven test panels, the database of each elastic property counts 240 by 7 elements. In the panel case, the discrepancy between the number of modal reference data (10 natural frequencies and 10 mode shapes) and the number of updating parameters (240 by 6) introduces uncertainty (Manan and Cooper 2010; Soize 2003; Schultz et al. 2007) of the updating results (elastic properties of skin and core). In order to quantify this uncertainty, stochastic model updating (Mares et al. 2002; Friswell and Mottershead 1995; Friswell et al. 2001; Chen and Guedes Soares 2008; Moens and Vandepitte 2006; Siemens) is needed. According to Zarate and Caicedo (2008), Ibrahim (1997) and Mares et al. (2006), there are mainly two ways of dealing with this problem.

The applied solution here is based on the idea of correlating multiple, perturbed numerical models to one set of experimental data (resonance frequencies and mode shapes in this case). In fact, this comes down to stochastic FE model updating.

This approach perturbes the initial values of the FEM updating parameters with respect to their nominal value. In this case, this perturbation is purely random; each parameter is consequently normally distributed in a certain interval. Each perturbed FE model is solved for the considered modal parameters and then updated. In fact, this comes down to the application of Monte Carlo stochastic FEM updating (Govers and Link 2010; Van Benthem 1976; Carmola and Chimowitz 1990; Kappagantu and Feeny 1999; Bultheel 2006). This procedure enables to study the sensitivity of the estimated optimized FE model to the set of initial values of the FEM updating parameters.

Table 5 compares the model updating uncertainty in case of the panel samples with the database variability for all considered elastic properties. Both are expressed as a COV.

Table 5 shows that the database variability is much higher than the model updating uncertainty. However, the updating uncertainty is not negligible and has to be considered in further stochastic processing of the databases.

6 Random Field Modelling of Panel Parameter Variability

In this research, the different considered elastic properties are treated as (independent) stochastic processes, modelled with the random field (RF) method (Loeve 1977; Ghanem and Doostan 2006; Ghanem et al. 2005; Desceliers et al. 2006, 2007b).

Figure 6 illustrates the obtained probability density functions (PDF) of the out-of-plane core shear modulus in case of the beam samples. For comparison the data originating from the finite element model updating process is shown. A third order RF is used here.

Since the amount of experimental validation data (natural frequencies and mode shapes, number of test samples) is very limited, the epistemic uncertainty involved has to be estimated. It is represented by the random character of the estimated coefficients of the polynomial chaos decomposition. In this study the method presented in (Soize 2010; Mehrez and Doostan 2012a, b) is followed. If enough test

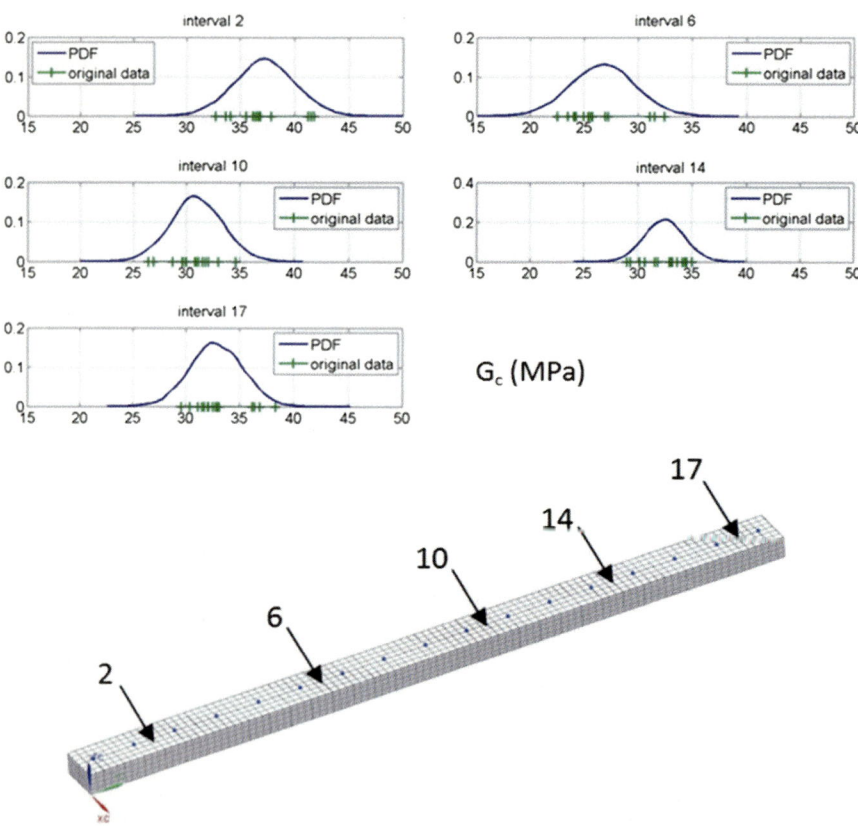

Fig. 6 Obtained PDF's for G_c (MPa) at zones 2, 6, 10, 14 and 17 (beam samples)

FE model zone	2	6	10	14	17
E_s mean stdev (GPa)	0.46	0.45	0.49	0.59	0.49
E_s stdev 2.5 % (GPa)	0.32	0.32	0.34	0.41	0.33
E_s stdev 97.5 % (GPa)	0.61	0.59	0.64	0.77	0.65
G_c stdev (MPa)	2.61	1.52	2.22	1.73	2.33
G_c stdev 2.5 % (MPa)	1.83	1.09	1.53	1.07	1.54
G_c stdev 97.5 % (MPa)	3.5	2.02	2.95	2.27	3.15

Table 6 95 % confidence intervals for the standard deviations of the estimated probability densities of E_s and G_c for all considered model zones (beam samples)

samples are observed, the distributions of the different estimated PC coefficients approach a multivariate normal distribution of which the covariance matrix is the inverse of the Fisher information matrix (Scott 2002).

Assuming normal distributions, Table 6 compares the mean standard deviations of the estimated PDF's with their corresponding epistemic uncertainty, expressed as a 95 % confidence intervals (CI) on the PDF standard deviations.

For E_s, the mean CI width caused by the epistemic uncertainty is approximately 3.3 % and for G_c approximately 4 %. The epistemic uncertainty is small, compared to the stated variability of the elastic parameters. The same methodology is applied in case of the panel samples. Due to the lower number of test samples here, the epistemic uncertainty is approximately five times higher than in case of the beam samples.

In this research, a specific point of interest is the characterization of the spatial variability of the considered elastic skin and core parameters. Due to the nature of the panel production process, one might expect significant patterns when studying spatial variability. The correlation function (5) defines a periodical spatial variability with period p and correlation length ξ, using the off-diagonal terms of the estimated correlation matrices. The 'noise' term σ^2 is a culmination of the uncertainty from the experimental and numerical modal analysis.

$$K\left(x_i, x_j\right) = \sigma^2 + \sigma_i \sigma_j e^{-\|x_i - x_j\|/\xi} \cos\left(\|x_i - x_j\|^* 2\pi/p\right) \quad (5)$$

In case of the beam samples, correlation lengths of 640 and 2,520 mm are estimated for parameters E_s and G_c respectively.

For the skin stiffness modulus E_s there is a clear periodicity (265 mm) which is smaller than the sample length; this is not the case for G_c.

7 General Conclusions

The goal of this research is to study the main elastic parameters of thermoplastic honeycomb sandwich structures by means of their experimentally determined dynamic behaviour. More specifically, the elastic parameter variability is the main issue of interest in this work. This research work is a mix of numerical simulations and experiments.

The variability of the most important elastic parameters of honeycomb beam and panel sandwich structures is estimated. The general random field method is successfully applied to estimate probability density functions of these parameters at different zones (intervals) in the structures. Apart from estimating the physically inherent (or aleatory) uncertainty, the epistemic uncertainty due to a lack of sufficient experimental data, has also been quantified. In case of the studied panel structures, of which only seven samples are available, this uncertainty is found to be high. The covariance matrices of all considered elastic parameters have been thoroughly studied to gather useful information on correlation length and periodicity of the observed and stated parameter variability.

However, this research is not an ending point. Some assumptions have been made for simplicity. For example, it is assumed that the different considered elastic parameters are independent. In reality they are not. Future work may focus on applying the necessary parameter relations. The associated parameter variability may then be described by multi-dimensional random fields. The ultimate sequel to this research could be the direct random field description of geometric parameters of the skin faces and the honeycomb core, starting from vibration measurements.

References

Arnst M, Ghanem R, Soize C (2010) Identification of Bayesian posteriors for coefficients of chaos expansions. J Comput Phys 229:3134–3154

Berthelot J-M (1996) Matériaux composites – Comportement mécanique et analyse des structures, 2nd edn. Masson. Lavoisier, Paris

Blevins RD (1984) Formulas for natural frequency and mode shape. Krieger Publishing Company, Malabar

Bultheel A (2006) Inleiding tot de numerieke wiskunde, Acco. Leuven

Carmola RE, Chimowitz EH (1990) Analysis of modal reduction techniques for the dynamics of tridiagonal systems. Comput Chem Eng 14(2):220–239

Cauberghe B (2004) Applied frequency-domain system identification in the field of experimental and operational modal analysis. Phd thesis, University of Brussels, Brussels

Chen N-Z, Guedes Soares C (2008) Spectral stochastic finite element analysis for laminated composite plates. Comput Method Appl Mech Eng 197:4830–4839

Cooker MJ (1990) A boundary-integral method for water wave motion over irregular beds. Eng Anal Bound Elem 7(4):205–213

Daniel O, Ishai IM (2006) Engineering mechanics of composite materials, 2nd edn. Oxford University Press, New York

De Gersem H, Moens D, Vandepitte D (2005) A fuzzy finite element procedure for the calculation of uncertain frequency response functions of damped structures: Part 2 – numerical case studies. J Sound Vib 288(3):463–486

Desceliers C, Ghanem RG, Soize C (2006) Maximum likelihood estimation of stochastic chaos representations from experimental data. Int J Numer Methods Eng 66(6):978–1001

Desceliers C, Soize C, Ghanem R (2007) Identification of chaos representations of elastic properties of random media using experimental vibration tests. Comput Mech 39:831–838

Ewins DJ (1986) Modal testing: theory and testing. Research Studies Press Ltd, London

Friswell MI, Mottershead JE (1995) Finite element model updating in structural dynamics. Kluwer Academic Publishers, Dordrecht/Boston, 286 pp. ISBN 0-7923-3431-0

Friswell MI, Mottershead JE, Ahmadian H (2001) Finite element model updating using experimental test data: parameterization and regularization. Trans R Soc Lond Ser A Spec Issue Exp Modal Anal 359(1778):169–186

Ghanem RG (1991) Stochastic finite elements, a spectral approach. Johns Hopkins University, Springer, New York

Ghanem RG (2006) On the construction and analysis of stochastic models: characterization and propagation of the errors associated with limited data. J Comput Phys 217:63–81

Ghanem RG, Doostan A, Red-Horse J (2005) A probabilistic construction of model validation. Comput Method Appl Mech Eng 197(29–32):2585–2595

Gibson LJ, Ashby MF (1988) Cellular solids. Pergamon Press, New York

Govers Y, Link M (2010) Stochastic model updating-covariance matrix adjustment form uncertain experimental data. Mech Syst Signal Process 24:696–706

Heylen W, Lammens S, Sas P (2003) Modal analysis: theory and testing. KU Leuven, Leuven

Ibrahim SR (1997) Multi-perturbed analytical models for updating and damage detection. In: Proceedings of IMAC XV conference, pp 127–141

Kappagantu R, Feeny BF (1999) An optimal modal reduction of a system with frictional excitation. J Sound Vib 224(5):863–877

Lauwagie T (2005) Vibration-based methods for the identification of the elastic properties of layered materials. Phd thesis, KU Leuven, Leuven. D/2005/7515/80

Leuven Measurement Systems: LMS International Siemens NX PLM Software

Liu Q (2001) Prediction of natural frequencies of a sandwich panel using thick plate theory. J Sandw Struct Mater 3(4):289

Liu Q (2002) Role of anisotropic core in vibration properties of honeycomb sandwich panels. J Thermoplast Compos Mater 15(1):23–32

Liu Q (2007) Effect of soft honeycomb core on flexural vibration of sandwich panel using low order and high order shear deformation models. J Sandw Struct Mater 9(1):95–108

Loeve M (1977) Probability theory, 4th edn. Springer, New York

Loeve M (1997) Probability theory, 4th edn. Springer, New York

Manan A, Cooper JE (2010) Prediction of uncertain frequency response function bounds using polynomial chaos expansion. J Sound Vib 329:3348–3358

Mares C, Friswell MI, Mottershead JE (2002) Model updating using Robust estimation. Mech Syst Signal Process 16(1):169–183

Mares C, Mottershead JE, Friswell MI (2006) Stochastic model updating: Part 1 – theory and simulated example. Mech Syst Signal Process 20(7):1674–1695

Mehrez L, Doostan A (2012a) Stochastic identification of composite material properties from limited experimental databases, Part I: Experimental database construction. Mech Syst Signal Process 27:471–483

Mehrez L, Doostan A (2012b) Stochastic identification of composite material properties from limited experimental databases, Part II: uncertainty modelling. Mech Syst Signal Process 27:484–498

Moens D, Vandepitte D (2004) An interval finite element approach for the calculation of envelope frequency response functions. Int J Numer Methods Eng 61(14):2480–2507

Moens D, Vandepitte D (2005a) A survey of non-probabilistic uncertainty treatment in finite element analysis. Comput Methods Appl Mech Eng 194(12–16):1527–1555

Moens D, Vandepitte D (2005b) A fuzzy finite element procedure for the calculation of uncertain frequency response functions of damped structures: Part 1 – procedure. J Sound Vib 288(3):431–462

Moens D, Vandepitte D (2006) Sensitivity analysis of frequency response function envelopes of mechanical structures. In: Proceedings of the international conference on noise and vibration engineering, Leuven. pp 4197–4212

Nilsson E, Nilsson AC (2002) Prediction and measurement of some dynamic properties of sandwich structures with honeycomb and foam cores. J Sound Vib 251(3):409–430

Pradlwarter HJ, Schuëller GI (1997) On advanced Monte Carlo simulation procedures in stochastic structural dynamics. Int J Non-Linear Mech 32(4):735–744

Randall RB B. Tech (1987) Frequency analysis. K. Larsen & Son. Glostrup

Robert CP, Casella G (2004) Monte Carlo statistical methods. Springer, New York

Schenk A, Schuëller GI (2005) Uncertainty assessment of large finite element systems. Springer, Innsbruck

Schuëller GI, Pradlwarter HJ (2009) Uncertain linear systems in dynamics: retrospective and recent developments by stochastic approaches. Eng Struct 31(11):2507–2517

Schultz T, Sheplak M, Louis N (2007) Application of multivariate uncertainty analysis to frequency response function estimates. J Sound Vib 305:116–133

Scott WA (1992) Multivariate density estimation: theory, practice and visualization. Wiley, New York

Scott WA (2002) Maximum likelihood estimation using the empirical Fisher information matrix. J Stat Comput Simul 72(8):599–611

Siemens NX Nastran, NX8, cast

Soize C (2003) Random matrix theory and non-parametric model of random uncertainties in vibration analysis. J Sound Vib 263:893–916

Soize C (2010) Identification of high-dimension polynomial chaos expansions with random coefficients for non-Gaussian tensor-valued random fields using partial and limited experimental data. Comput Method Appl Mech Eng 199(33–36):2150–2164

Soize C (2011) A computational inverse method for identification of non-Gaussian random fields using the Bayesian approach in very high dimension. Comput Method Appl Mech Eng 200:3083–3099

Topdar P (2003) Finite element analysis of composite and sandwich plates using a continuous inter-laminar shear stress model. J Sandw Struct Mater 5:207–231

Van Benthem JFAK (1976) Modal reduction principles. J Symb Log 42(2):301–312

Yang QW (2009) Model reduction by Neumann series expansion. Appl Math Model 33(12):4431–4434

Zarate BA, Caicedo JM (2008) Finite element model updating: multiple alternatives. Eng Struct 30(2008):3724–3730

Zenkert D (1997) An introduction to sandwich construction. Emas Publishing, London

Zhang EL, Feissel P, Antoni J (2011) A comprehensive Bayesian approach for model updating and quantification of modelling errors. Prob Eng Mech 26:550–560

Identification of a Mesoscale Model with Multiscale Experimental Observations

M.T. Nguyen, C. Desceliers, and C. Soize

Abstract This paper deals with a multiscale statistical inverse method for performing the experimental identification of the elastic properties of materials at macroscale and at mesoscale within the framework of a heterogeneous microstructure which is modeled by a random elastic media. New methods are required for carrying out such multiscale identification using experimental measurements of the displacement fields at macroscale and at mesoscale performed with only a single specimen submitted to a given external load at macroscale. In this paper, for a heterogeneous microstructure, a new identification method is presented and is formulated within the framework of the three dimensional linear elasticity. It permits the identification of the effective elasticity tensor at macroscale and the identification of the stochastic tensor field modeling the apparent elasticity at the mesoscale. A validation is presented with experimental measurements simulated with a numerical model with a 2D plane stresses hypothesis.

Keywords Multiscale identification • Heterogeneous microstructure • Random elasticity field • Mesoscale • Multi scale experiments

1 Introduction

The inverse methods for the experimental identification of the elastic properties of materials at the macroscale and/or mesoscale have been extensively studied. The experimental identification of microstructural morphology by image analysis began in the 1980s (see for instance Jeulin 1987, 1989, 2001) and it has led

M.T. Nguyen • C. Desceliers (✉) • C. Soize
Laboratoire Modélisation et Simulation Multi-Echelle, MSME UMR 8208 CNRS, Université Paris-Est, 5 bd Descartes, 77454 Marne-La-Vallée, Cedex 2, France
e-mail: christophe.desceliers@univ-paris-est.fr; christian.soize@univ-paris-est.fr

M. Papadrakakis and G. Stefanou (eds.), *Multiscale Modeling and Uncertainty Quantification of Materials and Structures*, DOI 10.1007/978-3-319-06331-7_8,
© Springer International Publishing Switzerland 2014

to significant advances in the identification of mechanical properties (see, for instance Avril and Pierron 2007; Avril et al. 2008a,b; Baxter and Graham 2000; Besnard et al. 2006; Bonnet and Constantinescu 2005; Bornert et al. 2009, 2010; Calloch et al. 2002; Chevalier et al. 2001; Constantinescu 1995; Geymonat et al. 2002; Geymonat and Pagano 2003; Graham et al. 2003; Hild 2002; Hild et al. 1999, 2002; Hild and Roux 2006, 2012; Madi et al. 2007; Rethore et al. 2008; Roux and Hild 2008; Roux et al. 2002, 2008). Concerning the identification of stochastic models, the methodologies for statistical inverse problems in finite and infinite dimension are numerous and have given rise to numerous studies and publications. These methods make extensive use of the formulations and the tools of the functional analysis of boundary value problems as well as those of probability theory, including mathematical statistics (finite and infinite dimensional cases). Concerning the mathematical statistics, one can refer to Lawson and Hanson (1974) and Serfling (1980) and Collins et al. (1974), Kaipio and Somersalo (2005), Walter and Pronzato (1997), and Spall (2003) for the general principles on the statistical inverse problems. Early work on the statistical inverse identification of stochastic fields for random elastic media, using partial and limited experimental data, have primarily be devoted to the identification of statistical parameters of prior stochastic models (such as the spatial correlation scales and the level of statistical fluctuations) (Arnst et al. 2008; Das et al. 2008, 2009; Desceliers et al. 2006, 2007; Guilleminot et al. 2009; Soize 2010; Ta et al. 2010). Those probabilistic/statistical methods are able to solve the statistical inverse problems related to the identification of prior stochastic models for the apparent elastic fields at mesoscale. Nevertheless, such experimental identification, which is carried out using measurements on a single specimen submitted to a given external load at macroscale and using measurements of the displacement fields at macroscale and mesoscale, requires new methods for identifying the statistical mean value of the random apparent elasticity tensor and the other parameters controlling its prior stochastic model as, for instance, the spatial correlation lengths and the parameters allowing the statistical fluctuations of the stochastic field to be controlled.

In this paper, a new identification method is presented. A statistical inverse multiscale method is formulated for a heterogeneous microstructure within the framework of the three-dimensional linear elasticity. This method permits both the identification of the effective elasticity tensor at macroscale and the identification of the stochastic tensor field which modelizes the apparent elasticity field at mesoscale. It is assumed that the experimental measurements of the displacement field are available at macroscale and at mesoscale. The prior stochastic model is a non-Gaussian tensor-valued random field adapted to the properties of the 3D-elasticity field and to the corresponding stochastic elliptic boundary value problem. The parameters of the prior stochastic model of the apparent elasticity random field at mesoscale, are its statistical mean value, its spatial correlation lengths and its level of statistical fluctuations. This identification of the stochastic model at mesoscale requires the knowledge of the effective elasticity tensor of macroscale and measurements of the displacements field at the two scales simultaneously for one given specimen submitted to a given static external loads. Thus, the proposed method is new. The theory will be presented for the 3D case and a numerical validation will be presented for the 2D plane stress in the framework of experimental measurements obtained by optical measurements (but, in the present paper, the validation will be performed with simulated experiments).

2   Multiscale Experimental Configuration

The specimen (whose microstructure is complex and heterogeneous at microscale) occupies a bounded macroscopic domain Ω^{macro} in $\mathbf{R}^3$. Surface forces, f^{macro}, are applied on a part Σ^{macro} of the boundary $\partial\Omega^{\text{macro}}$ of Ω^{macro}. The other part Γ^{macro} of $\partial\Omega^{\text{macro}}$ is fixed such that there is no rigid body displacement. At macroscale on Ω^{macro}, the measured displacement field is denoted as $u_{\text{exp}}^{\text{macro}}$ and its associated strain tensor is denoted as $\varepsilon_{\text{exp}}^{\text{macro}}$.

Let Ω^{meso} be a subdomain of the specimen at mesoscale (a REV) and let $\partial\Omega^{\text{meso}}$ be the boundary of Ω^{meso}. Let $u_{\text{exp}}^{\text{meso}}$ be the experimental measurement on Ω^{meso} of the displacement field at mesoscale. The associated strain tensor is denoted as $\mathcal{E}_{\text{exp}}^{\text{meso}}$. It is assumed that the experimental measurements of $u_{\text{exp}}^{\text{meso}}$ are obtained only for one subdomain Ω^{meso} related to one specimen. The volume average at mesoscale, $\underline{\mathcal{E}}_{\text{exp}}^{\text{meso}}$, of $\mathcal{E}_{\text{exp}}^{\text{meso}}$ is introduced such that

$$\underline{\mathcal{E}}_{\text{exp}}^{\text{meso}} = \frac{1}{|\Omega^{\text{meso}}|} \int_{\Omega^{\text{meso}}} \mathcal{E}_{\text{exp}}^{\text{meso}}(x) \, dx. \tag{1}$$

The statistical fluctuations level of the experimental linearized strain field at mesoscale around the volume average, $\underline{\mathcal{E}}_{\text{exp}}^{\text{meso}}$, is estimated by $\delta_{\text{exp}}^{\text{meso}}$ which is defined as

$$\delta_{\text{exp}}^{\text{meso}} = \frac{\sqrt{V_{\text{exp}}^{\text{meso}}}}{\|\underline{\mathcal{E}}_{\text{exp}}^{\text{meso}}\|_F}, \tag{2}$$

in which

$$V_{\text{exp}}^{\text{meso}} = \frac{1}{|\Omega^{\text{meso}}|} \int_{\Omega^{\text{meso}}} \|\mathcal{E}_{\text{exp}}^{\text{meso}}(x) - \underline{\mathcal{E}}_{\text{exp}}^{\text{meso}}\|_F^2 \, dx \tag{3}$$

and where $\|T\|_F$ is the Frobenius norm such that, for any second-order tensor $T = \{T_{ij}\}_{ij}$, one has

$$\|T\|_F^2 = \sum_{i=1}^{3} \sum_{j=1}^{3} T_{ij}^2. \tag{4}$$

3 Multiscale Statistical Inverse Problem

At macroscale, a deterministic boundary value problem is introduced for a 3D linear elastic medium, which modelizes the specimen in its experimental configuration (geometry, surface forces and Dirichlet conditions). At macroscale, the constitutive equation involves a prior model for the elasticity tensor $C^{\mathrm{macro}}(a)$ which is parameterized by a vector a. For the 3D anisotropic elasticity, a represents the 21 constants of the elasticity tensor. The boundary value problem is formulated in displacement and the solution is denoted as u^{macro} (deterministic macroscale displacement field). The linearized strain tensor associated with u^{macro} is denoted as $\varepsilon^{\mathrm{macro}}$. Tensor $C^{\mathrm{macro}}(a)$ is unknown and must experimentally be identified, which means that parameter a must be identified using the measurements of the displacement field at macroscale. Consequently, a first numerical indicator $\mathscr{I}_1(a)$ is introduced in order to quantify the distance between $\varepsilon_{\mathrm{exp}}^{\mathrm{macro}}$ and $\varepsilon^{\mathrm{macro}}$. For a fixed value of parameter a, this indicator is defined by

$$\mathscr{I}_1(a) = |||\varepsilon_{\mathrm{exp}}^{\mathrm{macro}} - \varepsilon^{\mathrm{macro}}(a)|||^2 , \tag{5}$$

in which

$$|||\varepsilon_{\mathrm{exp}}^{\mathrm{macro}} - \varepsilon^{\mathrm{macro}}(a)|||^2 = \int_{\Omega^{\mathrm{macro}}} \|\varepsilon_{\mathrm{exp}}^{\mathrm{macro}}(x) - \varepsilon^{\mathrm{macro}}(x; a)\|_F^2 \, dx . \tag{6}$$

At mesoscale, two additional numerical indicators, $\mathscr{I}_2(b)$ and $\mathscr{I}_3(a, b)$, are constructed to identify the parameters b involved in the prior stochastic model of the apparent elasticity random field $C^{\mathrm{meso}}(b)$ which is considered as the restriction to subdomain Ω^{meso} of a statistically homogeneous random field $\{C^{\mathrm{meso}}(x; b), x \in \mathbb{R}^3\}$.

Concerning the construction of the second numerical indicator $\mathscr{I}_2(b)$, a random boundary value problem is introduced for a 3D linear elastic random media occupying subdomain Ω^{meso} and for which the apparent elasticity random field is $C^{\mathrm{meso}}(b)$. This random boundary value problem is formulated in displacement and the solution is denoted as U^{meso} (displacement random field) with the Dirichlet condition $U^{\mathrm{meso}} = u_{\mathrm{exp}}^{\mathrm{meso}}$ on boundary $\partial\Omega^{\mathrm{meso}}$. The random linearized strain tensor field associated with U^{meso} is denoted as $\mathcal{E}^{\mathrm{meso}}$. For any given parameters b, numerical indicator $\mathscr{I}_2(b)$ is defined as

$$\mathscr{I}_2(b) = \int_{\Omega^{\mathrm{meso}}} (\delta^{\mathrm{meso}}(x; b) - \delta_{\mathrm{exp}}^{\mathrm{meso}})^2 \, dx , \tag{7}$$

in which

$$\delta^{\mathrm{meso}}(x; b) = \frac{\sqrt{V^{\mathrm{meso}}(x; b)}}{\|\mathcal{E}^{\mathrm{meso}}(b)\|_F} , \tag{8}$$

where

$$\underline{\mathfrak{C}}^{\text{meso}}(b) = \frac{1}{|\Omega^{\text{meso}}|} \int_{\Omega^{\text{meso}}} \underline{\mathfrak{C}}^{\text{meso}}(x; b) \, dx, \tag{9}$$

and

$$V^{\text{meso}}(x; b) = E\{\|\underline{\mathfrak{C}}^{\text{meso}}(x; b) - \underline{\mathfrak{C}}^{\text{meso}}(b)\|_F^2\}, \tag{10}$$

It should be noted that, for all b, $\underline{\mathfrak{C}}^{\text{meso}}(b) = \underline{\mathfrak{C}}^{\text{meso}}_{\text{exp}}$. The third numerical indicator $\mathscr{I}_3(a, b)$ depends on a and b since this numerical indicator quantifies the distance between the elasticity tensor $C^{\text{macro}}(a)$ used in boundary value problem at macroscale and the effective tensor $C^{\text{eff}}(b)$ calculated by homogenization of the stochastic model at mesoscale on the REV, which depends on b only. We then have

$$\mathscr{I}_3(a, b) = \|C^{\text{macro}}(a) - E\{C^{\text{eff}}(b)\}\|_F^2. \tag{11}$$

The identification of parameters a and b that describe the stochastic model of the apparent elasticity random field $C^{\text{meso}}(b)$ at mesoscale is obtained by solving a multi-objective optimization problem for the three indicators $\mathscr{I}_1(a)$, $\mathscr{I}_2(b)$ and $\mathscr{I}_3(a, b)$.

4 Validation of the Method in 2D Plane Stresses

The validation is performed within the framework of the linear elasticity in 2D plane stresses. It should be noted that the two directions are observed when the displacement fields are measured at macroscale and at mesoscale with a camera.

4.1 Prior Stochastic Model of the Apparent Elasticity Random Field in 2D Plane Stresses

At mesoscale, the prior stochastic model of the apparent elastic random field C^{meso} is indexed by subdomain Ω^{meso} which is assumed to be a REV. A representation of C^{meso} with a minimum of parameters and adapted to elliptic problems is used. Parametric stochastic models have been proposed for scalar-valued stochastic fields (Babuska et al. 2005, 2007; Desceliers et al. 2012; Graham et al. 2003) and for non-Gaussian tensor-valued stochastic fields in the framework of the heterogeneous anisotropic linear elasticity (Clouteau et al. 2013; Soize 2006, 2008; Ta et al. 2010), with important enhancements to take into account the material symmetry and the existence of elasticity bounds (Guilleminot and Soize 2011, 2012a,b, 2013). Hereinafter, the stochastic model is based on Soize (2006).

In using the Voigt notation, the fourth-order elasticity tensor $C^{\text{meso}}(x)$ can be represented by a (6×6) real matrix. The strain vector is then denoted as $(\varepsilon_{11}, \varepsilon_{22}, 2\varepsilon_{12},$ $\varepsilon_{33}, 2\varepsilon_{23}, 2\varepsilon_{13})$ and the associated stress vector is denoted as $(\sigma_{11}, \sigma_{22}, \sigma_{12}, \sigma_{33},$ $\sigma_{23}, \sigma_{13})$. Such numbering of those vectors, which is not usual, has been chosen for the sake of simplicity in 2D plane stresses, for which the (3×3) compliance matrix $[S^{2D}(x)]$ corresponds to the first (3×3) block of the (6×6) compliance matrix $[S^{\text{meso}}(x)] = [C^{\text{meso}}(x)]^{-1}$.

The *prior* stochastic model of C^{meso} is then constructed in choosing $[S^{\text{meso}}] = \{[S^{\text{meso}}(x)], x \in \Omega^{\text{meso}}\}$ in the set SFE^+ (Soize 2006, 2012) of non-Gaussian second-order stochastic fields with values in the set of all the positive-definite symmetric (6×6) real matrices, for which the mean value is a given matrix $[\underline{S}^{\text{meso}}] = E\{[S^{\text{meso}}(x)]\}$ for all x in Ω^{meso}. As a result, the matrix-valued random field $\{[S^{\text{meso}}(x)], x \in \Omega^{\text{meso}}\}$ is described as a function of the entries of matrix $[\underline{S}^{\text{meso}}]$, of three spatial correlation lengths ℓ_1, ℓ_2, ℓ_3 and of one parameter δ which controls the level of dispersion.

In the case of 2D plane stresses, random matrix $[S^{2D}(x)]$ (the left upper (3×3) block matrix of $[S^{\text{meso}}(x)]$) can be written as a function of the entries of matrix $[\underline{S}^{2D}]$ (left upper (3×3) block matrix of $[\underline{S}^{\text{meso}}]$), one spatial correlation length $\ell = \ell_1 = \ell_2$ and dispersion parameter δ. The prior model of the apparent elasticity random field $[C^{2D}] = \{[C^{2D}(x)], x \in \Omega^{\text{meso}}\}$ with values in the set of the (3×3) real matrices is then constructed, for all x in Ω^{meso}, as the inverse of the random matrix $[S^{2D}(x)]$. We then have

$$[C^{2D}(x)] = [S^{2D}(x)]^{-1} \quad , \quad a.s. \tag{12}$$

Consequently, the parameter b of the prior stochastic model of the apparent elasticity random field $[C^{2D}(b)]$ are $b = (\delta, \ell, \text{entries of } [\underline{S}^{2D}])$. It should be noted that random fields $[C^{2D}(b)]$ and $[S^{2D}(b)]$ do not belong to the set SFE^+ of non-Gaussian second-order stochastic fields with values in the set of all the positive-definite symmetric (3×3) real matrices.

4.2 Construction of a Simulated "Experimental" Database

To validate the methodology, "experimental" measurements at macroscale and at mesoscale are both simulated in using a computational model. The 2D domain Ω^{macro} in the plane (Ox_1x_2), is defined as a square whose dimensions are $h = 10^{-2}$ m. At mesoscale, the material is heterogeneous, anisotropic and linear elastic. A line force directed along $-x_2$, with an intensity of 5×10^{-2} N/m, is applied on the edge $x_2 = h$. The edge $x_2 = 0$ is fixed. A 2D plane stress state is assumed. At mesoscale, the 2D apparent elasticity field is constructed as a realization of the prior stochastic model of $[C^{2D}(b)]$ with $\ell = 1.25 \times 10^{-4}$ m, $\delta = 0.4$ and where the entries of $[\underline{S}^{2D}]$ are defined below. It is assumed that the elastic medium is transverse isotropic which yields

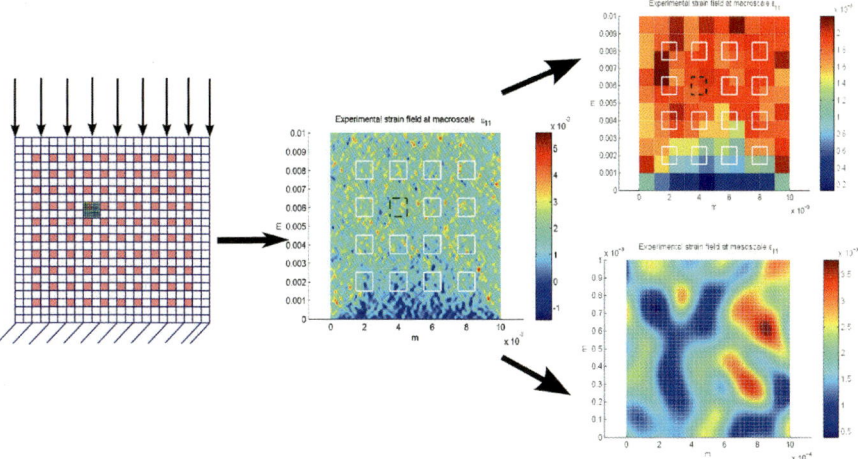

Fig. 1 Description of the methodology for the construction of the simulated experimental measurements in using the finite element method at macroscale and at mesoscale: FE model of the specimen at macroscale with a mesoscale resolution (*left*), component {11} of the strain field at macroscale with a mesoscale resolution (*center*), component {11} of the strain field at macroscale with a macroscale resolution (*upper right*) and component {11} of the strain field at mesoscale with a mesoscale resolution (*lower right*)

$$[\underline{S}^{\text{meso}}] = \begin{pmatrix} \frac{1}{E_T} & -\frac{\nu_T}{E_T} & 0 & -\frac{\nu_L}{E_L} & 0 & 0 \\ -\frac{\nu_T}{E_T} & \frac{1}{E_T} & 0 & 0 & 0 & 0 \\ 0 & 0 & \frac{2(1+\nu_T)}{E_T} & 0 & 0 & 0 \\ -\frac{\nu_L}{E_L} & -\frac{\nu_L}{E_L} & 0 & \frac{1}{E_L} & 0 & 0 \\ 0 & 0 & 0 & 0 & \frac{1}{G_L} & 0 \\ 0 & 0 & 0 & 0 & 0 & \frac{1}{G_L} \end{pmatrix}, \tag{13}$$

where $E_L = 15.8 \times 10^9 \, \text{Pa}$, $E_T = 9.9 \times 10^9 \, \text{Pa}$, $G_L = 5.2 \times 10^9 \, \text{Pa}$, $\nu_L = 0.31$ and $\nu_T = 0.38$. Consequently, we have

$$[\underline{S}^{2D}] = \begin{pmatrix} \frac{1}{E_T} & -\frac{\nu_T}{E_T} & 0 \\ -\frac{\nu_T}{E_T} & \frac{1}{E_T} & 0 \\ 0 & 0 & \frac{2(1+\nu_T)}{E_T} \end{pmatrix}. \tag{14}$$

Consequently, the vector b of parameters is then written as $b = (\delta, \ell, E_T, \nu_T)$. At mesoscale, the realization of the apparent elasticity random field is simulated on the whole domain Ω^{macro}.

A computational model is constructed with the finite element method and a regular finite element mesh with one million quadrangle elements (1,000 along x_1 and 1,000 along x_2, see Fig. 1, left). The strain field is numerically simulated in using a finite element interpolation in a regular grid of nodes with a mesoscale resolution on the whole domain Ω^{macro} (see Fig. 1, center). Measurements of

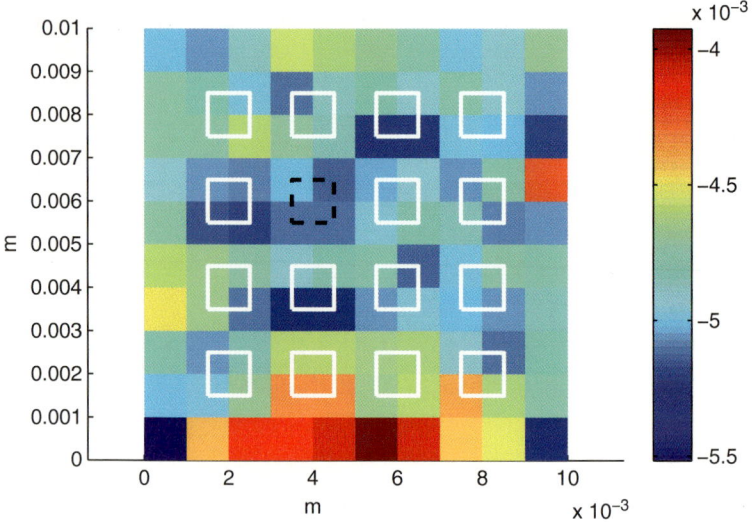

Fig. 2 Component $\{\varepsilon_{\text{exp}}^{\text{macro}}\}_{22}$ of the simulated experimental strain field at macroscale with a resolution 10×10

the strain field $\varepsilon_{\text{exp}}^{\text{macro}}$ is simulated at macroscale in extracting the values of the displacement field in a regular grid of 10×10 nodes and in using a finite element interpolation (see Fig. 1, upper right). In addition, in the subdomain defined as a square with dimension 10^{-3} m (mesoscale), the measurements of the strain field $\varepsilon_{\text{exp}}^{\text{meso}}$ are simulated at mesoscale in extracting the values of the displacement field in a regular grid of 100×100 nodes and in using a finite element interpolation (see Fig. 1, lower right).

Figure 2 shows the values of $\{\varepsilon_{\text{exp}}^{\text{macro}}\}_{22}$ for the simulated experimental strain field at macroscale with a resolution 10×10. The square in black dashed line represents the considered mesoscale subdomain. Figure 3 shows the values of $\{\varepsilon_{\text{exp}}^{\text{meso}}\}_{22}$ for the simulated experimental strain field at mesoscale with a resolution 100×100.

4.3 Multi-objective Optimization Problem

The identification of parameter b is carried out in searching for the optimal values a^{macro} and b^{meso} which solve the following multi-objective minimization problem

$$(a^{\text{macro}}, b^{\text{meso}}) = \underset{a \in \mathscr{A}^{\text{macro}}, b \in \mathscr{B}^{\text{meso}}}{\arg \min} \mathscr{I}(a,b), \qquad (15)$$

where $\mathscr{A}^{\text{macro}}$ and $\mathscr{B}^{\text{meso}}$ are the sets of the admissible values for a and b, and where the components $\mathscr{I}_1(a)$, $\mathscr{I}_2(b)$ and $\mathscr{I}_3(a,b)$ of vector $\mathscr{I}(a,b)$ are defined by Eqs. (5), (7) and (11).

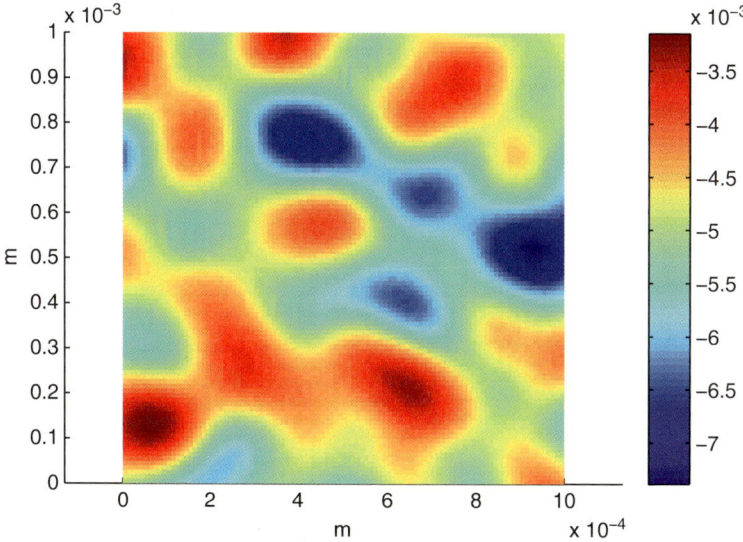

Fig. 3 Component $\{\varepsilon_{\text{exp}}^{\text{meso}}\}_{22}$ of the simulated experimental strain field at mesoscale with a resolution 100×100

The multi-objective optimization problem defined by Eq. (15) is solved in using a genetic algorithm with an initial population size of 50. Less than 100 generations has been enough for constructing the Pareto front which is iteratively constructed, at each generation of the genetic algorithm. The initial value of parameter a has been set to $a^{(0)}$ and corresponds to the solution of the following partial optimization problem: $a^{(0)} = \arg\min \mathscr{I}_1(a)$ for $a \in \mathscr{A}^{\text{macro}}$, which is solved with the simplex algorithm. Actually, the value of a^{macro} is almost unchanged through the iterations when the multi-objective problem is solved. The optimal value b^{meso} is chosen as the point on the Pareto front that minimizes the distance between the Pareto front and the origin.

4.4 Numerical Results and Validation

At macroscale, the prior model of the material is chosen as a transverse isotropic model. Consequently, in 2D plane stress, parameter $a = (E_T^{\text{macro}}, \nu_T^{\text{macro}})$ is made up of the transverse Young modulus and the transverse Poisson coefficient. The optimal value of $a = (E_T^{\text{macro}}, \nu_T^{\text{macro}})$ is $a^{\text{macro}} = (9.565 \times 10^9 \text{ Pa}, 0.3987)$.

Table 1 shows the values of $b = (\ell, \delta, E_T, \nu_T)$ for each point of the Pareto front displayed in Fig. 4. The optimal values correspond to the points 5, 6, 7, 8 and 9 where points 6 and 7 are close. The optimal value b^{meso} is such that $\ell^{\text{meso}} = 9.66 \times 10^{-5}$ m, $\delta^{\text{meso}} = 0.37$, $E_T^{\text{meso}} = 1.023 \times 10^{10}$ Pa, $\nu_T^{\text{meso}} = 0.376$. This result yields a

Table 1 Optimization in using the genetic algorithm

k	$\mathscr{I}_2(b)$	$\mathscr{I}_3(a,b)$	ℓ	δ	E_T	v_T
1	5.006529×10^{-9}	$2.311672\text{e-}01\times10^{-1}$	1.886667×10^{-4}	0.400000	1.023000×10^{10}	0.392667
2	5.006529×10^{-9}	9.477024×10^{-2}	2.500000×10^{-4}	0.400000	1.023000×10^{10}	0.392667
3	5.010827×10^{-9}	9.469903×10^{-2}	9.666667×10^{-5}	0.366667	1.023000×10^{10}	0.376200
4	5.132208×10^{-9}	9.201960×10^{-2}	1.273333×10^{-4}	0.383333	1.023000×10^{10}	0.392667
5	5.240100×10^{-9}	3.467300×10^{-2}	9.666667×10^{-5}	0.366667	1.023000×10^{10}	0.359733
6	5.259407×10^{-9}	2.455275×10^{-2}	5.066667×10^{-5}	0.350000	8.943000×10^{9}	0.293867
7	5.259407×10^{-9}	2.455275×10^{-2}	9.666667×10^{-5}	0.366667	1.023000×10^{10}	0.376200
8	5.386876×10^{-9}	2.064010×10^{-2}	5.066667×10^{-5}	0.350000	8.943000×10^{9}	0.310333
9	5.490529×10^{-9}	1.968774×10^{-2}	5.066667×10^{-5}	0.350000	1.237500×10^{10}	0.293867
10	6.57386×10^{-9}	1.962839×10^{-2}	2.193333×10^{-4}	0.400000	1.023000×10^{10}	0.392667
11	6.895467×10^{-9}	1.885624×10^{-2}	2.500000×10^{-4}	0.383333	1.023000×10^{10}	0.392667
12	7.254986×10^{-9}	1.759584×10^{-2}	2.500000×10^{-4}	0.333333	1.023000×10^{10}	0.392667
13	7.567184×10^{-9}	1.688894×10^{-2}	9.666667×10^{-5}	0.383333	1.023000×10^{10}	0.392667
14	7.996816×10^{-9}	1.623193×10^{-2}	2.000000×10^{-5}	0.350000	8.943000×10^{9}	0.310333
15	9.129340×10^{-9}	1.507042×10^{-2}	2.500000×10^{-4}	0.366667	1.023000×10^{10}	0.392667
16	9.368447×10^{-9}	1.333442×10^{-2}	1.273333×10^{-4}	0.266667	1.023000×10^{10}	0.392667

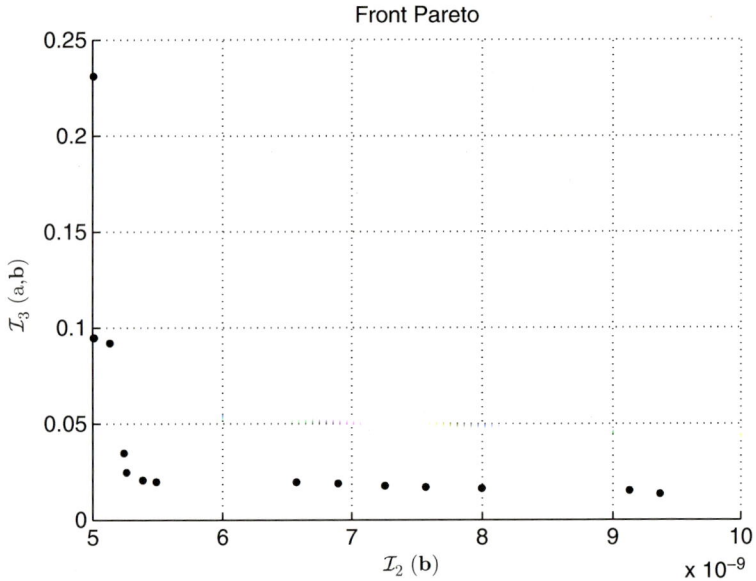

Fig. 4 Pareto front for the numerical indicators $\mathscr{I}_2(b)$ and $\mathscr{I}_3(a,b)$

validation of the proposed methodology since this identified optimal value b^{meso} is very close to the value b that has been used to construct the simulated experimental database and for which $\ell = 1.25 \times 10^{-4}$ m, $\delta = 0.4$, $E_T = 9.9 \times 10^9$ Pa, $\nu_T = 0.38$.

5 Conclusions

A multiscale inverse statistical method has been presented for the identification, in the framework of the 3D linear elasticity and in using experimental measurements at macroscale and at mesoscale, of the stochastic model of the apparent elasticity random field at mesoscale for a heterogeneous microstructure. A prior stochastic model depending of vector-valued parameter has been proposed for the apparent elasticity random field at mesoscale in the case of 2D plane stress. The identification has been formulated as a multi-objective minimization problem with respect to the parameter of the prior stochastic model. The optimal value of the parameter corresponds to the point that minimizes the distance of a Pareto front to the origin. The proposed statistical inverse method has been validated with a simulated experimental database.

Acknowledgements This research was supported by the "Agence Nationale de la Recherche", Contract TYCHE, ANR-2010-BLAN-0904.

References

Arnst M, Clouteau D, Bonnet M (2008) Inversion of probabilistic structural models using measured transfer functions. Comput Methods Appl Mech Eng 197:589–608

Avril S, Pierron F (2007) General framework for the identification of constitutive parameters from full-field measurements in linear elasticity. Int J Solids Struct 44:4978–5002

Avril S, Bonnet M, Bretelle AS, Grédiac M, Hild F, Ienny P, Latourte F, Lemosse D, Pagano S, Pagnacco E, Pierron F (2008) Overview of identification methods of mechanical parameters based on full-field measurements. Exp Mech 48:381–402

Avril S, Pierron F, Pannier Y, Rotinat R (2008) Stress reconstruction and constitutive parameter identification in plane-stress elastoplastic problems using surface measurements of deformation fields. Exp Mech 48(4):403–419

Babuska I, Tempone R, Zouraris GE (2005) Solving elliptic boundary value problems with uncertain coefficients by the finite element method: the stochastic formulation. Comput Methods Appl Mech Eng 194(12–16):1251–1294

Babuska I, Nobile F, Tempone R (2007) A stochastic collocation method for elliptic partial differential equations with random input data. SIAM J Numer Anal 45(3):1005–1034

Baxter SC, Graham LL (2000) Characterization of random composites using a moving window technique. ASCE J Eng Mech 126(4):389–404

Besnard G, Hild F, Roux S (2006) Finite-element displacement fields analysis from digital images: application to portevin-le chatelier bands. Exp Mech 46:789–803

Bonnet M, Constantinescu A (2005) Inverse problems in elasticity. Inverse Probl 21:R1–R50

Bornert M, Brémand F, Doumalin P, Dupré JC, Fazzini M, Grédiac M, Hild F, Mistou S, Molimard J, Orteu JJ, Robert L, Surrel Y, Vacher P, Wattrisse B (2009) Assessment of digital image correlation measurement errors: methodology and results. Exp Mech 49:353–370

Bornert M, Vales F, Gharbi H, Nguyen Minh D (2010) Multiscale full-field strain measurements for micromechanical investigations of the hydromechanical behaviour of clayey rocks. Strain 46:33–46

Calloch S, Dureisseix D, Hild F (2002) Identification de modèles de comportement de matériaux solides: utilisation d'essais et de calculs. Technol Form 100:36–41

Chevalier L, Calloch S, Hild F, Marco Y (2001) Digital image correlation used to analyze the multiaxial behavior of rubber-like materials. Eur J Mech A Solids 20:169–187

Clouteau D, Cottereau R, Lombaert G (2013) Dynamics of structures coupled with elastic media – a review of numerical models and methods. J Sound Vib 332:2415–2436

Collins JC, Hart GC, Kennedy B (1974) Statistical identification of structures. AIAA J 12:185–190

Constantinescu A (1995) On the identification of elastic moduli from displacement-force boundary measurements. Inverse Probl Eng 1:293–315

Das S, Ghanem R, Spall J (2008) Asymptotic sampling distribution for polynomial chaos representation of data: a maximum-entropy and fisher information approach. SIAM J Sci Comput 30(5):2207–2234

Das S, Ghanem R, Finette S (2009) Polynomial chaos representation of spatiotemporal random field from experimental measurements. J Comput Phys 228:8726–8751

Desceliers C, Ghanem R, Soize C (2006) Maximum likelihood estimation of stochastic chaos representations from experimental data. Int J Numer Methods Eng 66:978–1001

Desceliers C, Soize C, Ghanem R (2007) Identification of chaos representations of elastic properties of random media using experimental vibration tests. Comput Mech 39(6):831–838

Desceliers C, Soize C, Naili S, Haiat G (2012) Probabilistic model of the human cortical bone with mechanical alterations in ultrasonic range. Mech Syst Signal Process 32:170–177

Geymonat G, Pagano S (2003) Identification of mechanical properties by displacement field measurement: a variational approach. Meccanica 38:535–545

Geymonat G, Hild F, Pagano S (2002) Identification of elastic parameters by displacement field measurement. C R Méc 330:403–408

Graham LL, Gurley K, Masters F (2003) Non-Gaussian simulation of local material properties based on a moving-window technique. Probab Eng Mech 18:223–234

Guilleminot J, Soize C (2011) Non-Gaussian positive-definite matrix-valued random fields with constrained eigenvalues: application to random elasticity tensors with uncertain material symmetries. Int J Numer Methods Eng 88:1128–1151

Guilleminot J, Soize C (2012) Generalized stochastic approach for constitutive equation in linear elasticity: a random matrix model. Int J Numer Methods Eng 90:613–635

Guilleminot J, Soize C (2012) Probabilistic modeling of apparent tensors in elastostatics: a maxent approach under material symmetry and stochastic boundedness constraints. Probab Eng Mech 28:118–124

Guilleminot J, Soize C (2013) On the statistical dependence for the components of random elasticity tensors exhibiting material symmetry properties. J Elast 111:109–130

Guilleminot J, Soize C, Kondo D (2009) Mesoscale probabilistic models for the elasticity tensor of fiber reinforced composites: experimental identification and numerical aspects. Mech Mater 41:1309–1322

Hild F (2002) *CORRELI^{LMT}*: a software for displacement field measurements by digital image correlation. LMT-Cachan, Internal report 254

Hild F, Roux S (2006) Digital image correlation: from displacement measurement to identification of elastic properties – a review. Strain 42:69–80

Hild F, Roux S (2012) Comparison of local and global approaches to digital image correlation. Exp Mech 52:1503–1519

Hild F, Périé JN, Coret M (1999) Mesure de champs de déplacements 2D par intercorrélation d'images: CORRELI 2D. LMT-Cachan, Internal report 230

Hild F, Raka B, Baudequin M, Roux S, Cantelaube F (2002) Multiscale displacement field measurements of compressed mineral-wool samples by digital image correlation. Appl Opt 41:6815–6828

Jeulin D (1987) Microstructure modeling by random textures. Journal de Microscopie et de Spectroscopie Electroniques 12:133–140

Jeulin D (1989) Morphological modeling of images by sequential random functions. Signal Process 16:403–431

Jeulin D (2001) Caractérisation morphologique et modèles de structures aléatoires. In Homogénéisation en mécanique des matériaux 1. Hermès Science Publications

Kaipio J, Somersalo E (2005) Statistical and computational inverse problems. Springer, New York

Lawson CL, Hanson RJ (1974) Solving least squares problems. Prentice-Hall, Englewood Cliffs

Madi K, Forest S, Boussuge M, Gailliegue S, Lataste E, Buffiere JY, Bernard D, Jeulin D (2007) Finite element simulations of the deformation of fused-cast refractories based on x-ray computed tomography. Comput Mater Sci 39:224–229

Rethore J, Tinnes JP, Roux S, Buffiere JY, Hild F (2008) Extended three-dimensional digital image correlation (X3D-DIC). C R Méc 336:643–649

Roux S, Hild F (2008) Digital image mechanical identification (DIMI). Exp Mech 48:495–508

Roux S, Hild F, Berthaud Y (2002) Correlation image velocimetry: a spectral approach. Appl Opt 41:108–115

Roux S, Hild F, Viot P, Bernard D (2008) Three-dimensional image correlation from X-ray computed tomography of solid foam. Compos A Appl Sci Manuf 39:1253–1265

Serfling RJ (1980) Approximation theorems of mathematical statistics. Wiley, New York

Soize C (2006) Non-Gaussian positive-definite matrix-valued random fields for elliptic stochastic partial differential operators. Comput Methods Appl Mech Eng 195:26–64

Soize C (2008) Tensor-valued random fields for meso-scale stochastic model of anisotropic elastic microstructure and probabilistic analysis of representative volume element size. Probab Eng Mech 23:307–323

Soize C (2010) Identification of high-dimension polynomial chaos expansions with random coefficients for non-Gaussian tensor-valued random fields using partial and limited experimental data. Comput Methods Appl Mech Eng 199:2150–2164

Soize C (2012) Stochastic models of uncertainties in computational mechanics. Lecture notes in mechanics, vol 2. American Society of Civil Engineers, Reston

Spall JC (2003) Introduction to stochastic search and optimization. Wiley, Hoboken

Ta QA, Clouteau D, Cottereau R (2010) Modeling of random anisotropic elastic media and impact on wave propagation. Eur J Comput Mech 19:241–253

Walter E, Pronzato L (1997) Identification of parametric models from experimental data. Springer, Berlin

Part IV
Multiscale Stochastic Mechanics

Stochastic Multiscale Coupling of Inelastic Processes in Solid Mechanics

Hermann G. Matthies and Adnan Ibrahimbegović

Abstract Here we consider the inelastic nonlinear response of heterogeneous materials, possibly undergoing localised failure. We regard the material to be heterogeneous on many scales, but for simplicity we only look at one scale transition. The micro-scale is regarded as incompletely known and hence uncertain, therefore modelled probabilistically. Two alternative approaches are discussed: one for localised regions where a rather detailed micro-description is necessary to capture relevant effects, and the other in domains where it is accurate enough to define phenomenological models of 'generalised standard materials' on the macro-scale, which have to be identified via micro-scale computations. Apart from a proper transfer of mechanical quantities across scales, the same has to be achieved for the stochastic part of the model. Several main ingredients of the proposed approaches are discussed in detail, including micro-structure approximation with a structured mesh, random field representation, and Bayesian updating.

Keywords Uncertainty quantification • Multiscale • Inelastic deformation • Size effect

1 Introduction

We follow Ibrahimbegović and Matthies (2012) in addressing several issues related to the numerical analysis of irreversible or inelastic processes in heterogeneous materials, chief among them how to account for heterogeneities of real materials

H.G. Matthies (✉)
TU Braunschweig, Brunswick, Germany
e-mail: wire@tu-bs.de

A. Ibrahimbegović
ENS Cachan, Paris, France
e-mail: ai@lmt.ens-cachan.fr

M. Papadrakakis and G. Stefanou (eds.), *Multiscale Modeling and Uncertainty Quantification of Materials and Structures*, DOI 10.1007/978-3-319-06331-7_9,
© Springer International Publishing Switzerland 2014

and how to transfer the appropriate information provided from fine scales. Most examples and detailed illustrations of ideas pertaining to heterogeneities and related uncertainties are given for cement-based composite materials such as concrete or mortar, certainly the most widely used man-made materials.

The mechanical behaviour of heterogeneous materials can be represented at different scales, depending upon the objectives and the physical mechanisms that are important to account for. The choice of scale is also closely related to the corresponding uncertainty description.

Given computational resources for typical engineering applications, most frequently we ought to perform an analysis at the structure scale, or macro-scale. At this scale cement-based materials—in contrast to most situations involving geological materials—might be considered as homogeneous, and their properties obtained by using the key concept of Representative Volume Element (RVE, see Bornert et al. 2001; Kanit et al. 2003) to retrieve *homogeneous* phenomenological models of inelastic behaviour (e.g. see Ibrahimbegović 2009). Those models are well known for their robustness and lead to relatively moderate computational cost. Due to these two main points, phenomenological approaches are widely used. Such models are based on a set of 'material parameters' which need to be identified (e.g. see Kučerová et al. (2009) by minimising some kind of error measure between predicted and observed response) mainly from experiments providing unique load paths and boundary conditions. In simple idealised situations this *homogenisation* can be performed analytically.

Hence this methodology leads to a set of parameters which is linked to the chosen load-path. As they are not adapted to another path, the predictive features of those phenomenological macro-models is difficult to assess. The main reasons for this is that the macro-scale is not the right scale to consider with the aim to model failure of heterogeneous materials. Many authors have tried to overcome this major drawback by furnishing micro-mechanical bases to the macroscopic model set of parameters (see Markovič and Ibrahimbegović 2004; Ladevèze et al. 2001; Zohdi and Wriggers 2005) and provide more predictive macro-scale models. One possible way to achieve such a goal is to adopt homogenisation methods which lead to accurate results for linear problems. In the presence of non-linearities such methods are usually not capable to provide good estimates for the effective (macroscopic) properties (see Gilormini 1995). Moreover, such an approach does not take into account the inherent uncertainties attached to heterogeneous materials and structures.

The homogenisation process just alluded to aims at producing a spatially *homogeneous* material model, under the assumption that the small heterogeneous scales are *infinitesimally small* compared to the macro-scale, so that only the mean response is of interest and all response variability due to heterogeneity on small scales has been averaged away. There are several situations where this philosophy underlying the homogenisation process ceases to be valid. One is when the scales are not so well separated and significant variability due to small scale *uncertainty* is still present at the macro-scale. Another situation occurs when small scale variations cause large scale effects, like in the behaviour of brittle and quasi-brittle materials where cracks and local material failures may have severe macroscopic consequences.

At scales smaller than the macroscopic one, cement based materials appear to be heterogeneous, exhibiting an important variability on all relevant scales. For geological materials this is actually mostly the case also at the macro-scale, and the 'homogenisation process' described above is of questionable validity. This variability might be viewed from the geometrical point of view, considering the arrangement (positions, shapes) of the different phases, as well as the actual properties of the different phases. Here we propose to take account of the meso-scale variability in order to compute the macroscopic (i.e. effective) properties' statistics for a heterogeneous porous medium made of a non-linear matrix. Moreover, we show how simple statistics (mainly the correlation length through the covariance function) might be used at the macroscopic level to model particular features such as the size effects.

The material parameters at the meso-scale could be assumed to be deterministic, so that the variability is only related to the size and the positions of the voids in the porous media, or they can be considered as *uncertain* as the geometrical descriptors and modelled as random fields. In order to solve this stochastic problem and compute the statistical moments for the response quantities, the Monte Carlo method is the most widely known approach. Because of their promise of higher speed and accuracy, we explore approaches which solve the stochastic problem with a so-called *functional approximation* (e.g. Matthies (2007a,b, 2008)), which in many cases can lead to drastically lower computational requirements.

To lower the computational cost for the necessary evaluation of samples with different geometries and avoid repeated re-meshing, we propose a model based on a regular mesh which is not constrained by the physical interfaces. This model relies on either finite element or discrete element representations of the material micro-structure, whose kinematic description is enriched by the use of strain and displacements discontinuities in order to represent different phases (Hautefeuille et al. 2009).

The simultaneous approach is the first we propose (Ibrahimbegović and Matthies 2012)—it could also be referred to as an *online* procedure—, where the response of the meso-structure can not be precomputed as in the sequential approach (to be described in the following). Here the interaction between macro- and meso-scale is considered so strong that it is not possible, or rather not meaningful, to try and precompute all possible responses. This may happen in regions where severe irreversible material processes occur, such that the deformation path of the macro-scale, which is imposed onto the meso-scale, influences the meso-scale response in some profound way. Our idea is that this approach can actually be combined with the sequential approach described later. The sequential approach is regarded as the 'standard' way of transferring the meso-scale properties to the macro-scale. But in the circumstances just alluded to—severe meso-scale material irreversibility—the simultaneous approach could be switched on like a magnifying 'zoom lens' (Hautefeuille et al. 2008), and the meso-scale simulation can run simultaneously coupled with the macro-scale computation. As the meso-scale is modelled probabilistically, this probabilistic content or uncertainty has also to be

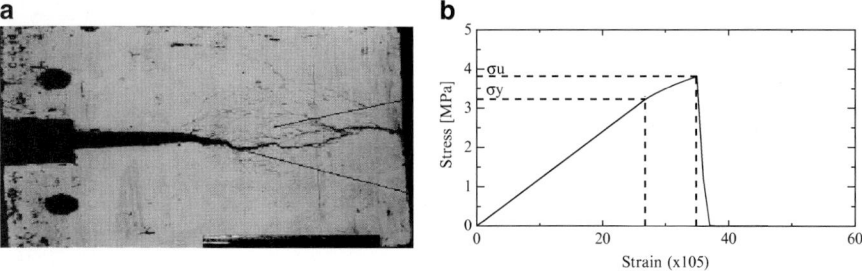

Fig. 1 Tensile test behaviour: (**a**) Typical quasi-brittle failure pattern, (**b**) Quasi-brittle material 1D model

transferred between the scales. The response at the meso-scale can then be used not only to identify the properties of macro-scale phenomenological models as in the sequential approach to be described next, but the macro-scale state of the system itself. Thus the *state* of the macro-scale is transferred or updated through a *Bayesian* procedure directly from the meso-scale.

The sequential approach is the second one we propose (Ibrahimbegović and Matthies 2012)—it could also be referred to as an *offline* procedure—, where the results obtained at the fine scale are used to define the probabilistic variation of the phenomenological model parameters used at the macro-scale. The key advantage of the sequential approach is to provide the appropriate probabilistic description in agreement with the given material microstructure.

Another very important issue is the ability to provide a sound explanation of the size effect (Colliat et al. 2007) encountered in failure phenomena of engineering structures built from quasi-brittle materials. The approach to failure analysis we propose (Ibrahimbegović and Matthies 2012) is placed within a stochastic framework, which provides a very good basis for taking into account the intrinsic randomness of the heterogeneities of real building materials: concrete, mortar, soils, or any other geo-material. Such materials have a particular mechanical behaviour, known as "quasi-brittle", which can be seen as a sub-category of softening materials (see Ibrahimbegović 2009). As a typical failure pattern we should be able to represent the fracture process zone (FPZ) along with the macro-crack that is a final threat for the structural integrity (see Fig. 1a). In the context of a simple 1D model interpretation, this behaviour can be described with four material parameters (e.g. see Ibrahimbegović and Brancherie 2003 or Brancherie and Ibrahimbegović 2009): Young's modulus E, the yield stress σ_y which induces micro-cracking or the FPZ creation, and the failure stress σ_u which induces macro-cracking after the sudden coalescence of the micro-cracks leading to a softening behaviour (see Fig. 1b). The last parameter is the fracture energy G_f, which represents the amount of energy necessary to create and open a macro-crack. Several theories exist on how to model failure in quasi-brittle materials, and most of them link the phenomenon of coalescence of micro-cracks to a size effect, a dependency on the size of a structure to its specific failure load. The aim of all those theories is to combine continuum damage mechanics (CDM), where the failure stress does not depend on the size of the structure, to linear fracture mechanics (LFM), where a size effect appears naturally, as the logarithm of the failure stress depends linearly on the logarithm of the size of the structure (see Fig. 10). It can be experimentally demonstrated that even if purely brittle materials follow LFM, quasi-brittle materials do not. They rather follow a non-linear relationship between the logarithm of the failure stress and the logarithm of the size of the structure. These materials exhibit a different size effect than the one encountered for purely brittle materials.

Both on the meso-scale level as well as on the macro-scale level, we see that mechanical models with a probabilistic description have to be dealt with both in a modelling aspect as well as numerically. As described here, on the meso-scale 'geometric' uncertainties as well as uncertainties in the values of properties have to be considered. These uncertainties, modelled probabilistically, induce uncertainties on the macro-scale. Hence on the macro-scale the continuum mechanics material description is probabilistic. For more details and a general overview of these general modelling and numerical aspects, see e.g. Matthies (2007a,b, 2008).

Due to considerations of space, in the following many of these issues can not be discussed in detail—these may be found in Ibrahimbegović and Matthies (2012) and in the references given. The outline of this chapter is as follows: Sect. 2 contains a short overview on numerical meso-structure modelling, the Sect. 2.1 is concerned with modelling random geometries with regular meshes, which can make computation of samples much easier. Some aspects of the probabilistic description of heterogeneous materials are given in Sect. 2.2. Turning to the actual computation, in Sect. 3 we present the simultaneous approach, whereas Sect. 4 is devoted to the description of the sequential approach. Concluding remarks are given in Sect. 5.

2 Meso-scale Model of Material Heterogeneities

Meshing is one of the major issues in modelling heterogeneous materials. The possibly high number of phases and their complex shapes frequently might lead to a quite high number of degrees-of-freedom and also quite distorted meshes. Moreover, the meshing process itself might consist in a complex and time-consuming algorithm.

2.1 Mechanical Meso-Scale Model

The objective of this first part is to indicate how to employ structured meshes in order to simplify the meshing process of heterogeneous materials. Hence this section presents the main ideas leading to regular meshes which are not constrained by the physical interfaces between the different phases. The key ingredients to provide such models are field discontinuities introduced inside the elements in which the

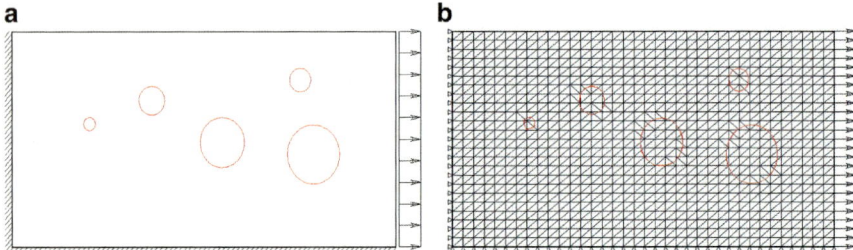

Fig. 2 Meso-structure geometry (**a**), corresponding structured mesh (**b**)

physical interfaces are present. These kinematic enhancements might be developed within the framework of the Incompatible Modes Method (see Wilson et al. 1973; Ibrahimbegović and Wilson 1991).

2.1.1   Structured Mesh and Element Kinematics Enhancements

As an example at the meso-scale consider a heterogeneous material in 2D built of different phases and assume that each of these phases is described by the inclusions positions and shapes. A typical situation is shown in Fig. 2a. In order to model such material with a structured mesh as in Fig. 2b, it is clear that one has to have the ability to represent different phases in one element. Those two phases are introduced through two types of discontinuities (see Ibrahimbegović and Melnyk 2007), namely a discontinuity of the strain field and a discontinuity of the displacement field, both of them lying at the same position (prescribed by the known physical interface between the two phases). The strain discontinuity permits the proper strain representation of two different sets of elastic properties corresponding to each phase. The displacement discontinuity leads to the possibility to model de-bonding or any failure mechanism at the interface. For the latter, two failure mechanisms are considered: one corresponding to the opening of the crack in the normal direction, and the second one to the sliding in the tangent direction (see Simo et al. 1993). Both of these discontinuities are introduced by using the Incompatible Modes Method (see Wilson et al. 1973; Ibrahimbegović and Wilson 1991). The key advantage of this method is to lead to a constant number of global degrees-of-freedom.

   Both of those kinematics enhancements are added on top of the standard constant strain triangle (CST) element. Hence this element is divided into two parts by introducing an interface whose position is obtained by the intersection of the chosen structured mesh with the inclusions placed within the structure. One of the most important and well-known features of strong (displacement field) discontinuity models is their capability to be independent from the mesh, even for *softening* laws (Ibrahimbegović and Brancherie 2003). This ability is based in the fact that the dissipation process occurs on a line (i.e. the interface) and not in the whole volume.

However, different elastic-plastic or elastic-damage behaviour laws with positive hardening might be chosen for each of the two sub-domains split by the interface with different elastic properties (see Ibrahimbegović and Markovič 2003).

2.1.2 Operator Split Solution Procedure

Deriving from the Incompatible Modes Method for the two kinds of discontinuities added on top of the classical CST element (strain field and displacement field), the total system to be solved consists of four equilibrium equations, the global equilibrium equation for the element nodal degrees of freedom (DOFs) u_e, and the local ones corresponding to normal α_I DOFs and tangential β_I DOFs for displacement discontinuities, and strain discontinuities DOFs α_{II}. The consistent linearisation (e.g. see Ibrahimbegović 2009) of this set of equations leads to a linear system, in incremental matrix form:

$$
\begin{bmatrix}
K^e & F_I^\alpha & F_I^\beta & F_{II} \\
F_I^{\alpha,T} & H_I^\alpha & F_H & F_{II}^\alpha \\
F_I^{\beta,T} & F_H^T & H_I^\beta & F_{II}^\beta \\
F_{II}^T & F_{II}^{\alpha,T} & F_{II}^{\beta,T} & H_{II}
\end{bmatrix}
\begin{pmatrix}
\Delta u_e \\
\Delta \alpha_I \\
\Delta \beta_I \\
\Delta \alpha_{II}
\end{pmatrix}
= -
\begin{pmatrix}
r \\
h_I^\alpha \\
h_I^\beta \\
h_{II}
\end{pmatrix}
\tag{1}
$$

The right hand side (rhs) are the corresponding residua. The expanded form for each block can be found in Hautefeuille et al. (2009). The operator split strategy consists in first solving the local equations of system Eq. (1) (namely the last three for $(\Delta \alpha_I, \Delta \beta_I, \Delta \alpha_{II})$ at each numerical integration point) for fixed global DOFs Δu_e. The second step is then to carry out a static condensations (e.g. see Wilson 1974). These static condensations leads to the effective stiffness matrix (see Simo and Taylor 1985; Hautefeuille et al. 2009)—the Schur complement—and residua for the global system of equations.

One of the key points to note is that the total number of global unknowns remains the same as with the standard CST element, which is a major advantage of the 'Incompatible Modes Method'. Simple illustrative examples dealing with the use of structured meshes might be found in Hautefeuille et al. (2009).

A Comparison between structured and unstructured mesh computations is shown in Fig. 3. For this we consider a porous material made of a perfectly plastic matrix with circular voids of different sizes. The first case (Fig. 3a) presents an adapted mesh obtained by using the software GMSH. Obviously, in this case each element contains only one phase (namely the matrix or the "voids"). Moreover several elements are strongly distorted and they exhibit quite different sizes. For these two reasons the stiffness matrix is poorly conditioned. The second case (Fig. 3b) relies on a structured mesh which is based on a regular grid as described above. Figure 3 shows the axial displacement contour plot (with an amplification factor of 100) for both unstructured and structured meshes. Note that both cases are providing very close results, but with a gain of computing time in favour

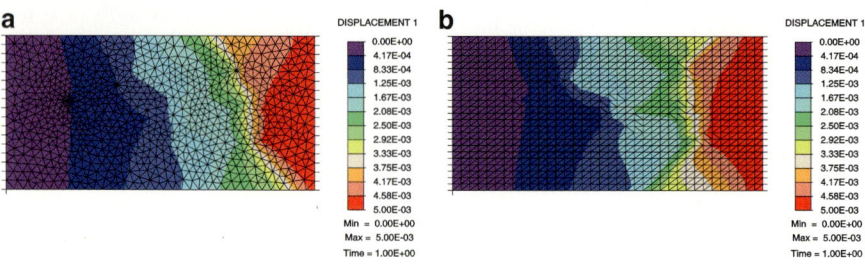

Fig. 3 Longitudinal displacement for adaptive mesh (**a**) and regular mesh (**b**)

of the structured mesh strategy (this point is mainly due to the tangent matrix optimal conditioning). Combined with a meshing process which is much easier, the structured mesh way appears to be a good and accurate method to model heterogeneous materials, especially in the context of many realisations that have to be analysed. This last point is one of the key issues considering probabilistic aspects for heterogeneous materials.

2.2 Probabilistic Aspects of Heterogenous Materials

At a finer scale than the macroscopic one, cement-based materials obviously appear to be heterogeneous. As an example, at this meso-scale mortars are made of at least three phases: two solid ones (the grains and the cement paste) and voids. It is well-known from experimental data that macroscopic properties of such materials are strongly linked to the (at least) meso-scale constituents. In Yaman et al. (2002) the authors gathered some experimental results showing the very important decrease of macroscopic mechanical strength (in tension or compression) along the increasing void volume fraction. Moreover, considering a constant porosity, the voids shapes and positions also have a major influence on the macroscopic properties, especially for small specimens. This key point is linked to the statistical RVE size (see e.g. Kanit et al. 2003), which has to be determined along a prescribed macroscopic error tolerance.

Here we only sketch briefly how heterogeneous materials may be modelled probabilistically. More details may be found in Ibrahimbegović and Matthies (2012) and the references quoted therein.

Probabilistic characterisation at the macro-scale is necessary for the sequential approach, and uses random fields as its main tool (Vanmarcke 1988; Matthies 2007b). Mechanical properties, be they scalar- or tensor-valued, are modelled as random fields. This covers both the elastic properties such as the bulk modulus, as well as the inelastic ones, which are also described by parameters like yield stress and ultimate stress in the simplest case (Rosić and Matthies 2008, 2011). One peculiarity is that most of these quantities have to be at each point and for each realisation positive scalars or positive definite tensors. In the usual approximations of random fields, which rely on interpolation or similar linear processes, this is not always easy to ascertain. One possibility is to approximate the logarithm of the quantity, this has also advantages in the identification process which has to be performed—and will be sketched later—to obtain the actual macro-values from the micro-scale (Rosić et al. 2013).

So let $\kappa(x, \omega)$ be a positive random field, where $x \in \mathcal{G}$ is a point in the spatial domain $\mathcal{G} \subseteq \mathbb{R}^n$, and $\omega \in \Omega$ is a sample in the probability space Ω of all realisations, equipped with a *probability measure* $\mathbb{P}$. One well-known approximation is the Karhunen-Loève expansion (KLE) $q(x, \omega) = \log \kappa(x, \omega) = \sum_m \xi_m(\omega)\phi_m(x)$, where $\int_\mathcal{G} \phi_k(x)\phi_m(x)\,\mathrm{d}x = \delta_{k,m}$ and $\int_\Omega \xi_k(\omega)\xi_m(\omega)\,\mathbb{P}(\mathrm{d}\omega) = \mathbb{E}(\xi_k\xi_m) = \delta_{k,m}$ are *orthonormal*, or, in probabilistic parlance, *uncorrelated* (see e.g. Matthies 2007b). To express everything in *independent* variables, we expand each $\xi_m(\omega) = \sum_i \zeta_m^i \psi_i(\boldsymbol{\theta}(\omega))$ (e.g. a polynomial chaos expansion, see e.g. Matthies 2007b) in known functions ψ_i of independent RVs $\boldsymbol{\theta}(\omega) = [\theta_1(\omega), \ldots, \theta_j(\omega), \ldots]$ to obtain

$$q(x, \omega) = \log \kappa(x, \omega) = \sum_{m,i} \zeta_m^i \psi_i(\boldsymbol{\theta}(\omega))\,\phi_m(x). \tag{2}$$

Probabilistic characterisation at the meso-scale has to be more detailed than Eq. (2). Usually it is not efficient to model the meso-scale variability just as a variability of mechanical properties. One has to try and model the different phases probabilistically fitting to the mechanical modelling in Sect. 2.1. This means that the geometric arrangement of the micro-structure has to be considered as uncertain, as hence random. In the simplest case the locations of the inclusions can be regarded as a *Gibbs process*, where a location is chosen randomly (Markovič and Ibrahimbegović 2004; Hautefeuille et al. 2008, 2012) according to a *Poisson* process, but the inclusion is only inserted if it 'fits', i.e. if there is no overlap with an inclusion which is already present.

A further refinement of this is to allow the sizes of the inclusions to vary according to some probability law, and also possibly the shapes from some parametrically defined family, like e.g. ellipsoids. As this gives only limited possibilities for the random shapes of inclusions, another, more general approach is to model each phase through a *phase field*. This is a random field $\varphi_A(x, \omega)$—and can be approximated as in Eq. (2)—which describes the presence or absence of a phase by saying that location $x \in \mathcal{G}$ is in phase 'A' if $\varphi_A(x, \omega) \in \mathcal{H}$, where $\mathcal{H}$ is the so-called 'hitting-set'. In the simplest case φ_A is a scalar field and $\mathcal{H} = [A, \infty[\subset \mathbb{R}$ is just a *level set*. The geometry of random sets thus defined can be very complex, but it can nevertheless be described mathematically through topological invariants (Adler 2008).

3 Simultaneous Approach to Multi-scale Analysis

This approach is given in Ibrahimbegović and Matthies (2012), and in more detail in Ibrahimbegović and Marković (2003), Marković and Ibrahimbegović (2004), and Hautefeuille et al. (2008, 2012); some theoretical background in Marković et al. (2005), and the computational setup including the software coupling in Niekamp et al. (2009).

We see the simultaneous approach as a way to fully resolve the meso-structural processes, without having to carry an excessive number of DOFs on the macroscopic level. In this respect this is a reduced order model. When the material experiences large strain or stress gradients, or if the pre-computed phenomenological responses on the macro-scale do not cover the material response adequately any more, e.g. when material instabilities, localisation, or cracks start developing, this is the case where the *simultaneous approach* comes into play.

3.1 Mechanical Two-Scale Coupling

A general mathematical approach for multi-scale coupling—but mainly aimed at homogenisation—has been described in Engquist (2003) and Abdulle and Nonnenmacher (2009). One approach is to have the meso-scale (or small-scale) evaluation each time a finite element on the macro-scale wants to evaluate a material response, i.e. in each Gauss-point. This has become known as the 'FE^2-method' (Feyel 1999; Feyel and Chaboche 2000), and has already been used extensively (e.g. Miehe and Koch 2002; Temizer and Zohdi 2007; Temizer and Wriggers 2008, 2011). As a Gauss-point has no extension, there is no way in this method to allow for scale-effects—the small scale is assumed to be infinitely smaller than the macro-scale—and hence it is only applicable when there is a really large separation of scales.

Here we want to allow for the scale effect and do not want to assume a separation of scales. The key idea is to have a finite element on the macro-scale as an 'empty hull' or 'window', to be filled with a meso- or small scale discretisation (see Fig. 4). Of course now the two meshes do not fit, and have to be coupled together. The simplest and most effective method is to just allow linear variations of the displacements along the meso-scale element boundaries (Ibrahimbegović and Marković 2003; Marković and Ibrahimbegović 2004; Marković et al. 2005) and use Lagrange multipliers. This is reminiscent of hybrid finite elements (Pian and Sumihara 1984), and can be performed in a completely analogous manner. The inner small-scale mesh is effectively under displacement control, which makes the computations simple and stable. The theory is described in Marković et al. (2005), and the computational coupling in Niekamp et al. (2009).

If one wants to allow higher order strain patterns to propagate through the small-scale boundaries, a more elaborate set-up is required. This problem also occurs when structures with different meshes are coupled, and one way to treat

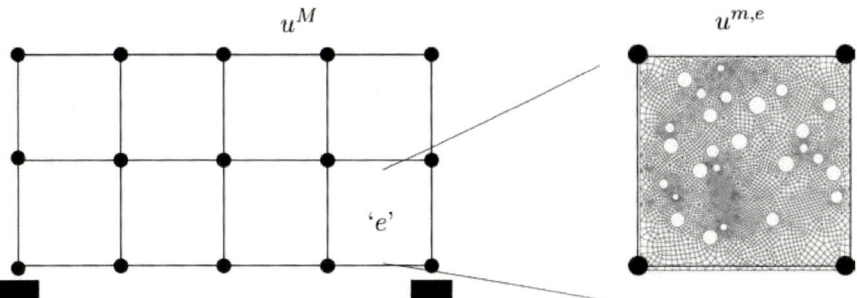

Notes: Each 'macro' finite element represents a mesoscale window containing the microstructure information, which is again modeled by a finite element mesh

Fig. 4 Macro- and meso-scale finite element model of a simple structure

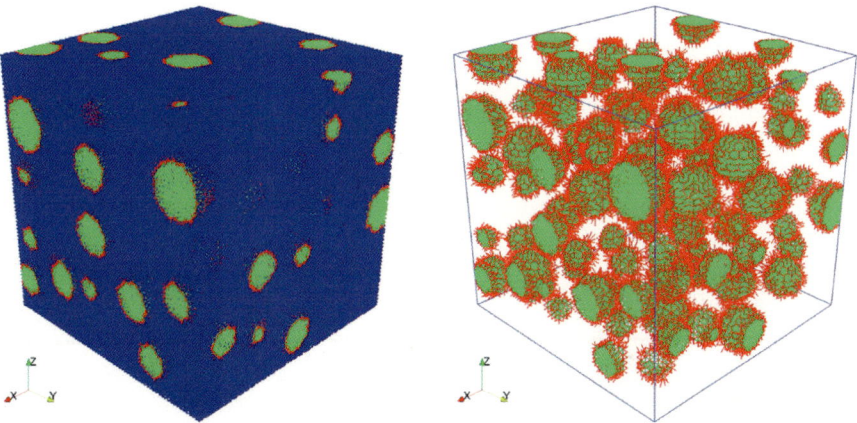

Fig. 5 Meso-scale discrete model with inclusions, shown with and without matrix and with discrete bonds

this is with localised Lagrange multipliers (Park and Felippa 2000; Park et al. 2002), and this approach has also been adopted here. A similar approach—but not with localised Lagrange multipliers—is the so-called 'Arlequin-method' (Ben Dhia 1998; Ben Dhia and Rateau 2005), and another multi-scale coupling approach may be found in Ladevèze et al. (2001) and Ladevèze and Nouy (2003). Our adaptation of the localised Lagrange multiplier method may be found in Hautefeuille et al. (2008, 2012), and a typical meso-model with a discrete—non-continuum mechanics—meso-structure is depicted in Fig. 5, as well as the de-bonding between the phases and a macro-crack developing across the macro-element in Fig. 6 in one of the tests computed.

The localised Lagrange multiplier method (Park and Felippa 2000; Park et al. 2002; Hautefeuille et al. 2008) produces upon discretisation and linearisation a system analogous to Eq. (1), just with a different interpretation of the quantities: u_e are again the macro-displacements to be retained in the global system, α_I are

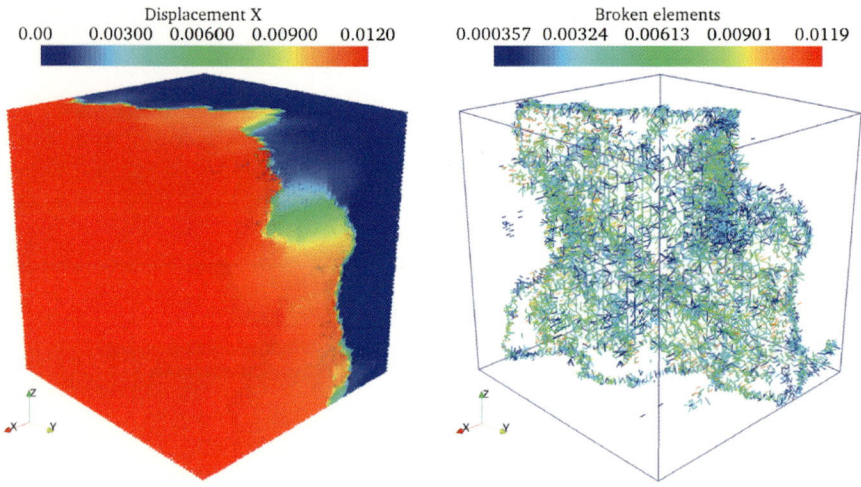

Fig. 6 Meso-model failure, showing displacements, and de-bonding (broken bonds)

the displacements of the meso-scale, $\boldsymbol{\beta}_I$ are the rigid body modes of the meso-scale, and $\boldsymbol{\alpha}_{II}$ are the localised Lagrange multipliers. And again the variables $(\Delta\boldsymbol{\alpha}_I, \Delta\boldsymbol{\beta}_I, \Delta\boldsymbol{\alpha}_{II})$ are condensed out for each meso-structure independently—and possibly in parallel—so that on the global level only the macro-displacements $\boldsymbol{u}_e$ appear. To the outside world this looks like a—maybe internally complicated—completely normal finite element which returns a macro-residual and a tangential macro-stiffness, and may be used as such in any code. The global system only carries the displacements of the 'window frame', and is hence relatively small—all the meso-scale details have been condensed away.

3.2 Probabilistic Scale Coupling

As was already mentioned, both the macro-scale as well as the small meso-scale are to be considered as uncertain, and hence are modelled in a probabilistic manner. The coupling described in the previous section describes really only the mechanical coupling, but the probabilistic information on both scales has to be coupled as well.

On both the meso-scale and the macro-scale the relevant quantities such as displacements, stresses or residua, or internal variables, are modelled as *random variables* (RVs), or rather *random fields* (RFs). In the functional approximation approach this means that, for example, the meso-displacements $\boldsymbol{\alpha}_I(\theta_1, \ldots, \theta_M) = \sum_i a_i \psi_i(\theta_1, \ldots, \theta_M)$ are expressed in some meso-structure related independent RVs $(\theta_1, \ldots, \theta_M)$ through a linear combination of some known functions ψ_i, e.g. a *polynomial chaos expansion* (PCE). The meso-structure variables $(\theta_1, \ldots, \theta_M)$ describe the uncertainty for an individual meso-patch.

If these are carried to the global level through the relations alluded to in the previous Sect. 3.1, so that we obtain $u_e = \sum_i u_i \psi_i(\theta_1, \ldots, \theta_M)$, there will be typically far too many RVs to describe just a few global variables, and some model reduction process should be carried out; the best known of these is the proper orthogonal decomposition (POD), which in the stochastic context is known under the name *Karhunen-Loève expansion* (KLE). But there is also another possibility, and this is the use of Bayes's theorem. This idea has already been suggested in Koutsourelakis and Bilionis (2010).

On each scale we regard the respective other scale as an essentially 'black box', which we try to approximate with the probabilistic description given on the scale on which we are sitting. It is then quite natural to use conditional expectation to perform the transfer (Rosić et al. 2013). Conditional expectation corresponds to a projection (Rosić et al. 2012), computed efficiently with the functional approximation, avoiding very time consuming Monte Carlo (MC) within Monte Carlo as, for example, in the Markov Chain Monte Carlo (MCMC) method, see Rosić et al. (2013) for a general description of computational possibilities. The quantities describing the *state* of the system are identified, i.e. displacements and forces or stresses and strains. Of course a tangent matrix may also be identified if needed in the computational procedure, e.g. in Newton's method.

3.3 Computational Coupling

After the mathematics of the coupling has been sketched, we turn to the code coupling required for the simulation. This was performed with the component framework already mentioned in Niekamp and Matthies (2004), Niekamp et al. (2009), Krosche et al. (2003), and Krosche and Matthies (2008), which is called 'Component Template Library (CTL)', and the computational structure as used in this application is depicted in Fig. 7.

In Markovič et al. (2005) on each of the scales a version of FEAP (or rather coFEAP) (Zienkiewicz et al. 2005; Kassiotis and Hautefeuille 2008) was used, and the small scale just appeared as a new kind of finite element on the macro-scale. This is particularly convenient as all the technology already present in FEAP (like iterative non-linear solvers, etc.) may be continued to be used.

Another aspect which can be handled on a coarse-grained basis by the component framework is parallelisation. From Fig. 7 one may easily glean that each component—using its own resources—may run concurrently with all the other components, while all the necessary synchronisation due to the information being passed around is being taken care of by the CTL (Niekamp and Matthies 2004; Niekamp et al. 2009). Each component may be a parallel code on its own, and also use a parallel processor in this coupled computations. This may be regarded as a fine-grained parallelisation, so that we have a multi-scale parallel computation to execute the multi-scale mechanical problem, a curious duplicity.

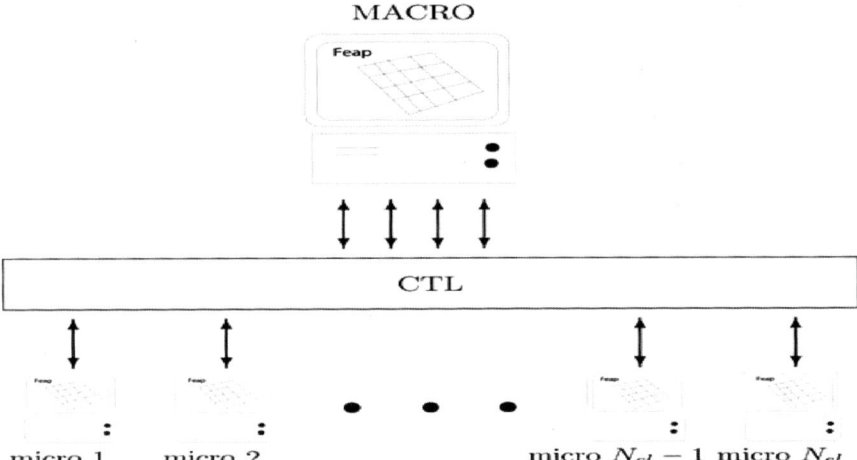

Fig. 7 Task distribution scheme on different processors. The component on top is the macro-scale component, and the 'satellites' at the bottom represent the meso-scale components. The codes which solve the macro-problem and the ones which solve the small-scale problem may be different, all that is required is that they look like a component

4 Sequential Approach to Multi-scale Analysis

At the macro-scale, the major mechanisms which are present at the meso-structure level have to be represented as well. On taking an 'energetic' look, one sees that there has to be at least an energy storage functionality for the reversible part of the material behaviour—represented through a stored energy function—and an energy dissipating functionality for the mechanically irreversible material behaviour, represented through a dissipation function. This corresponds to the description of 'generalised standard materials'. Here it is well-known that these functions can act as thermodynamic potentials resp. pseudo-potentials (see e.g. Lemaître and Chaboche 1988; Lubliner 1990; Ibrahimbegović 2009; Matthies 1991), and are thus sufficient to describe the behaviour alluded to.

4.1 Macro-scale Characterisation of Heterogeneous Materials

To achieve a probabilistic description for such a phenomenological model, it is then conceptually sufficient to model these two functions as random variables—or rather the parameters in these functions as random fields to take account of the spatial variation, as sketched in Sect. 2.2. In this short overview for the extension towards the stochastic situation we will follow the development as outlined in Matthies and Rosić (2008), Rosić and Matthies (2008, 2011), and Rosić et al. (2009).

On the macro-scale one wants to end up with—disregarding for a moment the probabilistic aspect—a conventional continuum mechanics description (Ibrahim-begović 2009). We look at an elasto-plastic material as the simplest instance of a rate-independent description. Rate dependent irreversible mechanisms can be dealt with similarly and are actually mathematically simpler. The elasto-plastic material may serve as a model problem, as other dissipative mechanisms can be formally treated in a completely analogous manner (Matthies 1991).

We look at a quadratic stored energy function (Matthies and Rosić 2008; Rosić and Matthies 2011). For the beginning it is sufficient to just consider a material point $x \in \mathcal{D}$. Denote the stress tensor by σ_x, the total strain due to some displacement u by $\epsilon_x(u)$, the plastic strain by ϵ_{px}, and the hardening variables by η_x. From these quantities we construct the generalised plastic strain $E_{px} = (0, \epsilon_{px}, \eta_x)$, and the generalised total strain $E_x = (\epsilon_x(u), \epsilon_{px}, \eta_x)$. Herewith we define the stored energy bilinear form

$$a_x(E_x, E'_x) := (\epsilon_x(u) - \epsilon_{px}) : C_x : (\epsilon'_x(u) - \epsilon'_{px}) + \langle \eta_x, H_x \eta'_x \rangle_x, \qquad (3)$$

where C_x is the fourth-order elasticity tensor at x, H_x a hardening modulus, and the bilinear form $\langle \cdot, \cdot \rangle_x$ an appropriate duality pairing depending on the specifics of the hardening variables (Matthies and Rosić 2008; Rosić and Matthies 2011). From this a Helmholtz free energy may be defined by $\psi_x(E_x) := \frac{1}{2} a_x(E_x, E_x)$.

Now if the macro-scale point $x \in \mathcal{D}$ corresponds to some meso-structure ensemble and RVE, all quantities in Eq. (3) have to be modelled as random quantities, effectively making the stored energy a random quantity. It can be shown (Matthies and Rosić 2008; Rosić and Matthies 2011), that a probabilistic plasticity problem at a material point may be formulated with the averaged quantities, i.e. the expected values of those in Eq. (3): $\mathbf{a}_x(E_x, E'_x) := \mathbb{E}\left(a_x(E_x, E'_x)\right)$. To develop an equation for the whole body, these quantities have to be integrated over the body, i.e. the domain $\mathcal{G}$, to give the probabilistic bilinear form for the whole body

$$\mathbf{a}(\mathbf{E}, \mathbf{E}') := \int_{\mathcal{G}} \mathbb{E}\left(a_x(E_x, E'_x)\right) \, \mathrm{d}x. \qquad (4)$$

A completely analogous extension has to be performed for the dissipation functional, see Matthies and Rosić (2008) and Rosić and Matthies (2011). Having obtained stochastic versions of the Helmholtz free energy and the dissipation function, one 'only' has to formally follow the normal derivation of evolution equations for generalised standard materials to obtain a stochastic version of the elasto-plastic problem with hardening (Matthies and Rosić 2008; Rosić and Matthies 2011). In Fig. 8 results of such a stochastic computation of a—in this case non-linear—elasto-plastic material are shown.

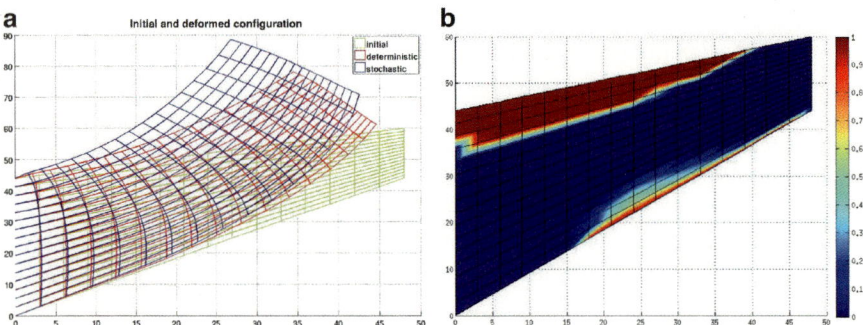

Fig. 8 Deformations (**a**) and exceedance probability of a shear stress level (**b**) for a stochastic elasto-plastic model

To actually compute the solution (Matthies and Rosić 2008; Rosić and Matthies 2011), developments which parallel the stochastic FEM (Matthies 2007a, 2008) have to be carried out, including a stochastic form of the closest point projection so well-known in computational plasticity. With the ability to perform such stochastic computations for inelastic materials, these may now serve as a macro-scale representations for heterogeneous media as explained before. What is needed additionally is the macro-scale identification of random material parameters.

4.2 Macro-scale Properties Identification

The probabilistic identification of macroscopic properties to represent the hetero-geneous random meso-structure will follow Bayes's rule, which is the preferred way of incorporating information into a stochastic model. In Rosić et al. (2012, 2013) this is described for the case where the identification is performed through measurements—although in that publication the measurements were 'virtual' ones.

The Bayesian update in its original form produces an updated, more precisely *conditional* (conditioned on the observation), probability measure. A closely con-nected alternative (Rosić et al. 2012, 2013) is to use the *conditional expectation* of that conditional probability measure to update the random variable directly. The Bayesian update for a random variable with finite variance boils down to an orthogonal projection onto the subspace of all (measurable) functions generated by a possible measurement. By sacrificing some information gained from the measurement, this can be approximated by a simpler linear—projecting only on the *linear* functions of the measurement—update which is reminiscent of the well-known Kalman filter—actually the Kalman filter can be shown (Rosić et al. 2012) to be the low-order part of this new linear update.

Here we use the example of Fig. 8 to try and identify the shear modulus from observations of the displacement under shear loading. Again no *real* experiments

Fig. 9 Identification of shear modulus through Bayesian update

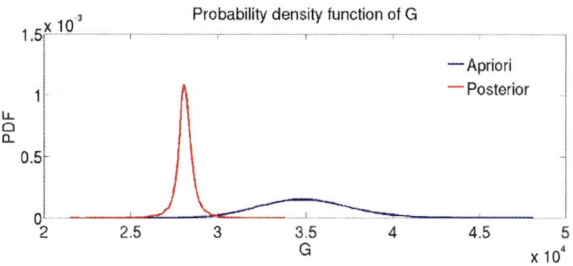

were performed, the experiments were 'virtual', i.e. computations. In Fig. 9 one may see the probability density functions prior to any identification and after a measurement—the posterior one.

For an identification of macro-scale properties we proceed similarly. We want to identify properties on the macro-scale, and we replace the measurement by a meso-structure computation, and the identification proceeds completely in the same manner. What was shown in Fig. 9 for a single parameter (the shear modulus, although we identify the logarithm of the shear modulus) can be performed for all random fields needed here, i.e. C_x and H_x which appear in the internal energy bilinear form Eq. (3)—or rather their logarithms. In the same manner the random fields describing the dissipation function or equivalently the yield criterion can be identified. This identification, as may be gleaned from Fig. 9, leaves some *residual* uncertainty in the macro-model, due to uncertainties in the micro-model, incomplete identification, and a possible inability of the macro-model to represent all detail of the micro-model.

4.3   Size Effect Representation

Size effects for quasi-brittle materials can be experimentally demonstrated at macro-scale, and several ways exist of dealing with its modelling. Most of them are linking the micro-cracks coalescence phenomenon, which consists in the failure process as a first step, to such a size effect. An extensive literature exists on that topic—it has already been noted by Leonardo da Vinci and Galileo Galilei (e.g. Jaramillo and Héctor 2011)—from the early studies of Weibull (see Weibull 1951) dealing with infinite chains built from brittle links (theory of the weakest link), to the current two concurrent theories of Bažant on the one side and of Carpinteri on another. The first one tends to describe the size effect as a deterministic theory of strength redistribution in a Fracture Process Zone (FPZ), the size of which is proportional to a characteristic length, that leads to energetic dissipation. At some level the micro-cracks coalesce and that induces both heterogeneous behaviour and some kind of localisation, and is strongly intricate to the size effect. Hence, a way to study the fracture of quasi-brittle materials is to study the size effect. Recently, Bažant has developed a new theory as a combination of this previous theory with Weibull's one leading to the so called energetic-statistical size effect (see Bažant 2004). Another theory combining a non-local model and a stochastic approach has been developed in Sab and Lalaai (1993). On the other hand, Carpenteri's theory is based on the study of quasi-brittle materials seen as materials with a fractal micro-structure (see Carpenteri 2003).

Our goal is to stress the possibility to model the size effect, taking place at the macro-scale, with the use of correlated random fields for macroscopic properties. With some basic assumptions, such macroscopic random fields are in the simplest case characterised by their marginal (point-wise) distributions as well as their spatial mean and covariance function, which of course have to match the first two moments of the marginal distribution. Adding an isotropy condition, this covariance function might be parameterised using, for example, a unique scalar value: the correlation length L_c. This length plays a key role in the context of size effects. Contrary to classical macroscopic models which are based on the RVE concept only, L_c actually defines a scale to which the whole structure size is compared. In that sense such correlated fields naturally incorporate size effects. Moreover, to some extent such a correlation length L_c might be considered as the 'characteristic length' which needs to be defined when using well-known macroscopic non-local models (Pijaudier-Cabot and Bažant 1987). Contrary to this characteristic length for which there is a lack of physical interpretation, the correlation length L_c as well as the marginal distribution necessary to characterise random fields for the macroscopic properties might be retrieved from a two-scale analysis as the one presented in the previous section.

The macro-model we consider here is based on a strong discontinuity model (see Ibrahimbegović and Melnyk 2007) which leads to the possibility to couple diffuse plasticity or damage (describing the volumetric dissipation due to the homogeneous micro-cracking which takes place in the FPZ) with surface dissipation at the macro cracks. The latter drives the stress to zero without any mesh dependency.

Considering tensile tests as shown in Fig. 10, three different lengths have been treated under displacement control (0.01, 0.1 and 1 m truss), keeping the correlation length equal to $L_c = 0.01$ m. These three cases will be called respectively small, medium and large. It is worth to note that for the small case, the bar is the same size as the correlation length (see Fig. 10a).

Figure 10b presents the cumulative density functions for the maximum load. Considering a given percentile of broken bars, it is worth to note that the smaller the bar is, the higher is its ultimate stress, e.g. 3.68 MPa for the small truss, 3.3 MPa for the medium one, and upto 2.87 MPa for the large one. In other words, the strength of the structure is directly linked to its size. The larger the structure is compared to the correlation length, the weaker it is. Hence, this stochastic way of modelling quasi-brittle failure naturally reveals the size effect.

Clearly, the correlation length here plays the key role. Comparable to the characteristic length which appears in the non-local theory, it can be linked to the size of the Fracture Process Zone (FPZ) where micro-cracking occurs. The more the size of the FPZ prevails relative to the global size of the structure (which is the case for the small bar), the more similar to continuum damage mechanics (CDM)

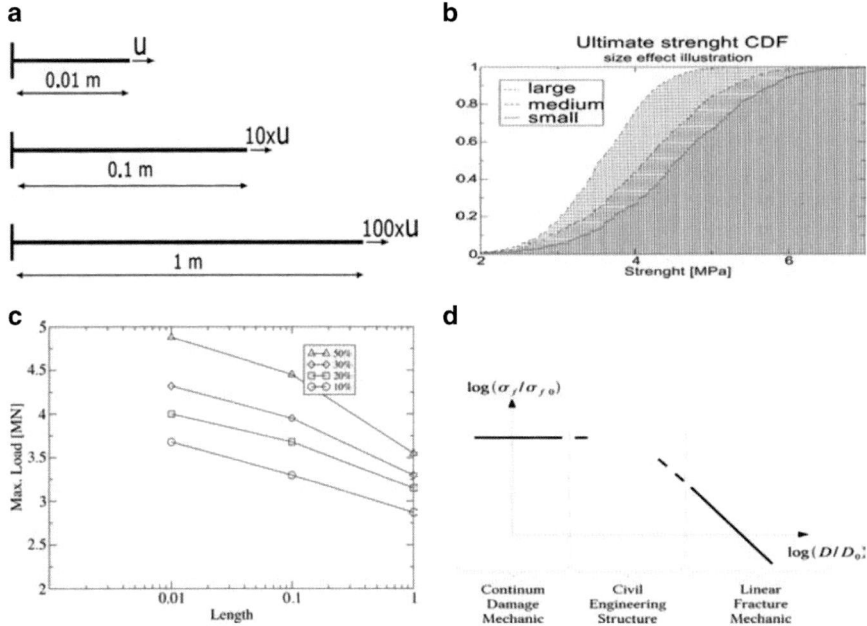

Fig. 10 (**a**) Short, medium and large bar, (**b**) Ultimate stress cumulative distribution (**c**) Probabilistic size effect diagram, (**d**) Standard size effect illustration

is the structure's behaviour. On the opposite, if the FPZ size is negligible with respect to the size of the structure (i.e. the large bar), its influence on the global behaviour of the structure is small. Thus a macro-crack occurs following linear fracture mechanics (LFM). Modelling the behaviour of quasi-brittle materials is an attempt to link these two limiting behaviours (LFM and CDM).

5   Concluding Remarks

In order to improve predictive modelling of failure of quasi-brittle materials, the meso-scale has been chosen here as the one to describe failure mechanisms. At this scale, cement-based composite materials (concrete, mortar etc.) are properly interpreted as heterogeneous and a special structured mesh methodology has been developed. Any such structured mesh relies on a regular grid where elements are not constrained to the physical interfaces between the different phases and can also contain phase interface.

With such an efficient failure modelling tool in hand we presented how to take into account the variability of the geometrical description of a heterogeneous material at the meso-scale level, as well as the modelling of material properties as random fields.

Finally, based on the macroscopic properties, we showed that the proposed macro-model might provide a straight-forward modelling for the size effect. Such size effects are a major issue in modelling quasi-brittle failure like for cement-based materials, and bridge two limited cases, the continuum damage mechanic one and linear fracture mechanics such as in Weibull's theory. In between, several authors have proposed size effects laws corresponding to different kind of structures and loading paths, or tried to model this particular feature. One attempt consists in using nonlocal models and retrieve size effects through their characteristic lengths, although this length has no physical basis. Here, we showed that the use of macroscopic correlated random fields naturally leads to size effects.

Another key ingredient of our development pertains to a phenomenological model that can account for both the fracture process zone (FPZ) and the localised failure introducing the displacement discontinuity and softening. This particular feature leads to the possibility to retrieve the size effect governed response that remains valid anywhere between the two limit cases: the one described by continuum damage mechanics, where the FPZ is the dominant failure mechanism, and another defined by linear fracture mechanics, where the displacement discontinuity quickly takes over failure leaving a negligible FPZ. This method might be viewed as an extension of Weibull's theory which can be retrieved considering uncorrelated random field.

Acknowledgements This work was partially supported by the French Ministry of Research, the Franco-German University (DFH-UFA), and the German research foundation "Deutsche Forschungsgemeinschaft" (DFG).

References

Abdulle A, Nonnenmacher A (2009) A short and versatile finite element multiscale code for homogenization problems. Comput Methods Appl Mech Eng 198:2839–2859

Adler R (2008) Some new random field tools for spatial analysis. Stoch Environ Res Risk Assess 22:809–822

Bažant ZP (2004) Probability distribution of energetic-statistical size effect in quasibrittle fracture. Probab Eng Mech 19:307–319

Ben Dhia H (1998) Problèmes mécaniques multi-échelles: la méthode Arlequin. C R Acad Sci Ser IIb 326:899–904

Ben Dhia H, Rateau G (2005) The Arlequin method as a flexible engineering design tool. Int J Numer Methods Eng 62:1442–1462

Bornert M, Bretheau T, Gilormini P (2001) Homogénéisation en mécanique des matériaux. Hermes-Science, Paris

Brancherie D, Ibrahimbegović A (2009) Novel anisotropic continuum-damage model representing localized failure. Part I: theoretical formulation and numerical implementation. Eng Comput 26(1–2):100–127

Carpenteri A (2003) On the mechanics of quasi-brittle materials with a fractal microstructure. Eng Fract Mech 70:2321–2349

Colliat JB, Hautefeuille M, Ibrahimbegović A, Matthies H (2007) Stochastic approach to quasi-brittle failure and size effect. C R Acad Sci Mech (CRAS) 335:430–435

Engquist WEB (2003) The heterogeneous multiscale methods. Commun Math Sci 1:87–133

Feyel F (1999) Multiscale FE^2 elastoviscoplastic analysis of composite structures. Comput Mater Sci 16:344–354

Feyel F, Chaboche J-L (2000) FE^2 multiscale approach for modelling the elastoviscoplastic behaviour of long fibre SiC/Ti composite materials. Comput Methods Appl Mech Eng 183: 309–330

Gilormini P (1995) A shortcoming of the classical non-linear extension of the self-consistent model. C R Acad Sci 320(116):115–122

Hautefeuille M, Colliat J-B, Ibrahimbegović A, Matthies HG (2008) Multiscale zoom capabilities for damage assessment in structures. In: Ibrahimbegović A, Zlatar M (eds) Damage assessment and reconstruction after natural desasters and previous military activities. NATO-ARW series. Springer, Dordrecht

Hautefeuille M, Melnyk S, Colliat J-B, Ibrahimbegović A (2009) Failure model for heterogeneous structures using structured meshes and accounting for probability aspects. Eng Comput 26(1–2):166–184

Hautefeuille M, Colliat J-B, Ibrahimbegović A, Matthies HG, Villon P (2012) A multi-scale approach to model localized failure with softening. Comput Struct 94–95:83–95

Ibrahimbegović A (2009) Nonlinear solid mechanics: theoretical formulations and finite element solution methods. Springer, Berlin

Ibrahimbegović A, Brancherie D (2003) Combined hardening and softening constitutive model of plasticity: precursor to shear slip line failure. Comput Mech 31:88–100

Ibrahimbegović A, Markovič D (2003) Strong coupling methods in multiphase and multiscale modeling of inelastic behavior of heterogeneous structures. Comput Methods Appl Mech Eng 192:3089–3107

Ibrahimbegović A, Matthies HG (2012) Probabilistic multiscale analysis of inelastic localized failure in solid mechanics. Comput Assist Methods Eng Sci 19:277–304

Ibrahimbegović A, Melnyk S (2007) Embedded discontinuity finite element method for modeling of localized failure in heterogeneous materials with structured mesh: an alternative to extended finite element method. Comput Mech 40:149–155

Ibrahimbegović A, Wilson EL (1991) A modified method of incompatible modes. Commun Appl Numer Methods 7:187–194

Jaramillo S, Héctor E (2011) An analysis of strength of materials from postulates of "discourses and mathematical demonstrations relating to two new sciences" of Galileo Galilei. Lámpsakos 5:53–59 (in Spanish)

Kassiotis C, Hautefeuille M (2008) coFeap's manual. LMT-Cachan internal report, vol 2

Kanit T, Forest S, Galliet I, Mounoury V, Jeulin D (2003) Determination of the size of the representative volume element for random composites: statistical and numerical approach. Int J Solid Struct 40:3647–3679

Krosche M, Matthies HG (2008) Component-based software realisations of Monte Carlo and stochastic Galerkin methods. Proc Appl Math Mech (PAMM) 8:10765–10766

Krosche M, Niekamp R, Matthies HG (2003) PLATON: a problem solving environment for computational steering of evolutionary optimisation on the grid. In: Bugeda G, Désidéri JA, Periaux J, Schoenauer M, Winter G (eds) Proceedings of the international conference on evolutionary methods for design, optimisation, and control with application to industrial problems (EUROGEN 2003), Barcelona. CIMNE

Koutsourelakis P-S, Bilionis E (2010) Scalable Bayesian reduced-order models for high-dimensional multiscale dynamical systems. The Smithsonian/NASA Astrophysics Data System. arXiv: 1001.2753v2 [stat.ML]

Kučerová A, Brancherie D, Ibrahimbegović A, Zeman J, Bitnar Z (2009) Novel anisotropic continuum-damage model representing localized failure. Part II: identification from tests under heterogeneous test fields. Eng Comput 26(1–2):128–144

Ladevèze P, Nouy A (2003) On a multiscale computational strategy with time and space homogenization for structural mechanics. Comput Methods Appl Mech Eng 192:3061–3087

Ladevèze P, Loiseau O, Dureisseix D (2001) A micro-macro and parallel computational strategy for highly heterogeneous structures. Int J Numer Methods Eng 52:121–138

Lemaître J, Chaboche JL (1988) Mécanique des Matériaux solides. Dunod, Paris

Lubliner J (1990) Plasticity theory. Macmillan, New York

Markovič D, Ibrahimbegović A (2004) On micro-macro interface conditions for micro-scale based FEM for inelastic behavior of heterogeneous material. Comput Methods Appl Mech Eng 193:5503–5523

Markovič D, Niekamp R, Ibrahimbegović A, Matthies HG, Taylor RL (2005) Multi-scale modelling of heterogeneous structures with inelastic constitutive behaviour: part I – physical and mathematical aspects. Eng Comput 22:664–683

Matthies HG (1991) Computation of constitutive response. In: Wriggers P, Wagner W (eds) Nonlinear computational mechanics—state of the art. Springer, Berlin/New York

Matthies HG (2007a) Quantifying uncertainty: modern computional representation of probability and applications. In: Ibrahimbegović A, Kožar I (eds) Extreme man-made and natural hazards in dynamic of structures. Springer, Dordrecht

Matthies HG (2007b) Uncertainty quantification with stochastic finite elements. In: Stein E, de Borst R, Hughes TRJ (eds) Encyclopedia of computational mechanics. Wiley, Chichester. doi:10.1002/0470091355.ecm071/pdf. http://dx.doi.org/10.1002/0470091355.ecm071/pdf

Matthies HG (2008) Stochastic finite elements: computational approaches to stochastic partial differential equations. Zeitschrift für Angewandte Mathematik und Mechanik (ZAMM) 88:849–873

Matthies HG, Rosić B (2008) Inelastic media under uncertainty: stochastic models and computational approaches. In: Reddy BD (ed) IUTAM symposium on theoretical, computational, and modelling aspects of inelastic media, Cape Town. IUTAM bookseries. Springer, Dordrecht

Miehe C, Koch A (2002) Computational micro-to-macro transitions of discretized microstructures undergoing small strains. Arch Appl Mech 72:300–317

Niekamp R, Matthies HG (2004) CTL: a C++ communication template library. GAMM Jahreshauptversammlung in Dresden, 21–27 Mar 2004

Niekamp R, Markovič D, Ibrahimbegović A, Matthies HG, Taylor RL (2009) Multi-scale modelling of heterogeneous structures with inelastic constitutive behaviour: part II – software coupling implementation aspects. Eng Comput 26:6–28

Park KC, Felippa CA (2000) A variational principle for the formulation of partitioned structural systems. Int J Numer Methods Eng 47:395–418

Park KC, Felippa CA, Rebel G (2002) A simple algorithm for localized construction of nonmatching structural interfaces. Int J Numer Methods Eng 53:2117–2142

Pian THH, Sumihara K (1984) Rational approach for assumed stress finite elements. Int J Numer Methods Eng 20:1638–1685

Pijaudier-Cabot G, Bažant ZP (1987) Nonlocal damage theory. J Eng Mech 113:1512–1533

Rosić B, Matthies HG (2008) Computational approaches to inelastic media with uncertain parameters. J Serbian Soc Comput Mech 2:28–43

Rosić BV, Matthies HG (2011) Stochastic Galerkin method for the elastoplasticity problem. In: Müller-Hoppe D, Löhr S, Reese S (eds) Recent developments and innovative applications in computational mechanics. Springer, Berlin, pp 303–310

Rosić B, Matthies HG, Živković M, Ibrahimbegović A (2009) Formulation and computational application of inelastic media with uncertain parameters. In: Oñate E, Owen DRJ, Suárez B (eds) Proceedings of the Xth conference on computational plasticity (COMPLAS), Barcelona. CIMNE

Rosić BV, Litvinenko A, Pajonk O, Matthies HG (2012) Sampling-free linear Bayesian update of polynomial chaos representations. J Comput Phys 231:5761–5787

Rosić BV, Kučerová A, Sýkora J, Pajonk O, Litvinenko A, Matthies HG (2013) Parameter identification in a probabilistic setting. Eng Struct 50:179–196

Sab K, Lalaai I (1993) Une approche unifiée des effects d'échelle dans les matériaux quasi fragiles. C R Acad Sci Paris II 316(9):1187–1192

Simo JC, Taylor RL (1985) Consistent tangent operators for rate-independent elastoplasticity. Comput Methods Appl Mech Eng 48:101–118

Simo JC, Oliver J, Armero F (1993) An analysis of strong discontinuity induced by strain softening solution in rate independent solids. Comput Mech 12:277–296

Temizer I, Wriggers P (2008) On the computation of the macroscopic tangent for multiscale volumetric homogenization problems. Comput Method Appl Mech Eng 198:495–510

Temizer I, Wriggers P (2011) An adaptive multiscale resolution strategy for the finite deformation analysis of microheterogeneous structures. Comput Method Appl Mech Eng 200:2639–2661

Temizer I, Zohdi TI (2007) A numerical method for homogenization in non-linear elasticity. Comput Mech 40:281–298

Vanmarcke E (1988) Random fields: analysis and synthesis. MIT, Cambridge, MA

Weibull W (1951) A statistical distribution function of wide applicability. J Appl Mech 18: 293–297

Wilson EL (1974) The static condensation algorithm. Int J Numer Method Eng 8:199–203

Wilson EL, Taylor RL, Doherty WP, Ghaboussi J (1973) Incompatible displacement models. In: Fenves S (ed) Numerical and computer models in structural analysis. Academic, New York, pp 43–57

Yaman IO, Hearn N, Aktan HM (2002) Active and non-active porosity in concrete. Part I: experimental evidence. Mater Struct 15:102–109

Zienkiewicz OC, Taylor RL, Zhu JZ (2005) The finite element method, vols 1 and 2, 6th edn. Elsevier Butterworth Heinemann, Oxford

Zohdi TI, Wriggers P (2005) Introduction to computational micromechanics. Springer, Berlin

A Note on Scale-Coupling Mechanics

X. Frank Xu, Guansuo Dui, and Qingwen Ren

Abstract After two decades of rapid growth, multiscale research is currently undergoing a transition from an initial surge of excitement to a more rational stage. The major hindrance in multiscale modeling is continuous use of classical scale-separation theories in conflict with scale-coupling phenomena, and often a price to pay is we end up with some physically inconsistent models or parameters, e.g. non-local theory in damage mechanics, negative mass or moduli in elastodynamics. In this note we propose development of new scale-coupling theories as the most important direction of multiscale research. In Sect. 2, a conceptual distinction is made between scale-separation and scale-coupling strategies. In Sect. 3, a scale-coupling mechanics theory is introduced in a format that leads to derivation of non-local and gradient theories. Conclusion is made at the end.

Keywords Scale coupling • Green function • Multiscale • Non-local • Strain gradient

1 Introduction

Multiscale research began to formally emerge in science and engineering fields in early 1990s, not by coincidence, around the time when the first carbon nanotubes were discovered leading to the birth of nanotechnology. After two decades of rapid growth, multiscale research is currently undergoing a transition from an initial surge of excitement to a more rational stage.

X.F. Xu (✉) • G. Dui
School of Civil Engineering, Beijing Jiaotong University, Beijing 100044, China
e-mail: xixu@bjtu.edu.cn

Q. Ren
College of Mechanics and Materials, Hohai University, Nanjing 210098, China

M. Papadrakakis and G. Stefanou (eds.), *Multiscale Modeling and Uncertainty Quantification of Materials and Structures*, DOI 10.1007/978-3-319-06331-7__10,
© Springer International Publishing Switzerland 2014

In this note, we emphasize lack of scale-coupling mechanics theories as the most critical issue. When classical scale separation theories are obstinately or incautiously applied to scale-coupling phenomena, a price to pay is we often end up with some physically inconsistent models or parameters, e.g. non-local theory in damage mechanics, negative mass or moduli in elastodynamics, etc. In Sect. 2, a conceptual distinction is made between two fundamental multiscale strategies, i.e. scale-separation (or homogenization) and scale-coupling. In Sect. 3, a scale-coupling mechanics theory is introduced in a format that leads to derivation of non-local and gradient theories. Conclusion is made at the end.

2   Scale Coupling or Decoupling

2.1   Classical Modeling Methodologies

Before proceeding to a discussion about multiscale strategies, let us have a brief overview of modeling methodologies in mechanics of materials. There are two classical and fundamental modeling methodologies, i.e. phenomenological and micro-macro, as shown in Fig. 1. A phenomenological modelling process typically consists of the following basic steps:

1. Define a number of state variables and parameters (e.g. constitutive parameters);
2. Formulate a mathematical model using the defined variables and parameters to describe the observed phenomenon;
3. Find the parameters and validate the model experimentally.

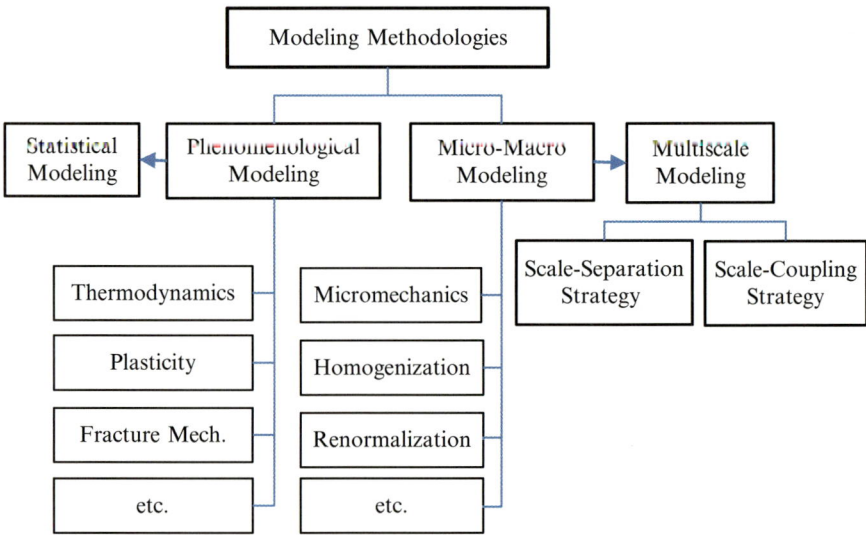

Fig. 1 Diagram of modeling methodologies

Most models developed in science and engineering are phenomenological ones, e.g. thermodynamics, plasticity, fracture mechanics, etc. A major drawback of the methodology lies in its lack of incorporation of micro-mechanisms and thereby of in-depth physical insights. When dealing with a complex phenomenon involving nonlinearity, multiple scales, high dimension, or uncertainty, a typical pitfall in phenomenological modelling is the number of variables and parameters can rise to a level that is hardly amenable theoretically. In an extreme case, by purely using curve fitting parameters a phenomenological model degenerates to an entirely statistical model with no any input or understanding of underlying physical mechanisms. For example, when a constitutive model of an unsaturated soil contains many parameters with unclear physical meaning, the applicability and generality of the model can be severely restricted, and the model is practically close to a statistical one. On the other hand, it should be noted that statistical modeling remains to be the most essential approach to tackle highly complex phenomena of which we have primitive understanding, e.g. in the field of life science.

Micro-macro methodology pertains to analytical or numerical derivation of macro-scale laws or properties based on certain validated micro-scale laws and properties. In general there are no special assumptions or experimentally fitted parameters, and in this case a micro-macro model can also be called a first principle or ab initio model. For example, given the Newton's laws of motion and the inter-atomic potential, macroscopic thermodynamics properties of an ideal gas can be numerically predicted by using molecular dynamics. Compared with phenomenological modelling, micro-macro modelling has two major advantages:

1. An explicit micro-macro relation becomes known, thereby allowing optimization of micro-scale (geometric, physical, chemical) properties to achieve certain desired macro-scale laws or properties;
2. Analytical or numerical implementation can be frequently used to substitute more expensive physical testing, and to resolve many otherwise experimentally challenging or prohibitive problems.

The so-called micro-scale and macro-scale are two scales relative to each other with a wide scale separation in either space or time. For example, the grain scale is a micro-scale with respect to a specimen of steel bar, but a macro-scale with respect to the iron atoms. Two famous micro-macro models in early stage of modern science are statistical mechanics and Brownian motion. Due to technological limitation in small scale physical measurement, early micro-macro models such as statistical mechanics and Brownian motion, adopt the top-down strategy, i.e. derivation of micro-scale properties from macro-scale measurement. A simple example is we can estimate the variance of molecular velocity in the air by measuring a mercury thermometer. Since 1950s, the bottom-up strategy emerged, with three representative micro-macro theories developed in mechanics, mathematics, and physics, respectively:

- Micromechanics theory based on a Representative Volume Element concept
- Homogenization theory based on a periodic unit cell concept
- Renormalization theory based on a scale-invariance or self-similarity concept

2.2   Multiscale Methodology

With the fast growth of computing technology and rising of nanotechnology, multiscale modeling has emerged as a new methodology since 1990s. While multiscale methodology can be alternatively treated as an extension of classical micro-macro methodology (Fig. 1), it does contain several important new features as follows.

1. Classical micro-macro models deal with scale separation or scale-invariance problems, with the results normally obtained analytically. In a scale-coupling problem whereas the scales are not well separated, in general there is no analytical solution available and the problem has to be resolved numerically by using a certain multiscale computational model. It is therefore not surprising that most of complicated unsolved problems in mechanics are scale-coupling ones, from classical problems such as turbulence, boundary layer, fatigue cracks to novel applications like MEMS, metamaterials, etc.;
2. Classical micro-macro models deal with two scales only with no consideration of intermediate scales, except for renormalization theory treating an infinite number of intermediate scales self-similarly. Multiscale methodology enables modeling of microscopic effects propagating across multiple discrete scales explicitly. For example, a micro-crack initiated from atomistic defects gradually extends into a macro-crack with its length comparable to the width of a beam specimen. In this example there are three scales involved, namely, atomistic scale, crack length scale, and specimen size scale. A multiscale model based on hybrid of molecular dynamics and finite element method can be used to simulate the above crack propagation;
3. Classical micro-macro models have been developed in individual traditional scientific and engineering disciplines. Many such models in different research fields are actually quite similar, overlapping, or complementary to each other, simply due to disciplinary barriers. New multiscale models have been developed with synergistic efforts crossing disciplinary boundaries.

According to the condition of scale separation or scale coupling, there are two distinctive multiscale strategies, i.e. scale-decoupling and scale-coupling. Classical mechanics theories are formulated based on scale separation, e.g. elasticity, thermodynamics, and density functional theory, in which the effect of uncertainty is minimized in the process of energy summation according to the central limit theorem. There are however many unsolved challenging issues, e.g. turbulence, fractal fracture, fatigue failure, critical phenomena, etc., which are typically characterized with complexity of amplified uncertainty amid scale-coupling interactions, even subjected to questioning of computability (Belytschko and Mish 2001). The major hindrance, in our opinion, is continuous use of classical scale-separation theories in conflict with scale-coupling phenomena that dominate the above unsolved problems. Lack of rigorous scale-coupling mechanics theories is considered to be

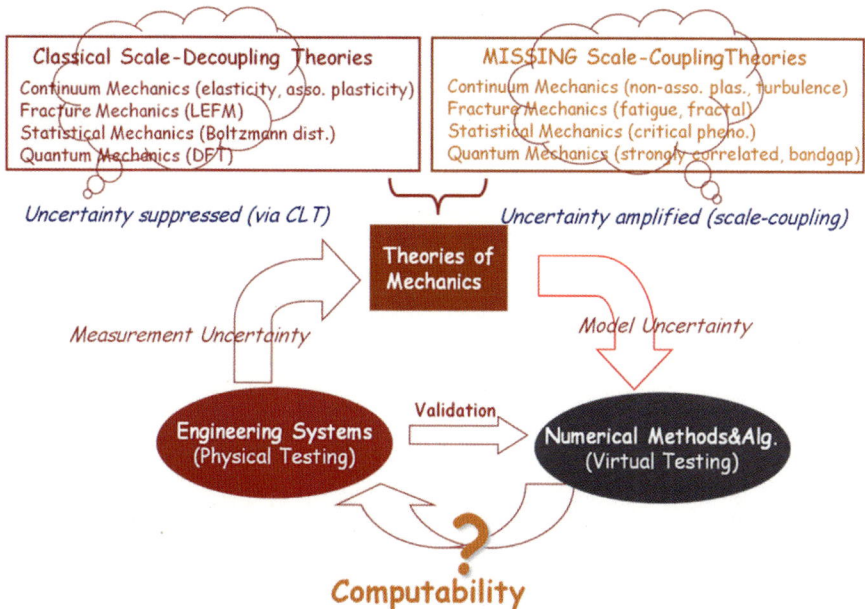

Fig. 2 Missing scale-coupling theories across scales

the most critical issue in the field of multiscale modeling. In Fig. 2, this critical issue is illustrated along with uncertainty propagating through theory, physical testing, and virtual testing of engineering systems.

3   A Scale-Coupling Mechanics Theory

3.1   Existing Methods to Address Scale Coupling Effect

Little theoretical work has been specifically devoted to scale-coupling problems. The only existing relevant theoretical framework is generalized continua theory (or mixtures theory) originated from the idea of Cosserat continua and developed by Eringen (1999), Toupin (1962), Mindlin (1964), Germain (1973) from different perspectives. Generalized continua theory attempts to explicitly account for scale-coupling effects of microstructure by introducing additional degrees-of-freedom or internal variables (e.g. a spin field in micropolar continua, a micro-displacement field in micromorphic continua) into a continuum framework.

While generalized continua theory is mathematically sound, its application to engineering problems remains very limited and the theory itself has been subjected to considerable debates. Main reasons are attributed to lack of clarity and experimental methods surrounding unconventional higher order variables, issues related to boundary conditions for higher order tensors such as couple forces and couple stresses, and difficulties of resolving boundary layer effects (see e.g. Kunin 1983). Many relevant works on generalized continua factually degenerate into purely mathematical endeavors, and the specific area becomes largely occupied by mathematicians, with one exception on the strain gradient plasticity. The strain gradient plasticity theory, with its fundamental roots in generalized continua theory, has been developed with a close tie to phenomenological size effect of plasticity. Following the first proposal of the theory (Aifantis 1984), further development has been made with phenomenological modeling (Fleck and Hutchinson 1997; Nix and Gao 1998, etc.). As microstructure information have not yet been directly taken into account, the major issues of the theory center on the interpretation of the internal length scale (Evans and Hutchinson 2009).

Relevant to the above strain gradient theory, non-local constitutive laws have been proposed in elasticity, damage, and plasticity (Rogula 1982; Eringen 1983; Bazant and Pijaudier-Cabot 1988, etc.). As shown next, the non-local and strain gradient formulations are convertible from each other. A major issue for the non-local theory is lack of physical ground in determination of the non-local kernel, while the choice of the latter affects the modeling results significantly.

3.2 Scale Coupling Mechanics Formulation

In (Xu 2009) variational principles are formulated for scale-coupling elastic composites, based on which a Green-function-based multiscale method is developed to compute scale-coupling boundary value problems (BVPs) (Xu et al. 2009). Below the proposed scale coupling mechanics (SCM) theory is presented in a form leading to derivation of non-local and strain gradient theories.

3.2.1 Derivation of the Non-local Constitutive Law

An elasticity BVP characterized with spatially heterogeneous moduli $\mathbf{L}(\mathbf{x})$ in domain D reads

$$\nabla \cdot \boldsymbol{\sigma} + \mathbf{f} = 0 \tag{1a}$$

$$\boldsymbol{\sigma} = \mathbf{L}(\mathbf{x})\mathbf{e} \tag{1b}$$

$$\text{BC } \mathbf{u} = \tilde{\mathbf{u}} \quad \partial D_u \tag{1c}$$

$$\text{BC } \boldsymbol{\sigma} \cdot \mathbf{n} = \tilde{\mathbf{t}} \quad \partial D_t \tag{1d}$$

which can be decomposed into a slow-scale BVP with spatially homogeneous moduli $\mathbf{L}_0$

$$\nabla \cdot \boldsymbol{\sigma}_0 + \mathbf{f} = 0 \tag{2a}$$

$$\boldsymbol{\sigma}_0 = \mathbf{L}_0 \mathbf{e}_0 \tag{2b}$$

$$\text{BC } \mathbf{u}_0 = \tilde{\mathbf{u}} \ \ \partial D_u \tag{2c}$$

$$\text{BC } \boldsymbol{\sigma}_0 \cdot \mathbf{n} = \tilde{\mathbf{t}} \ \ \partial D_t \tag{2d}$$

and a fast scale BVP

$$\nabla \cdot \boldsymbol{\sigma}^* + \nabla \cdot \mathbf{p} = 0 \tag{3a}$$

$$\boldsymbol{\sigma}^* = \mathbf{L}_0 \mathbf{e}^* \tag{3b}$$

$$\text{BC } \mathbf{u}^* = 0 \ \ \partial D_u \tag{3c}$$

$$\text{BC } \boldsymbol{\sigma}^* \cdot \mathbf{n} = 0 \ \ \partial D_t \tag{3d}$$

with the stress polarization

$$\mathbf{p}(\mathbf{x}) = [\mathbf{L}(\mathbf{x}) - \mathbf{L}_0] \mathbf{e}(\mathbf{x}) \tag{4}$$

The fast scale BVP solution is given in terms of Green function $\mathbf{G}$

$$\mathbf{u}^*(\mathbf{x}) = \int_D \mathbf{G}(\mathbf{x}, \mathbf{x}') \nabla \cdot \mathbf{p}(\mathbf{x}') d\mathbf{x}' \tag{5}$$

which yields

$$\mathbf{e}^*(\mathbf{x}) = -\int_D \boldsymbol{\Gamma}(\mathbf{x}, \mathbf{x}') \mathbf{p}(\mathbf{x}') d\mathbf{x}' \tag{6}$$

with the modified Green function

$$\Gamma_{ijkl}(\mathbf{x}, \mathbf{x}') = \frac{1}{2} \left(\frac{\partial^2 G_{ik}(\mathbf{x}, \mathbf{x}')}{\partial x_l \partial x_j{}'} + \frac{\partial^2 G_{jk}(\mathbf{x}, \mathbf{x}')}{\partial x_l \partial x_i{}'} \right).$$

By substituting (6) into (4), the stress polarization (4) thus can be expanded as an infinite series

$$\mathbf{p} = (\Delta \mathbf{L} - \Delta \mathbf{L} \boldsymbol{\Gamma} \Delta \mathbf{L} + \Delta \mathbf{L} \boldsymbol{\Gamma} \Delta \mathbf{L} \boldsymbol{\Gamma} \Delta \mathbf{L} - \cdots) \mathbf{e}_0 \tag{7}$$

where a contracted notation is used and $\Delta L = L - L_0$. The stress therefore can be expressed in terms of the slow-scale strain e_0

$$
\begin{aligned}
\sigma &= L e_0 + L e^* \\
&= L e_0 - L \Gamma p \\
&= (L - L \Gamma \Delta L + L \Gamma \Delta L \Gamma \Delta L - L \Gamma \Delta L \Gamma \Delta L \Gamma \Delta L - \cdots) e_0
\end{aligned}
\tag{8}
$$

which shows that the stress is non-local with respect to e_0. In the non-local theory formulated as

$$
\sigma(x) = \int_D K(x, x') e_0(x') dx'
\tag{9}
$$

the non-local kernel K is always assumed empirically, e.g. an exponential function. This kernel issue is considered to be the major drawback of the non-local theory. The above result (8) provides the rigorous mathematical expression for the non-local kernel as

$$
K = L - L \Gamma \Delta L + L \Gamma \Delta L \Gamma \Delta L - L \Gamma \Delta L \Gamma \Delta L \Gamma \Delta L - \cdots
\tag{10}
$$

In engineering practice, since the microstructure information in domain D is always incomplete, the strain in a BVP domain is conventionally obtained by using certain "effective" moduli, corresponding to the slow scale strain e_0. With such an "apparent" strain to represent the true strain at a particular location, as Eq. (8) shows, the constitutive law at this location becomes non-local.

Remark The above result also explains clearly why certain regularization, such as the non-local or gradient formulation, should be necessarily taken to prevent spurious mesh dependence of material softening problems. For a BVP even initially characterized with homogeneous moduli, increase of the loading upon it will eventually yield microcracks leading to a heterogeneous field $\Delta L(x)$, and the consequent non-local kernel K Eq. (10).

3.2.2 Representative Volume Element

Consider a subdomain D_0 in which the slow scale strain e_0 is approximately constant, e.g. a linear finite element domain. If the subdomain size $l \ll L$ the slow scale strain wave length and the boundary effect is negligible, the modified Green function can be approximated as the free space modified Green function Γ^∞ with its closed-form solution available in Fourier space

$$
\widehat{\Gamma}^\infty_{ijkl}(\xi) = \frac{1}{\mu_0 |\xi|^4} \left[\frac{|\xi|^2}{4} \left(\delta_{ik} \xi_j \xi_l + \delta_{il} \xi_j \xi_k + \delta_{jk} \xi_i \xi_l + \delta_{jl} \xi_i \xi_k \right) - \frac{3\kappa_0 + \mu_0}{3\kappa_0 + 4\mu_0} \xi_i \xi_j \xi_k \xi_l \right]
\tag{11}
$$

where κ_0, μ_0 denote the bulk and shear moduli, respectively, and $\boldsymbol{\xi}$ the vector of the wave number. The volume average of the stress in the subdomain D_0 is therefore expressed as

$$\langle \boldsymbol{\sigma} \rangle = (\langle \mathbf{L} \rangle - \langle \mathbf{L}\boldsymbol{\Gamma}^\infty \Delta \mathbf{L} \rangle + \langle \mathbf{L}\boldsymbol{\Gamma}^\infty \Delta \mathbf{L}\boldsymbol{\Gamma}^\infty \Delta \mathbf{L} \rangle - \langle \mathbf{L}\boldsymbol{\Gamma}^\infty \Delta \mathbf{L}\boldsymbol{\Gamma}^\infty \Delta \mathbf{L}\boldsymbol{\Gamma}^\infty \Delta \mathbf{L} \rangle - \cdots) \, \mathbf{e}_0 \tag{12}$$

where the bracket denotes the volume average. When the subdomain size l, is much greater than the characteristic length ℓ of the microstructure, the $(n+1)$-th term on the right hand side of Eq. (8) involves n-fold free-space modified Green function and $(n+1)$-th order correlation function, which is convergent to a tensor result independent of the ratio between ℓ and l, as long as the scale separation condition $\ell \ll l$ is satisfied. For example, by invoking the ergodicity assumption and denoting by the overbar the statistical average, the 3rd term $\langle \mathbf{L}\boldsymbol{\Gamma}^\infty \Delta \mathbf{L}\boldsymbol{\Gamma}^\infty \Delta \mathbf{L} \rangle = \overline{\mathbf{L}\boldsymbol{\Gamma}^\infty \Delta \mathbf{L}\boldsymbol{\Gamma}^\infty \Delta \mathbf{L}}$ involves the 3rd-order correlation function and can be analytically simplified into explicit results in terms of a bulk parameter and a shear parameter between 0 and 1 (see Milton and Phan-Thien 1982; Xu 2011).

In the scale separation case, a well-known name for the subdomain D_0 in micromechanics is Representative Volume Element (RVE). When the scales are coupling, i.e. ℓ is not significantly smaller than l, the results for all the terms on the right hand side of Eq. (8) become size-dependent. In this scale coupling case, size-dependence of so-called stochastic or statistical RVE can be numerically assessed, e.g. in (Xu and Chen 2009).

3.2.3 Derivation of Strain Gradient Formulation

By employing Taylor expansion for the slow scale strain

$$\mathbf{e}_0 \left(\mathbf{x}' \right) = \mathbf{e}_0 \left(\mathbf{x} \right) + \left(\mathbf{x}' - \mathbf{x} \right) \cdot \nabla \mathbf{e}_0 \left(\mathbf{x} \right) + \frac{1}{2} \left(\mathbf{x}' - \mathbf{x} \right) \left(\mathbf{x}' - \mathbf{x} \right) \nabla^2 \mathbf{e}_0 \left(\mathbf{x} \right) + \cdots \tag{13}$$

and substituting it into the non-local formulation Eq. (9), it leads to a strain gradient form

$$\boldsymbol{\sigma} \left(\mathbf{x} \right) = \mathbf{L} \left(\mathbf{x} \right) \mathbf{e}_0 \left(\mathbf{x} \right) + \mathbf{L}^{(3)} \left(\mathbf{x} \right) \nabla \mathbf{e}_0 \left(\mathbf{x} \right) + \mathbf{L}^{(4)} \left(\mathbf{x} \right) \nabla^2 \mathbf{e}_0 \left(\mathbf{x} \right) + \cdots \tag{14}$$

where the two higher order constitutive tensors

$$\mathbf{L}^{(3)} \left(\mathbf{x} \right) = \int_D \mathbf{K} \left(\mathbf{x}, \mathbf{x}' \right) \left(\mathbf{x}' - \mathbf{x} \right) d\mathbf{x}' \tag{15}$$

$$\mathbf{L}^{(4)} \left(\mathbf{x} \right) = \frac{1}{2} \int_D \mathbf{K} \left(\mathbf{x}, \mathbf{x}' \right) \left(\mathbf{x}' - \mathbf{x} \right) \left(\mathbf{x}' - \mathbf{x} \right) d\mathbf{x} \tag{16}$$

The above strain gradient formulation is first derived by Drugan and Willis (1996) by using the statistical average of stress and strain, while the statistical average concept is not directly linked to the existing strain gradient theory. Hereby Eq. (14) shows that, by using the slow-scale strain, no statistical average is needed and all the higher order constitutive tensors of the strain gradient theory are exactly derivable from the SCM formulation.

Remark As shown in Eq. (14) the main ingredient of generalized continua theory is also recovered from the SCM formulation. In fact the fast-scale and slow-scale variables employed in the SCM theory correspond to micro- and macro-volume defined in Mindlin's work (1964). It is thought that the decomposition made in Eqs. (1, 2 and 3) and the use of Green function are theoretically clearer and simpler than the original formulation developed in generalized continua theory.

4   Conclusion

In this note the scale-coupling issue is emphasized as the most critical research direction of multiscale research. By using an elasticity scale-coupling BVP as a benchmark model, three types of scale coupling/decoupling conditions are demonstrated:

1. The RVE size should be much smaller than the slow scale strain wave length, i.e. $l \ll L$;
2. The characteristic length of microstructure should be much smaller than the RVE size, i.e. $\ell \ll l$;
3. The scale-coupling effect is always present within a boundary layer with its thickness about the minimum RVE size as quantified in (Xu and Chen 2009).

The above conditions based on spatial length scales equally apply to wave propagation in complex media. There are also scale coupling problems in time domain, e.g. in stochastic dynamics the use of white noise input is an important scale separation assumption with respect to the relaxation time or the natural period of the system, which leads to nice analytical results. When the characterization time length of an excitation is close to the natural period of the system, a so-called "probabilistic resonance" phenomenon occurs, as noted in a recent report (Xu 2014). The scale-coupling effect in one-dimensional time domain is expected to be much more amplified than in three-dimensional spatial domain, especially with regard to propagation of uncertainty, e.g. in chaotic systems.

Acknowledgement This material was supported by National Science Foundation of China (11132003).

References

Aifantis EC (1984) On the microstructural origin of certain inelastic models. Trans ASME J Eng Mater Tech 106:326–330

Bazant ZP, Pijaudier-Cabot G (1988) Nonlocal continuum damage, localization instability and convergence. J Appl Mech 55:287–293

Belytschko T, Mish K (2001) Computability in non-linear solid mechanics. Int J Numer Methods Eng 52:3–21

Drugan WJ, Willis JR (1996) A micromechanics-based nonlocal constitutive equation and estimates of representative volume element size for elastic composites. J Mech Phys Solid 44:497–524

Eringen AC (1983) Theories of nonlocal plasticity. Int J Eng Sci 21:741–751

Eringen AC (1999) Microcontinuum field theories. I. Foundations and solids. Springer, New York

Evans AG, Hutchinson JW (2009) A critical assessment of theories of strain gradient plasticity. Acta Mater 57:1675–1688

Fleck NA, Hutchinson JW (1997) Strain gradient plasticity. Adv Appl Mech 33:295–361

Germain P (1973) Method of virtual power in continuum mechanics. 2. Microstructure. SIAM J Appl Math 25(3):556–575

Kunin IA (1983) Elastic media with microstructure, vol 2. Three-dimensional models. Springer, Berlin

Milton GW, Phan-Thien N (1982) New bounds on effective elastic moduli of two-component materials. Proc R Soc Lond Ser A 380:305–331

Mindlin RD (1964) Micro-structure in linear elasticity. Arch Ration Mech Anal 16:51–78

Nix D, Gao H (1998) Indentation size effects in crystalline materials: a law for strain plasticity. J Mech Phys Solid 46(3):411–425

Rogula D (1982) Introduction to nonlocal theory of material media. In: Rogula D (ed) Nonlocal theory of material media, CISM courses and lectures, vol 268. Springer, Wien, pp 125–222

Toupin RA (1962) Elastic materials with couple-stresses. Arch Ration Mech Anal 11:385–414

Xu XF (2009) Generalized variational principles for uncertainty quantification of boundary value problems of random heterogeneous materials. J Eng Mech 135(10):1180–1188

Xu XF (2011) On the third-order bounds of effective shear modulus of two-phase composites. Mech Mater 43(5):269–275

Xu XF (2014) Probabilistic resonance and variance spectra. In: The proceedings of second international conference on vulnerability and risk analysis and management, in press

Xu XF, Chen X (2009) Stochastic homogenization of random multi-phase composites and size quantification of representative volume element. Mech Mater 41(2):174–186

Xu XF, Chen X, Shen L (2009) A green-function-based multiscale method for uncertainty quantification of finite body random heterogeneous materials. Comput Struct 87:1416–1426

Statistical Volume Elements for Metal Foams

M. Geißendörfer, A. Liebscher, C. Proppe, C. Redenbach, and D. Schwarzer

Abstract Open cell metal foams can be represented by a network of beams. Due to the heterogeneity of the geometry, the length scale of the representative volume element is often nearly of the same order as the length scale of structures made of metal foam. Therefore, classical homogenization techniques for the computation of effective properties can not be applied. Statistical volume elements lead to apparent material properties that depend on the boundary conditions. Here, we introduce a model for structures made of metal foam that consists of two domains, an interior region and a boundary region. For both regions, unique random fields are identified by simulations of the microstructure. The model is validated by comparison with Finite Element simulations and experiments.

Keywords Metal foam • Stochastic analysis • Multiscale method • Modal analysis

1 Introduction

For heterogeneous materials, the size of the representative volume element can be quite large (Dirrenberger et al. 2014). In this case, the assumption of scale separation is not valid anymore. For metal foams, the representative volume element is estimated to consist of about 1,000 cells (Kanaun and Tkachenko 2007).

M. Geißendörfer (✉) • C. Proppe • D. Schwarzer
Institute of Engineering Mechanics, Karlsruhe Institute of Technology, Karlsruhe, Germany
e-mail: maximilian.geissendoerfer@kit.edu; proppe@kit.edu; schwarzer@kit.edu

A. Liebscher • C. Redenbach
Mathematics Department, University of Kaiserslautern, Kaiserslautern, Germany
e-mail: liebscher@mathematik.uni-kl.de; redenbach@mathematik.uni-kl.de

M. Papadrakakis and G. Stefanou (eds.), *Multiscale Modeling and Uncertainty Quantification of Materials and Structures*, DOI 10.1007/978-3-319-06331-7_11,
© Springer International Publishing Switzerland 2014

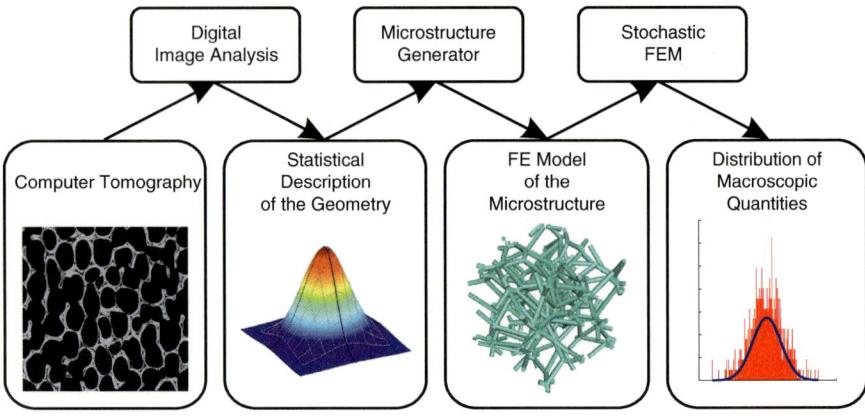

Fig. 1 Overview of the proposed computational procedure

If scale separation can not be assumed, it is still possible to compute apparent material properties (Huet 1990) and to provide a macroscopic description by random fields (Guilleminot et al. 2011). However, the apparent material properties depend on the boundary conditions. Therefore, a unique random field for the apparent material properties does not exist (Ostoja-Starzewski 2011). Recently, Di Paola (2011) proposed an averaging method that leads to unique boundary effect independent apparent properties. In this method, material properties are obtained by averaging over a volume that is smaller than the volume element on which the boundary conditions are applied.

Since the basic deformation mechanism is bending dominated, the mechanical model of metal foam is a three dimensional network of connected Timoshenko type beams with the material properties of the solid structure (Gibson and Ashby 1997). Averaging over a volume inside a beam network would predict a stiffer behavior than averaging over the whole volume element, due to the free, unconnected ends of the beams at the boundary of the volume element. Therefore, averaging over volumes smaller than the volume element will be valid only in the interior of a structure. For these reasons, a model for structures made of metal foam that consists of two domains – an interior region and a boundary region – is introduced and investigated here. In the interior region, a unique random field is identified that represents boundary effect independent apparent properties. For the boundary region, a random field is obtained by applying appropriate boundary conditions. The proposed approach allows to work with a uniquely defined random field by introducing a slightly more complex structural model.

For structures made of metal foam, uncertainties are mainly due to the heterogeneous geometry. By introducing suitable statistical volume elements, applying boundary conditions and an averaging procedure, uncertainties are propagated from the geometry to the material properties and to macroscopic structural behavior. This propagation process is sketched in Fig. 1. It applies also to other types of heterogeneous materials and macroscopic properties than those studied in this paper. This paper is organized along the propagation process as follows: the next section discusses the microstructure model, Sects. 3 and 4 treat the generation and analysis of mesoscale volume elements. In Sect. 5, predictions are compared with experiments for an open cell copper foam and in Sect. 6, conclusions are drawn.

2   Microstructure

2.1   CT Analysis

As an example of application we investigate a Duocel® copper foam. A stochastic microstructure model is developed and adapted to the geometric characteristics estimated from three-dimensional μCT images. For that purpose, ten cubes of 25 mm side length were imaged by CT with a voxel edge length of 38.15 μm. A visualization of one of the samples is shown in Fig. 2. The volume fraction of copper was on average 12.6 %. Foam cells have been reconstructed by the following procedure (Ohser and Schladitz 2009): First, the images were binarized using a global threshold. On the resulting binary images, the Euclidean distance transform was applied which assigns to each cell pixel its distance to the nearest strut. Ideally, the resulting image has local maxima at the cell centers. In practice, however, additional maxima may appear due to discretization effects and irregular cell shapes. These were removed using an adaptable h-maxima transform. Finally, the watershed algorithm was applied to the inverted distance images to separate the single cells. All image processing steps were performed using the MAVI software package (Fraunhofer ITWM, Department Image Processing (Hrsg.) 2006).

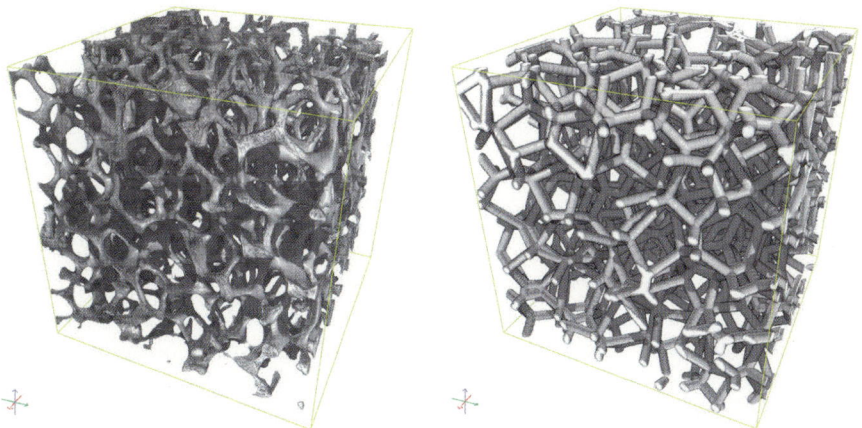

Fig. 2 Visualizations of a CT image of the Cu Duocel® foam (*left*) and the model (*right*). Visualized are 500^3 voxels

Table 1 Cell properties obtained from CT analysis

Property	Mean	Standard deviation
Diameter	5.09 mm	0.30 mm
Surface area	80.19 mm^2	9.58 mm^2
Volume	49.64 mm^3	9.12 mm^3
Facets	13.90	1.48

From the reconstructed cell systems, the geometric quantities summarized in Table 1 were estimated using minus-sampling edge correction (Ohser and Schladitz 2009). The mean number of cells per 1,000 mm^3 was 20.15. The mean cell diameters in the coordinate directions indicated an anisotropy in the cell structure. Although this can be included in the microstructure model (Redenbach 2009), a simplified model assuming isotropy of the microstructure was used here.

2.2 Microstructure Generation

Solid foams show a high variability in cell sizes and shapes, which influences their elastic properties (Zhu et al. 2000). This variability cannot be represented by deterministic models. Random tessellations (Stoyan et al. 1995) proved to be a suitable model class in this regard.

An important class of random tessellation models are Voronoi tessellations generated from realizations of random point processes. Voronoi tessellations generated by hard-core point processes are of particular interest for the modelling of foam cells due to their relatively regular cell shapes. However, the adaptability of the size distribution of the tessellation cells to the estimated size distribution is limited. To overcome this problem, weighted Voronoi tessellations can be considered. For modelling foam cells, Laguerre or power-tessellations (Aurenhammer 1987) are a promising model class. This model is defined as follows: given a set S of spheres, the Laguerre cell $C(s(x, r), S)$ of a sphere $s(x, r)$ in S with center x and radius r is defined as

$$C(s(x,r), S) = \{y \in \mathbb{R}^3 : \|y - x\|^2 - r^2 \le \|y - x'\|^2 - r'^2, \forall s(x',r') \in S\}, \quad (1)$$

where $\|\cdot\|$ denotes the Euclidean norm. The Laguerre tessellation is the set of all non-empty Laguerre cells of spheres in S. It forms a space-filling system of convex polytopes. If all spheres have equal radii, the Voronoi tessellation is obtained.

If the set S forms a system of non-overlapping spheres, each sphere is completely contained in its Laguerre cells. Consequently, the volume distribution of the spheres can, to a certain degree, be used to control the volume distribution of the cells. A method for adapting Laguerre tessellations generated by hard sphere packings to real foams based on the statistical analysis of CT images is presented in Redenbach (2009). The superiority of Laguerre tessellations over Poisson and hard-core

Voronoi tessellations has been shown in Lautensack (2008) and Hardenacke and Hohe (2010). Laguerre tessellations were used to determine the elastic properties of a representative volume element of metal foam in Kanaun and Tkachenko (2008) and Hardenacke and Hohe (2009).

Based on the data shown in Table 1, a Laguerre tessellation was fit to the foam structure using the procedure introduced in Redenbach (2009). A system of non-overlapping spheres simulated by the force-biased algorithm was chosen to reproduce the regularity of the observed cell shapes. The lognormal distribution provided a good fit to the cell volume distribution of the foam. Therefore, it was also chosen for the volume distribution of the generating spheres. The probability density function of this distribution family is given by

$$p(r) = \frac{\exp\left(-\frac{(\log r - m)^2}{2\sigma^2}\right)}{\sqrt{2\pi}\sigma r}, \quad r \geq 0, \tag{2}$$

with parameters $m \in \mathbb{R}$ and $\sigma > 0$.

To determine the model parameters, the geometric characteristics of the foam cells were compared to the characteristics of the tessellation cells using the following distance measure. Denote by $\hat{c}_i, i = 1, \ldots, 8$, the eight quantities given in Table 1 and let $c_i(m, \sigma), i = 1, \ldots, 8$, be estimates of these quantities obtained from Laguerre tessellations with parameters m and σ for the sphere volume distribution. The optimal parameters are those, for which the relative distance

$$d_{m,\sigma} = \sqrt{\sum_{i=1}^{8} \left(\frac{\hat{c}_i - c_i(m, \sigma)}{\hat{c}_i}\right)^2} \tag{3}$$

is minimized. In the application, the optimal parameters for the volume distribution were found to be $m = 1.0508$ and $\sigma = 0.2849$. Visualizations of one of the CT images and of the fitted model are shown in Fig. 2.

Until now, we only considered the cell system of the foam. In a second modeling step, the actual open- or closed-cell foam model is derived from the edges or facets of the tessellation model by morphological operations (Soille 1999). When modeling open-cell foams, the cross-section thickness along the strut is usually kept constant.

Locally variable strut thickness was considered by Kanaun and Tkachenko (2008) and Liebscher and Redenbach (2013). Here, the strut thickness is modeled as a polynomial of the distance x from the strut center to the adjacent nodes. In Liebscher and Redenbach (2013) it turned out that a polynomial of the form

$$p(x) = ax^8 + bx^4 + cx^2 + d, \tag{4}$$

$a, b, c, d \in \mathbb{R}$ results in the best model for the strut thickness.

3   Determination of Linear Elastic Properties

3.1   Stiffness Tensor

In order to compute the linear elastic properties of metal foam, mesoscopic volume elements were created with the microstructure generator and boundary conditions yielding an upper (kinematic uniform boundary conditions, KUBC) and a lower bound (static uniform boundary conditions, SUBC) for the compliance tensor (Hazanov and Huet 1994) were applied. This procedure is often utilized in the context of homogenization techniques (see e.g. Kanit et al. 2003; Ostoja-Starzewski 2007), but mainly for the determination of the size of a statistically representative volume element (RVE). The size of the RVE is defined by the element size, for which these two bounds converge against the same value. Therefore, the mechanical properties of the RVE are theoretically deterministic – in the sense of being accurate enough to represent the mean constitutive response (Drugan and Willis 1996).

It is well known that these homogenization techniques are based on the condition that the scale of the microstructure and the scale of the observed mechanical properties can be separated due to a large difference in their characteristic lengths. Unfortunately, the characteristic length scale of metal foam is in many applications not much smaller than the characteristic length scale of the structure to be investigated. For these reasons, homogenization schemes can not be applied. One has to consider stochastic volume elements (SVE) instead. In this case the above mentioned method of loading different boundary conditions can still be adopted yielding so-called apparent properties (Huet 1990)

$$\sigma_0 = \mathbf{C}_{apparent}^{SUBC} <\epsilon>,$$
$$<\sigma> = \mathbf{C}_{apparent}^{KUBC}\epsilon_0, \tag{5}$$

where $<.>$ denotes the volume average, σ and ϵ the stress and strain tensor and σ_0 and ϵ_0 the imposed stresses and strains according to SUBC and KUBC, respectively. When ensemble averaged, the apparent properties yield bounds for the effective material properties of interest. For a larger SVE the bounds become closer and their scatter smaller (Ostoja-Starzewski 2007). Here, the aim is not to compute effective properties, but to model the scatter in the stiffness tensor and to predict its consequences on the natural frequencies. Under assumption of the Hill condition, the following inequality holds for the apparent stiffness tensor $\mathbf{C}_{app}$ (Hazanov and Huet 1994):

$$\mathbf{C}_{apparent}^{SUBC} \leq \mathbf{C}_{app} \leq \mathbf{C}_{apparent}^{KUBC}, \tag{6}$$

100 SVEs with a side length of 25 mm are generated and each SVE is loaded by different load cases for the boundary conditions mentioned above and solved with the help of the finite element method. After that, the apparent material parameters are calculated from the results.

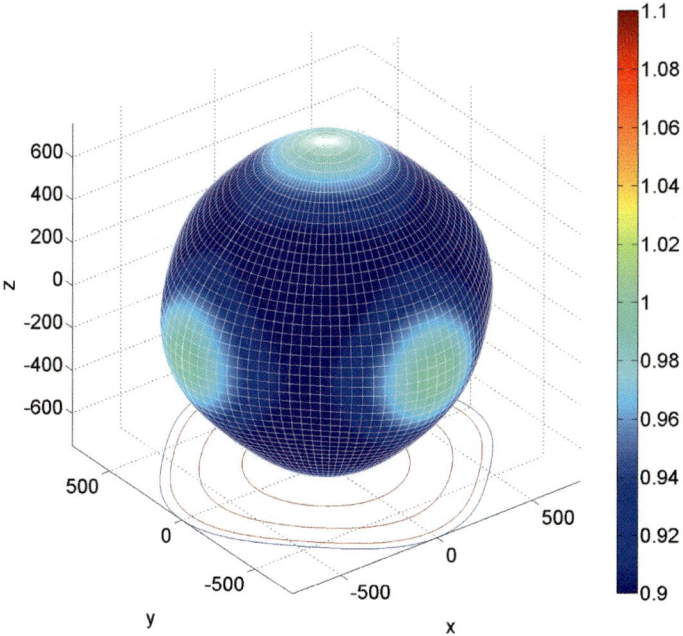

Fig. 3 Isotropic-orthotropic behavior: Young's modulus in different spatial directions

3.2 *Material Symmetry*

Additionally, the symmetry of the elastic properties can be investigated. It turns out that the symmetry properties depend on the size of the SVE: for a small size, cubic symmetry with three independent linear elastic material parameter is obtained, while for a larger SVE size, isotropic behavior is found. This can be illustrated graphically by projecting the ensemble averaged compliance tensor on space directions $d = [x, y, z]^T$ and inverting in order to obtain a directional dependent Young's modulus:

$$E(d) = [(d \otimes d) : S : (d \otimes d)]^{-1}. \tag{7}$$

While a cube represents cubic symmetry, a sphere means isotropic symmetry. Figure 3 shows the relative deviation of the directional dependent Young's modulus from the average value in the three space directions,

$$\frac{1}{3} \left(E([1, 0, 0]^T) + E([0, 1, 0]^T) + E([0, 0, 1]^T) \right). \tag{8}$$

It indicates that there is a mixture of both symmetries for the mentioned SVEs with a side length of 25 mm.

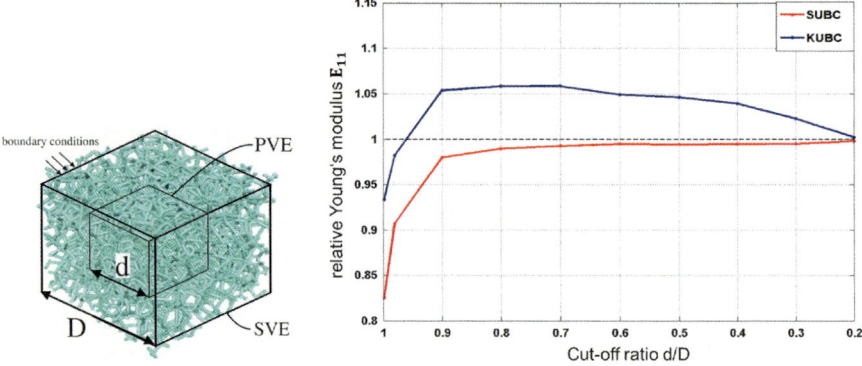

Fig. 4 Definition of the partial volume element (PVE) and the behavior of the relative Young's modulus as a function of ratio d/D

3.3 Partial Volume Averaging

As discussed in Sect. 3.1 we obtain a lower and upper bound of the stiffness tensor. For the determination of an apparent boundary effect free stiffness tensor, the influence of the boundary conditions has to be eliminated. In order to omit the border areas of the SVE, which are mainly influenced by the boundary conditions, strains and stresses are averaged over an inner partial volume of the SVE. The center point of the partial volume element (PVE) coincides with that of the SVE.

Defining the edge length D of the SVE and d of the PVE, the behavior of the relative Young's modulus as a function of the ratio d/D is illustrated in Fig. 4. By reducing the ratio $r_{dD} = d/D$, the influence of the boundary conditions is minimized and as a result the stiffness tensors for SUBC and KUBC converge against the same value.

Contrary to the expectation that the upper bound should become lower by increasing r_{dD}, both bounds first increase and then approach for $r_{dD} < 0.9$. The reason for this characteristic curve lies in the microstructure of the foam. The SVE is cut out from a surrounding network causing cut struts in the border area. This reduces the stiffness in that area. In the inner part of the SVE struts are not cut and the interior is stiffer than the exterior. This leads us to a paradox: On the one hand we want to determine the material parameters without influence of the boundary conditions and on the other hand we have to consider the boundary effect because of the difference of the stiffness tensor in the inner and outer part of a SVE.

The same 100 SVEs of Sect. 3.1 are used to interpolate the results from the finite element method to the predefined side surfaces of the PVE. The interpolated results are used to calculate the material parameters inside the PVE.

3.4 Two Section Model

Due to the difference in stiffness shown in the interior and exterior of a SVE, it is not completely correct to determine material parameters by averaging over the entire volume. This would be associated with the assumption that the stiffness is constant over the volume.

For a more accurate mapping of real foams, the apparent stiffness tensor $\mathbf{C}_{app}$ which is assumed as the mean value of $\mathbf{C}_{apparent}^{SUBC}$ and $\mathbf{C}_{apparent}^{KUBC}$ is interpreted as the mean value of the stiffness tensor of the outer and inner part. For the interior the stiffness is calculated at $r_{dD} = 0.2$ in Fig. 4. The boundary between the interior and exterior is determined by the average length of the cut struts. The stiffness in the border area is calculated with the model of springs connected in series. The Young's modulus is then described by the formula

$$\frac{1}{E_{SVE}} = \frac{p_i}{E_{interior}} + \frac{1-p_i}{E_{exterior}}, \tag{9}$$

where p_i is the percentage of the inner edge length and E_{SVE} is Young's modulus obtained from $\mathbf{C}_{app}$.

3.5 Variable Strut Thickness

In reality, the thickness of each strut varies and can be described by a polynomial. Therefore every strut is now divided in several beams. The thickness of each beam is adapted to the polynomial in Eq. (4).

For the Duocel© copper foam eight beams for every strut are used to find a compromise between the accuracy of modeling and calculation time.

4 Statistical Evaluation of Material Properties

4.1 Determination of the Distribution Function

Relative frequencies for the boundary effect free apparent material properties are shown in Fig. 5. From the relative frequencies, empirical distribution functions can be obtained.

4.2 Determination of the Correlation Functions

As the linear-elastic material parameters and the mass density will serve as input parameters for structural computations, they are represented as random fields. The

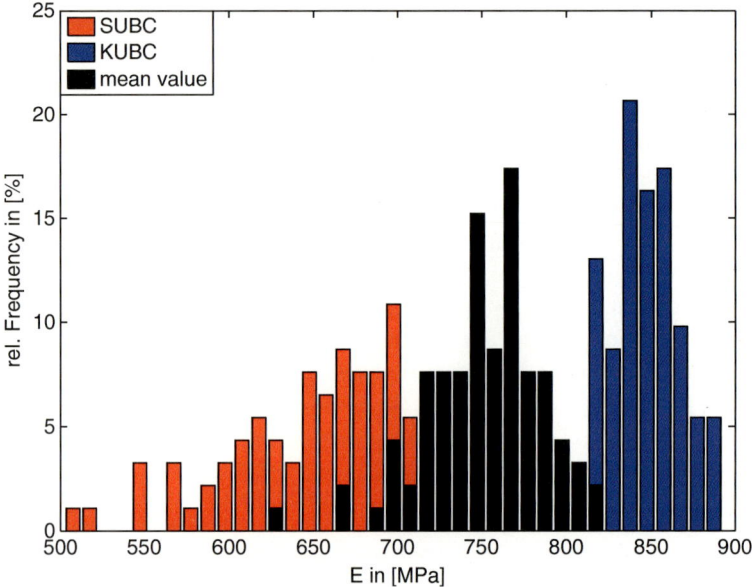

Fig. 5 Histogram of the boundary effect free Young's modulus

random fields are assumed to be homogeneous as a consequence of the homogeneity of the generated microstructure geometry. In order to find the correlation functions for the linear-elastic material parameters, 15 beam structures ($100 \times 10 \times 10$ mm) made of foam are analyzed by a method of moving SVEs: SVEs of the same size are cut out of each of these beams at different positions along the longitudinal axis. For each SVE the material parameters are calculated as functions of the center point coordinate x on the longitudinal axis.

For the computation of the autocorrelation function, the 15 realizations are made mean free and scaled to unit variance. After that, the autocorrelation function is obtained by taking the mean value over all 15 realizations at each distance Δ. The results for the Young's modulus E, shear modulus G, bulk modulus K and mass density ρ are shown in Fig. 6. It can be seen from these results that the autocorrelation functions approach zero with increasing distance. Moreover, the correlation length is rather small.

Figure 6 also indicates that the autocorrelation functions reveal a similar behavior. Therefore, the autocorrelation functions have been fitted to the expression

$$C(x) = e^{-c|x|} \left(1 - c|x|\right),\tag{10}$$

which has been proposed in Liebscher et al. (2012). In Fig. 7 the fitted autocorrelation function is plotted.

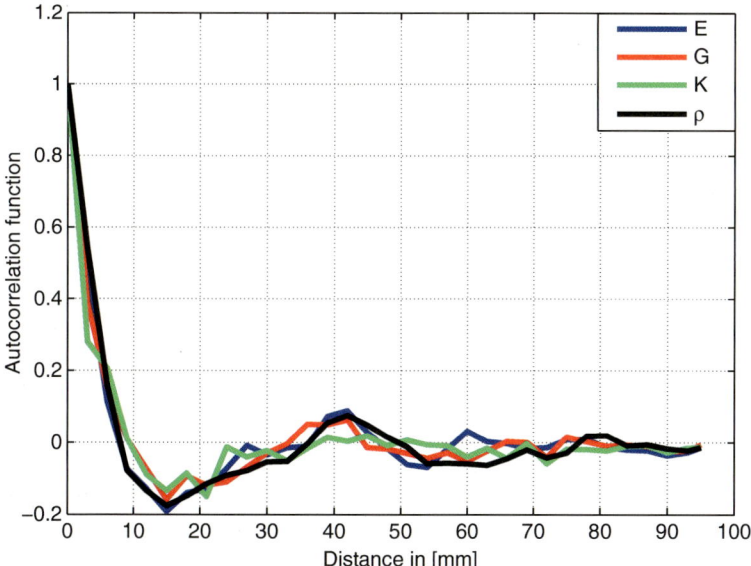

Fig. 6 Estimated autocorrelation functions

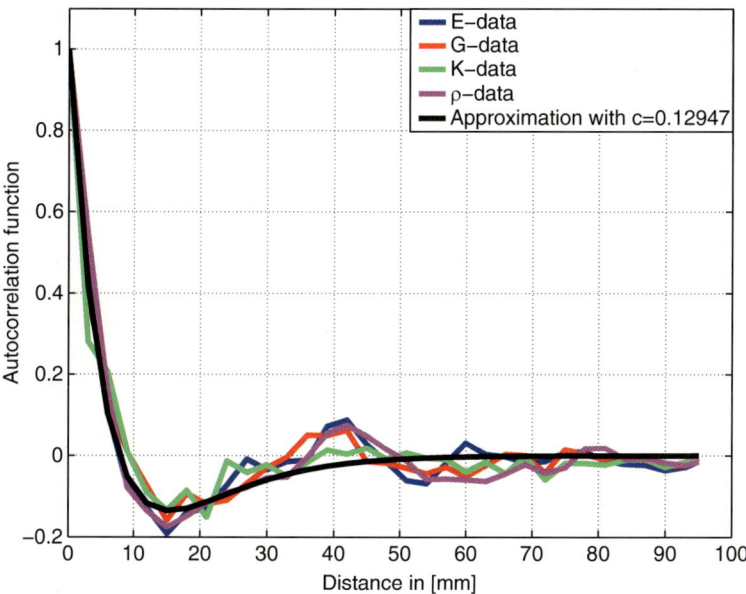

Fig. 7 Approximation of the autocorrelation functions

In the same manner, the crosscorrelation functions can be obtained. It turns out that the material parameters are almost uncorrelated.

The random fields for the material parameters and the mass density are described by the empirical distribution functions and the autocorrelation functions. They are discretized by a truncated KLE. Samples of the random variables involved in the KLE are generated iteratively by adapting the empirical marginal distribution. For this, a procedure described in Phoon et al. (2005) is applied.

5 Validation of the Implemented Model

5.1 Comparison with Finite Element Method

The bending eigenfrequencies from foam beams calculated by the finite element method (FEM) are compared with the results obtained from the method presented in this article. 100 Duocel© copper foam beams are generated with the dimensions $255 \times 15 \times 15$ mm. Their eigenfrequencies are calculated with the help of the FEM-software ABAQUS©. The mean values and the coefficient of variation of the results are listed in Table 2.

As the next step, a SVE with side length 15 mm is cut out of each generated copper foam beam, so that from 100 SVEs the material parameters can be calculated using the presented method in Sect. 3.1. With the obtained material parameters the eigenfrequencies of beams are calculated using Timoshenko theory and Monte Carlo simulation. This procedure is named *one section model* (OSM), because stresses and strains are averaged over the whole volume of a SVE.

To investigate the effects of the *two section model* (TSM) from Sect. 3.4, the material parameters are calculated using the PVE. The percentage of the inner edge length p_i from Eq. (9) is 0.8613. In a Monte Carlo simulation the bending stiffness EI is split for the interior and exterior of the foam beam. The results of these three methods are summarized in Table 2.

OSM and FEM yield similar results for the first two bending eigenfrequencies. The deviation between these two methods becomes larger for higher bending modes.

Table 2 Young's modulus and Eigenfrequencies using FEM, OSM and TSM

	E_{SVE}	$E_{interior}$	$E_{exterior}$
MPa (c.o.v.)	615 (10.2 %)	977 (7.9 %)	193 (22.9 %)
	Bending frequencies [Hz] (c.o.v.)		
	FEM	OSM	TSM
1st bending mode	173 (1.6 %)	173 (1.4 %)	175 (1.1 %)
2nd bending mode	468 (1.4 %)	465 (1.7 %)	474 (1.1 %)
3rd bending mode	897 (1.1 %)	881 (1.7 %)	900 (1.5 %)
4th bending mode	1,432 (1.4 %)	1,398 (1.7 %)	1,444 (1.5 %)

Table 3 Material parameters for Cu Duocel®

Property	μ	σ	COV (%)
Young's modulus	737 MPa	42 MPa	5.7
Shear modulus	239 MPa	13 MPa	5.3
Mass density	1,047 kg/m³	53 kg/m³	5

Table 4 Comparison for the beam of Cu Duocel® (250 × 25 × 25 mm)

	Bending frequencies [Hz] (c.o.v.)			
	OSM	TSM	VST	Experiments
1st bending mode	334 (1.4 %)	332 (1.5 %)	320 (1.3 %)	314 (2.8 %)
2nd bending mode	861 (1.9 %)	866 (2.0 %)	827 (2.2 %)	761 (3.4 %)

Nevertheless the error remains less than 4 %. Also the coefficient of variation becomes larger for the OSM. The TSM is consistent to FEM even for higher frequencies with an error less than 2 %. It behaves stiffer than the OSM which is due to the stiff interior region. The coefficient of variation is relatively low for all three methods, because the deviations of the material parameter average out along the longitudinal axis of the beam.

Obviously the disctinction between the different stiffnesses in the foam beam become more important for higher bending modes. The remaining error between FEM and TSM may attributed to the inaccuracy of the determination of the transition between the interior and exterior of the beam.

5.2   Comparison with Experiments

In this section, the natural frequencies of beams made of Cu Duocel® are predicted by OSM, TSM, a model using the variable strut thickness (VST) and compared with experimental values. To this end, 25 beams of size 25 × 25 × 250 mm are investigated experimentally in two ways. First, the density is determined via optical measurements and second, experimental modal analysis was performed.

The linear-elastic material properties of Cu Duocel® and the mass density were calculated with the proposed mesoscopic model. The input parameter to this model were

- The material data of copper,
- The geometric characteristics estimated from the CT data of 10 Cu Duocel® cubes of length 25 mm,
- The cross section shape of the beam network.

The result of the mesoscopic modeling is given in Table 3. Table 4 compares the first two bending frequencies obtained from Monte Carlo simulations and experiments. The mean values obtained by OSM and TSM are larger than the experimentally determined mean values. This is due to the constant strut thickness. The results between OSM and TSM does not differ significantly because the influence of the border area from the TSM is low compared to the cross-sectional area from the beam. The VST using the polynomial from Sect. 2.2 results in a better accordance.

The remaining difference between VST and experiments could be related to experimental conditions and the simplified microstructure geometry of the model (e.g. ignoring the anisotropy).

6 Conclusions

In this paper, a novel model for structures made of metal foam is developed. It consists of an interior region and a boundary region. For both regions, non-Gaussian random fields are identified by averaging stresses and strains on statistical volume elements that represent the heterogeneous network of struts.

Comparisons of simulations with the novel two region model and numerical as well as experimental results demonstrate that highly accurate dynamical properties can be obtained with the proposed model. Moreover, it has been shown that the random variation of the strut thickness constitutes an important parameter that has to be taken into account in order to produce accurate predictions of macroscopic properties.

The proposed model can be refined to take the anisotropy of the network geometry into account. It can be applied to the study of other macroscopic properties of metal foams, notably their damping and crushing behavior. Finally, the proposed model can be applied to other classes of materials with heterogeneous microstructure as well.

Acknowledgements This work was supported in part by the German Research Foundation (DFG) under grants RE 3002/1-1 and PR 1114/10-1.

References

Aurenhammer F (1987) Power diagrams: properties, algorithms and applications. SIAM J Comput 16:78–96

Di Paola F (2011) Modélisation multi-échelles du comportement thermo-élastique de composites à particules sphériques. Dissertation, Ecole Centrale de Paris

Dirrenberger J, Forest S, Jeulin D (2014) Towards gigantic RVE sizes for 3D stochastic fibrous networks. Solids Struct 51:359–376

Drugan WJ, Willis JR (1996) A micromechanics-based nonlocal constitutive equation and estimates of representative volume element size for elastic composites. J Mech Phys Solids 44(4):497–524

Fraunhofer ITWM, Department Image Processing (Hrsg.) (2006) MAVI – modular algorithms for volume images. Department Image Processing, Fraunhofer ITWM, Kaiserslautern. http://www.mavi-3d.de

Gibson LJ, Ashby MF (1997) Cellular solids – structure and properties. Cambridge University Press, Cambridge

Guilleminot J, Noshadravan A, Soize C, Ghanem RG (2011) A probabilistic model for bounded elasticity tensor random fields with application to polycristalline microstructures. Comput Methods Appl Mech Eng 200:1637–1648

Hardenacke V, Hohe J (2009) Local probabilistic homogenization of two-dimensional model foams accounting for micro structural disorder. Int J Solids Struct 46:989–1006

Hardenacke V, Hohe J (2010) Assessment of space division strategies for generation of adequate computational models for solid foams. Int J Mech Sci 52:1772–1782

Hazanov S, Huet C (1994) Order relationships for boundary conditions effect in heterogeneous bodies smaller than the representative volume. J Mech Phys Solids 42(12):1995–2011

Huet C (1990) Application of variational concepts to size effects in elastic heterogeneous bodies. J Mech Phys Solids 38(6):813–841

Kanaun S, Tkachenko O (2007) Representative volume element and effective elastic properties of open cell foam materials with random microstructues. J Mech Mater Struct 2(6):1607–1628

Kanaun S, Tkachenko O (2008) Effective conductive properties of open-cell foams. Int J Eng Sci 46:551–571

Kanit T, Forest S, Galliet I, Mounoury V, Jeulin D (2003) Determination of the size of the representative volume element for random composites: statistical and numerical approach. Int J Solids Struct 40:3647–3679

Lautensack C (2008) Fitting three-dimensional Laguerre tessellations to foam structures. J Appl Stat 35(9):985–995

Liebscher A, Redenbach C (2013) Statistical analysis of the local strut thickness of open cell foams. Image Anal Stereol. Accepted for publication 32:1–12

Liebscher A, Proppe C, Redenbach C, Schwarzer D (2012) Uncertainty quantification for metal foam structures by means of digital image based analysis. Probab Eng Mech 28:143–152

Ohser J, Schladitz K (2009) 3D images of materials structures – processing and analysis. Wiley, Heidelberg

Ostoja-Starzewski M (2007) Microstructural randomness and scaling in mechanics of materials. Chapman & Hall/CRC, Boca Raton

Ostoja-Starzewski M (2011) Stochastic finite elements: where is the physics? Theor Appl Mech 38:379–396

Phoon KK, Huang SP, Quek ST (2005) Simulation of strongly non-Gaussian processes using Karhunen-Loève expansion. Probab Eng Mech 20:188–198

Redenbach C (2009) Microstructure models for cellular materials. Comput Mater Sci 44:1397–1407

Soille P (1999) Morphological image analysis. Springer, Berlin

Stoyan D, Kendall, WS, Mecke J (1995) Stochastic geometry and its applications, 2nd edn. Wiley, Chichester

Zhu HX, Hobdell JR, Windle AH (2000) Effects of cell irregularity on the elastic properties of open-cell foams. Acta Mater 48(20):4893–4900

Stochastic Characterisation of the In-Plane Tow Centroid in Textile Composites to Quantify the Multi-scale Variation in Geometry

Andy Vanaerschot, Brian N. Cox, Stepan V. Lomov, and Dirk Vandepitte

Abstract Optical imaging is performed to quantify the long-range behaviour of the in-plane tow centroid of a 2/2 twill woven textile composite produced by resin transfer moulding. The position of the carbon fibre tow paths is inspected over a square region of ten unit cells and characterised by decomposing the centroid data into a non-periodic non-stochastic handling effect and non-periodic stochastic fluctuations. A significantly different stochastic behaviour is observed for warp and weft direction. Variability of the in-plane coordinate, identified by the standard deviation, is found to be six times higher in weft direction. The spatial dependency of deviations along the tow demonstrates a correlation length of ten unit cells for warp tows, which is twice the length computed for weft tows. The observed bundling behaviour of neighbouring tows of the same type is quantified by a cross-correlation length. Warp tow deviations affect neighbouring centroid values within the unit cell dimension, while this effect exceeds the unit cell size for weft tows. The stochastic information reflects the difference in tow tensions during the weaving of the fabric.

Keywords Textile composites • Non-determinism • Probabilistic methods

A. Vanaerschot (✉) • D. Vandepitte
Department of Mechanical Engineering, Katholieke Universiteit Leuven, Leuven, Belgium
e-mail: Andy.Vanaerschot@mech.kuleuven.be

B.N. Cox
Teledyne Scientific Co. LLC, Thousand Oaks, CA, USA

S.V. Lomov
Department of Metallurgy and Materials Engineering, Katholieke Universiteit Leuven, Leuven, Belgium

M. Papadrakakis and G. Stefanou (eds.), *Multiscale Modeling and Uncertainty Quantification of Materials and Structures*, DOI 10.1007/978-3-319-06331-7_12,
© Springer International Publishing Switzerland 2014

187

1 Introduction

Fibre reinforced composite materials are subjected to a significant amount of scatter in the geometrical structure, leading to a remarkable variability in performance. The nominal periodicity in the tow reinforcement of a textile composite, prescribed by the manufacturer, is only approximated in real samples. Different work already demonstrated that the tow paths in textile composites should not be represented as deterministic, but as stochastic entities where deviations are fluctuating around a mean trend (Desplentere et al. 2005; Endruweit et al. 2006; Gan et al. 2012). Mapping the variation in geometry and material properties will support material design and certification of structural composites (Rousseau et al. 2012; Zhu 1993). It increases the reliability of numerical analyses of composite structures.

Almost all published research deals with randomness of local properties without considering the correlation of a property at different positions along a tow, or correlation between different properties at the same position on a tow. Although, experimental work Mehrez et al. (2012) already has demonstrated that spatial variability must be considered to achieve an accurate description of the material. Also the sources of variability remain poorly understood and the inadequacy of experimental data (Charmpis et al. 2007; Vandepitte and Moens 2009) result in assumptions for the input probability density functions of numerical modelling techniques. Further, only a few tools are available to partially model the geometrical variation of textile reinforcements (Cox and Yang 2006). Significant advances in realistic material modelling can be achieved by Charmpis et al. (2007): (i) collecting sufficient experimental data on the spatially correlated random fluctuations of the uncertain tow path parameters and (ii) deriving probabilistic information for the macroscopic properties from the lower scale mechanical characteristics. This work is part of a series of papers following the approach of Charmpis et al. (2007). The objective is to create virtual specimen of polymer reinforced composites possessing the same statistical information as observed in experimental samples. Such random composite structures are subsequently used to define the spatial variability in the mechanical properties caused by geometrical variation in the tow path.

Variation in the geometry is a multi-scale phenomenon in textiles: geometrical scatter should be investigated on the short range, i.e. deviations correlated over distances less than or compared to the size of the unit cell, complemented with long range information, i.e. spanning several unit cells. The methodology is tested on a carbon-epoxy 2/2 twill woven composite produced with resin transfer moulding (RTM). Short range tow path data are already quantified in Vanaerschot et al. (2013a), while this paper reports and analyses the collection of the long range deviations of the in-plane position.

While data about the out-of-plane centroid and cross-sectional variations demand the investigation of the internal geometry, in-plane centroid information can be deduced from scans performed of the top view of the composite. It does not require sectioning or need a full three-dimensional representation. Optical imaging of the surface of textile composites has already been applied to characterise the in-plane geometrical variations for several woven and stitched composites. Endruweit et al. (2006) links fabric irregularities with permeability variations. Optical images are taken of several woven fabrics to inspect the scatter in tow width, tow spacing and inter-tow angle. Depending on the structure of the fabric, higher or limited fibre tow mobility is allowed. Skordos and Sutcliffe (2008) investigated the influence of fibre architectural parameters on the forming of woven composites. Variability in tow directions and unit cell size are quantified for a pre-impregnated carbon-epoxy satin weave textile using the Fourier transform of a grey-scale image. A two-dimensional spectrum with directional structure is obtained which corresponds to the physical tow directions. This methodology is found to be effective, but does not permit to analyse the local centroid coordinate along the tow. Gan et al. (2012) used an optical technique to quantify the variability of three different glass reinforcement structures. The translucent property of glass fibres is exploited to set up an automated characterisation procedure using Matlab. Samples spanning several unit cells are quantified in areal weight variations, with additional local tow orientations, tow spacings and widths for the periodic reinforcements. This automated procedure can however not be applied for carbon fibres due to its opacity.

A full characterisation of the in-plane centroid of the 2/2 twill woven textile is obtained by scanning the top surface of the impregnated composite. Local and correlated information of the tow centroids are investigated over a region spanning multiple unit cells. The objectives of the paper are to (i) perform optical imaging with derivation of the in-plane tow centroid of a one-ply 2/2 twill woven carbon epoxy composite, (ii) develop a procedure to define the in-plane centroid deviations (iii) compute the mean trend and statistical information of the deviations in terms of standard deviation and correlation length. The statistical information is prepared to be used as input in a stochastic multiple unit cell modelling technique.

2 Material

The inspected tow reinforcement is a 2/2 twill woven Hexcel fabric (G0986 injectex) (HexForce 2014). The unit cell topology is given in Fig. 1 with λ_x and λ_y, respectively the periodic length in warp (x-axis) and weft (y-axis) direction. Nominal areal density measures $285\,g/m^2$ with an ends/picks count of 3.5 resulting in unit cell dimensions of 11.4 by 11.4 mm. Two one-ply reinforcements of this fabric, spanning a region of 13 unit cells by 13 unit cells, are impregnated with epoxy resin in a RTM process. The production of one-ply samples is more challenging than multiple-ply samples but offers the advantage to obtain a high contrast between tow and resin regions for the image processing step. Characterisation of the in-plane centroid on these multiple unit cell samples provides new information of the geometrical scatter on the long range. Deviations correlated over distances less than or compared to the size of the unit cell are already quantified using laboratory micro-computed tomography in Vanaerschot et al. (2013a). The tow path is statistically characterised for the centroid location (in- and out-of-plane), area, aspect ratio and

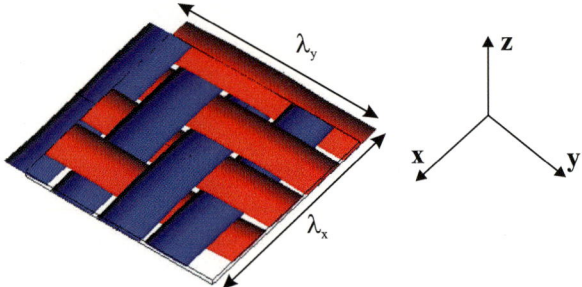

Fig. 1 WiseTex model of a 2/2 twill woven reinforcement. The x-axis and y-axis of the coordinate system are respectively parallel to the warp and weft direction

orientation in cross-section. The reference period collation method (Bale et al. 2012) is applied, where each tow parameter is decomposed in non-stochastic, periodic systematic trends and non-periodic stochastic fluctuations. Average behaviour of the tow parameter is represented by the systematic trend, while the stochastic characteristics are given in terms of the standard deviation and correlation length. The procedure and statistical information is described in Vanaerschot et al. (2013a).

The investigation of short range variations pointed out that only the in-plane centroid component of the tow path possesses a long range effect, indicated by the correlation length along the tow which exceeds the unit cell dimensions. The out-of-plane centroid and tow cross-sectional properties vary within the unit cell dimensions.

3   Image Processing and Analysis

An optical scan of the in-plane dimension of both the samples (sample 1 & sample 2) is performed with a resolution of 1,200 dots per inch (DPI). The obtained image of the first sample is shown in Fig. 2. A region of 10 unit cells by 10 unit cells is indicated where the in-plane tow data are analysed. This area of interest is chosen away from the edges to minimise possible edge effects and large enough, roughly one magnitude larger than the short range data.

The freeware image processing tool GIMP is used to extract the centroid line of the tow reinforcement in order to quantify the in-plane position of warp and weft tows. In a first step, boundaries of the tows are marked based on visual recognition for prescribed grid spacings of 125 pixels in x (warp) and y (weft) direction. These distances correspond to the nominal tow spacing of a 2/2 twill weave. In a second step, the centroid locations are computed as half the tow width at each grid location. Typical patterns in the centroid positions are further investigated to quantify the global deformation of the fabric.

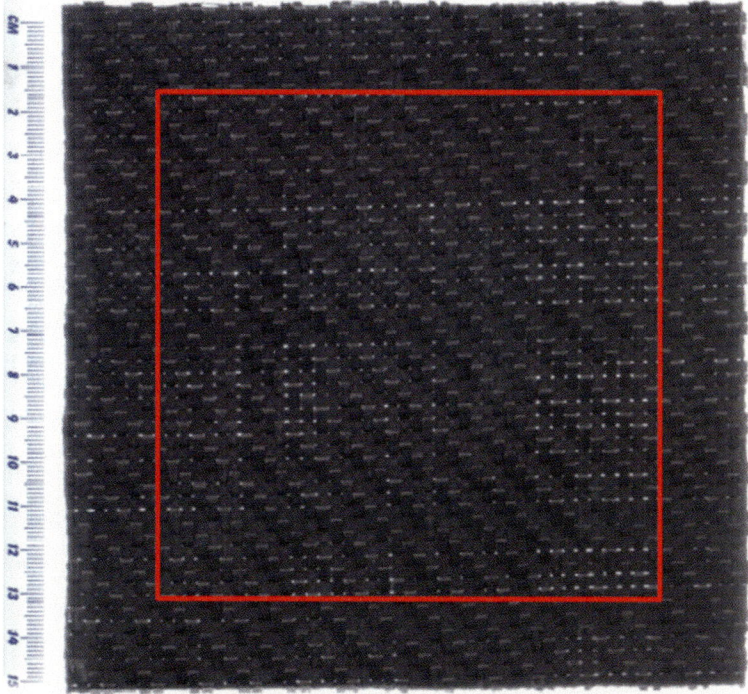

Fig. 2 Optical scan of a one-ply 2/2 twill woven carbon fibre fabric impregnated with epoxy resin. Warp tows are oriented horizontally, while weft tows are positioned in vertical direction

Table 1 Unit cell periods obtained from long range data, short range data and manufacturer's data

	λ (mm) sample 1	λ (mm) sample 2	λ (mm) combined	λ (mm) short range (Vanaerschot et al. 2013a)	λ (mm) manufacturer (HexForce 2014)
Warp direction	11.43	11.53	11.48	11.55	11.40
Weft direction	11.38	11.37	11.38	11.48	11.40

The discrete representations of the tows are given as input to Matlab. Before further inspection, data are translated to a global axis system and rotated. Average warp and weft angles are computed to verify if a rotation of the entire data set is required to compensate a possible shift in the sample data due to manual placement in the scanning device. A continuous representation of the discrete tow data is afterwards obtained by cubic spline interpolation.

Using the same approach as for the unit cell sample in Vanaerschot et al. (2013a), the unit cell periods can be defined by a minimisation algorithm. Table 1 compares the periodic lengths (i) obtained from the considered long range samples, (ii) derived

Table 2 Tow width and spacing of the one-ply 2/2 twill woven fabric

	Warp tows	Weft tows
w_{tow} – mean (mm)	2.64	2.49
w_{tow} – COV (%)	4.96	7.17
sp_{tow} – mean (mm)	2.86	2.90
sp_{tow} – COV (%)	3.62	8.73

from the unit cell sample and (iii) given by the manufacturer. The experimental data for the short and long range demonstrate that the unit cell period of the warp tows is slightly longer than the weft tows.

Geometrical characteristics of the single ply 2/2 twill woven fabric, such as tow spacing sp_{tow} and width w_{tow}, can be defined from the in-plane dimensional image. These parameters are derived from the boundary points and centroid locations of the tows and presented in Table 2. Weft tows have on average a smaller width, while the tow spacing is similar for warp and weft tows. The variation is higher for the weft direction.

The digital image also allows to quantify the open gaps between neighbouring tows. Pattern of these gaps originates from the fluctuating in-plane centroid of the tow path over the experimental sample and represent regions fully occupied by resin. After thresholding of the digital image, characterisation of the gaps is further performed automatically in Matlab to define the location and shape. The approximated rectangular shape of the gaps is represented by the width and height at each gap location. To eliminate disturbances due to e.g. dust particles in the image, a gap is considered to be significant if it has at least an area of 16 pixels^2 with a width and/or height of at least 4 pixels. Maps of the significant gaps located over the sample dimension can be constructed such as presented in Fig. 3 for sample 1. Each gap size is categorised in one of the five considered intervals of the gaps area as indicated by the legend. Such mapping of gaps demonstrates bundling behaviour of the tows, which is more noticeable for the weft tows. Local shifts in a tow affects neighbouring tows over a certain distance. The fluctuations along a single tow spans several unit cells, but do not persist over the entire length of the tow. This is reflected by the open gaps which occur in bands over the entire sample.

Statistics of the gaps (width w_{gap}, height h_{gap} and area A_{gap}) are described in Table 3 in terms of sample mean value and coefficient of variation (COV). Individual gap dimensions are Weibull distributed with a scale parameter approximating the zero value and shape parameter between 1.2 and 1.9. The width of the gaps w_{gap} (x- or warp direction) is 2–2.5 times the height of the gaps h_{gap} (y- or weft direction). This reflects the larger variability in the weft in-plane tow path compared to the warp tows, as will be discussed in Sect. 4. A higher width leads to significant gaps between neighbouring weft tows in the warp direction. The amount of gaps in the one-ply sample is less than 1 % of the entire area, which dominates the through-thickness permeability. It can be used to optimise flow simulation in accurately describing the local flow and minimising the number of voids.

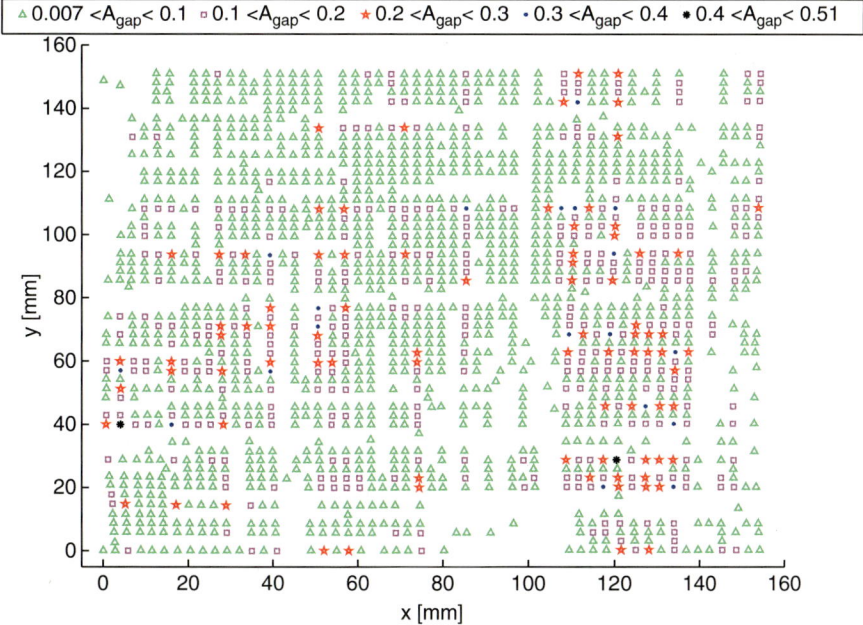

Fig. 3 Map of gaps distributed over sample 1. The area of the significant gaps (mm²) are indicated and categorised in five intervals

Table 3 Sample mean and coefficient of variation of the gaps in sample 1 (N = 1,762) and sample 2 (N = 1,862)

	Sample 1	Sample 2
w_{gap} – mean (mm)	0.418	0.451
w_{gap} – COV (%)	61.46	53.49
h_{gap} – mean (mm)	0.184	0.177
h_{gap} – COV (%)	56.85	59.18
A_{gap} – mean (mm²)	0.075	0.080
A_{gap} – COV (%)	84.93	83.35
Porosity full sample (%)	0.56	0.65

4 Statistical Characterisation of the In-Plane Centroid

Figures 2 and 3 demonstrate that the in-plane tow centroid does not follow straight paths with equal tow spacing. These in-plane undulations and shifts in tow spacing are quantified by comparing the experimental tow paths with an ideal lattice description. Tows of this lattice are represented as straight lines, with nominal spacing equal to $\delta = \lambda_y/4$ and $\delta' = \lambda_x/4$ respectively for the warp and weft spacing (λ_x, λ_y taken from the experimental unit cell periods in Table 1). A best-fit

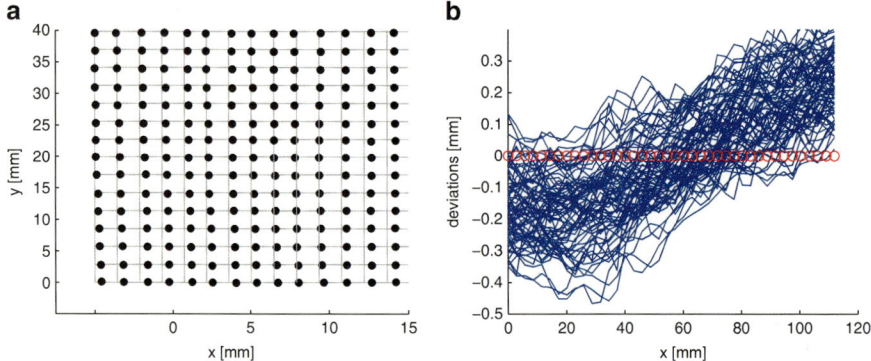

Fig. 4 Procedure to define the in-plane centroid deviations applied on sample 1. (**a**) Detail image of best-fit grid to the experimental cross-over locations. (**b**) Deviations pattern of all warp tows after subtraction of the experimental cross-over data from the grid values

of this grid with the experimental cross-over locations of the tows is searched by a minimisation algorithm reducing the overall standard deviation of the fluctuations from the grid. This procedure is shown in detail in Fig. 4a, where the lattice is fitted to the centroid data of the left bottom side.

4.1 Analysis of the In-Plane Deviations

Deviations from the nominal architecture are determined by computing the difference between the experimental tow path and the lattice at each grid location. The in-plane warp and weft fluctuations are considered respectively in y- and x-direction. This procedure results in a deviations pattern for the warp tows, combination of sample 1 and 2, as given in Fig. 4b.

The obtained deviations are further represented as $c_i^{(j,t,s)}$, with i the grid location ($i = 1 \ldots N_i$ and $N_i = 40$), j the tow index ($j = 1 \ldots N_f$ and $N_f = 40$) in each direction, $t =$ warp or weft tows and $s = 1$ or 2 referring to the sample. The in-plane centroid deviations along the different tows have a particular non-periodic trend. This pattern is shown for the warp tows in Fig. 5a, by considering the mean value per grid point $< \varepsilon_i^{(j,t,s)} >$. The lack of periodicity signifies that this tendency should not be interpreted as a systematic trend, representing the repetitive mean behaviour of the tow path, but as an effect due to handling. Variability already originates in the in-plane centroid before production due to storage and handling of dry fabrics, e.g. unwinding of the fabric from the pulley and preparing the stacking sequence in the RTM mould. This kind of variation should not be considered as stochastic, but as an added deterministic effect. Subtracting the handling pattern per sample

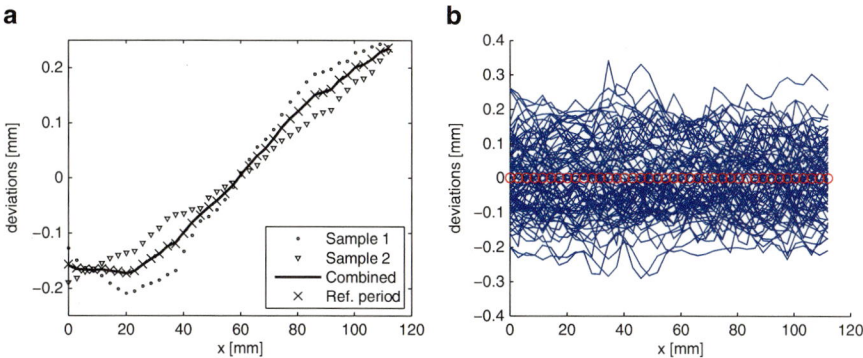

Fig. 5 Deviations trend of warp tows decomposed in (**a**) handling effect and (**b**) stochastic deviations

$< \varepsilon_i^{(j,t,s)} >$ in Fig. 4a from the deviation values $\varepsilon_i^{(j,t,s)}$ in Fig. 4b, results in stochastic variations $\zeta_i^{(j,t,s)}$ which are attributed only to the loom itself. Decomposition of in-plane deviations in a deterministic and stochastic part is summarised as

$$\varepsilon_i^{(j,t,s)} = < \varepsilon_i^{(j,t,s)} > + \zeta_i^{(j,t,s)} \qquad (1)$$

The deviations $\zeta_i^{(j,t,s)}$ are presented in Fig. 5b for the warp tows.

Equal results can also be obtained following the reference period method (Bale et al. 2012; Vanaerschot et al. 2013a). The proposed procedure uses a similar approach by defining the nominal tow spacing, expressed as δ in the ideal lattice, equal to $\lambda/4$. The periodic lengths λ_x and λ_y are experimentally obtained using the same minimisation algorithm. The similarity is indicated in Fig. 5a where the cross marks indicate the systematic trend obtained using the reference period technique. However, this systematic curve should be interpreted as a handling effect.

The transformed deviations $\zeta_i^{(j,t,s)}$ are further analysed to obtain stochastic information about the geometrical scatter of the in-plane tow centroid. Both sample deviations are grouped in one data set, which is permitted since no special relationship is expected between the different samples after subtraction of the mean trend. The warp and weft deviations approximately follow a normal distribution. The normal probability plot of the in-plane deviations of the warp tows (Fig. 6a) show good agreement, except for the tails where a lower frequency of deviations is present. The weft deviations on the other hand do show significant differences from normality (Fig. 6b), with a higher frequency of values in the left tail of the distribution and a higher frequency of deviations values around the zero mean. The weft fluctuations behaviour could be depicted as a combination of two normal distributions: one with low standard deviation (peaked curve) and one with high standard deviation (wide curve). However, there is no physical reason why these

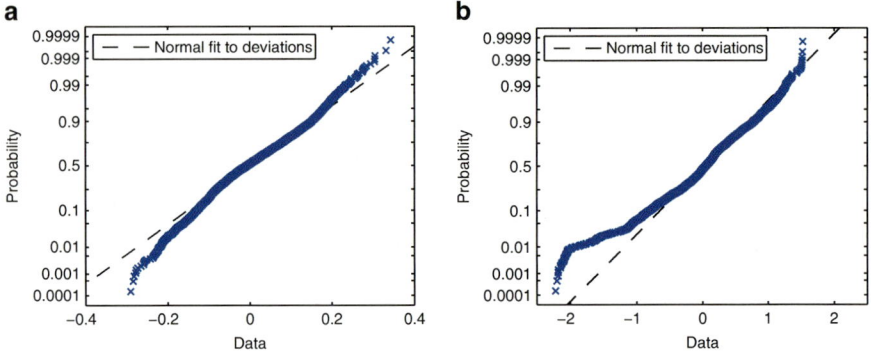

Fig. 6 Normal probability plot for (**a**) warp and (**b**) weft tow deviations showing approximately normal behaviour of the in-plane deviations

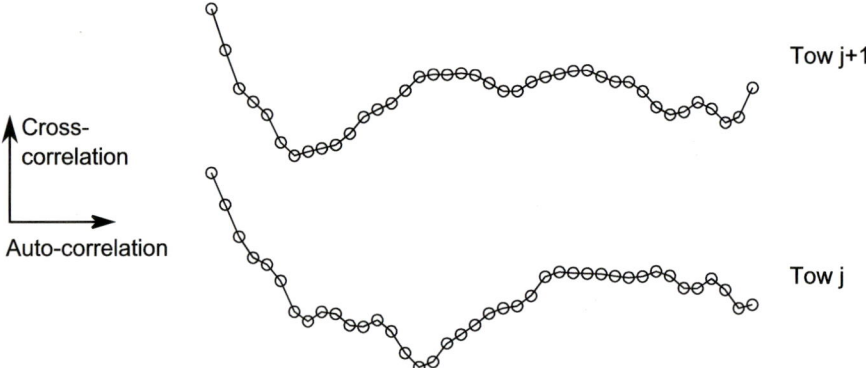

Fig. 7 Definition of spatial dependencies of deviations demonstrated for two weft tows: auto-correlation (along the tow) and cross-correlation (between neighbouring tows)

fluctuations would follow such a distribution. Therefore, the experimental data are considered for now as if they are normally distributed.

The random behaviour of the in-plane position is described in terms of standard deviation and correlation length. These statistics are required to generate virtual random specimen possessing the same statistical information as the experimental samples. Correlation of the centroid is considered along a single tow, further called auto-correlation, but also between neighbouring tows of the same type, named cross-correlation (Fig. 7). The latter correlation type is not observed for the other tow properties in the short range data (Vanaerschot et al. 2013a), but is significantly present for the in-plane centroid as indicated by the bundling of tow positions in Figs. 2 and 3.

Table 4 Standard deviation of warp and weft tows for the combined data set

	Sample 1	Sample 2	Combined
σ_{warp} (mm)	0.108	0.105	0.106
σ_{weft} (mm)	0.701	0.515	0.615

4.2 Standard Deviation

The standard deviation σ for the combined data set of warp and weft tows are presented in Table 4. Weft in-plane centroids are subjected to a much larger variation which can be explained by the production process. Warp tows are put under tension during fabrication of the fabric, while weft tows are inserted. Weft tows are therefore less restricted in their in-plane movement. Similar results are obtained by Skordos and Sutcliffe (2008) for a carbon epoxy five harness satin weave where the variability in local tow orientations, which can be related to the in-plane deviations, is higher for the weft direction.

4.3 Correlation Information

Correlation information is summarised by evaluation of the Pearson's moment correlation parameter for pairs of data taken at distinct locations on a single tow, spaced by $k\delta$ (auto-correlation C_{auto}), and pairs of data on neighbouring tows but fixed at the same grid location on a tow, spaced by $k\delta'$ (cross-correlation C_{cross}). The correlation parameter for computing the auto-correlation is given by:

$$C_{auto}^{(j,t,s)}(k) = \frac{\sum_{i=1}^{n-k} \zeta_i^{(j,t,s)} \zeta_{i+k}^{(j,t,s)}}{\sqrt{\sum_{i=1}^{n-k} (\zeta_i^{(j,t,s)})^2}\sqrt{\sum_{i=1}^{n-k} (\zeta_{i+k}^{(j,t,s)})^2}} \tag{2}$$

with k the lag index ($k = 1 \ldots N_i - 1$ in warp and weft direction), and δ, δ' the grid spacings in warp and weft direction.

Next, exponential functions are fitted to represent the computed correlation information and to estimate the correlation length ξ. The objective is not to find the optimal function perfectly representing the experimental correlation information, but to consider conventional functions which are physically reasonable and give a good estimate of the centroid behaviour. However, when more data are collected an optimal correlation function can be searched. In this work, only two types of exponential functions are considered which approximately represent the correlation information. These are functions of $\tau = |x_2 - x_1| = k\delta$ (or replace by $k\delta'$ in case of cross-correlation):

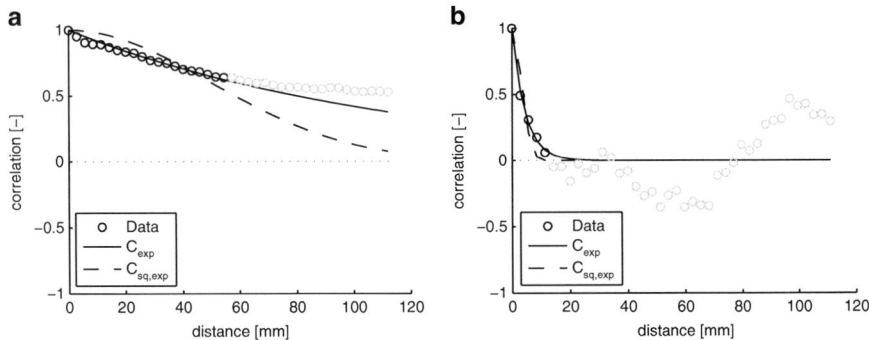

Fig. 8 Correlation graphs of the warp tows for the (**a**) auto-correlation and (**b**) cross-correlation. The data points in *lighter colour* are not considered for the fitting procedure

$$C_{exp}(\tau) = e^{-\frac{|\tau|}{\xi}} = e^{-\frac{k\delta}{\xi}} \tag{3}$$

$$C_{sq,exp}(\tau) = e^{-\frac{|\tau|^2}{\xi^2}} = e^{-\frac{(k\delta)^2}{\xi^2}} \tag{4}$$

Both functions are fitted in a least-square sense to the correlation graphs, also called correlograms, which represent the correlation values in function of the lag $k\delta$. For this procedure, a maximum of 20 data points are considered, corresponding to the first 20 lags or a length of five unit cells. Correlation information of larger point spacings are not used for fitting since these correlation data are based on a smaller data set size leading to a larger variability. In the case that the correlation data cross the zero-correlation before 20 lags are reached, no further data points are considered since negative correlation is not expected but can be present in the data due to a larger variation. To obtain the optimal fit, the sum of squares of the residuals E_{res} at each lag between the experimental correlation data $c_{data,i}$ and the fitted correlation data $c_{fit,i}$ should be minimised:

$$E_{res} = \sum_{i=1}^{n}(c_{data,i} - c_{fit,i})^2 \tag{5}$$

Figures 8 and 9 respectively present the correlation information of the warp and weft tows. The most appropriate function is evaluated by considering the sum of squares error estimate E_{res} and observation of the correlation graphs. The deduced correlation lengths are described in Table 5. It is demonstrated that either the exponential or square exponential function is preferred with an error E_{res} less than or equal to 1 %.

The warp auto- and cross-correlation information are well represented by an exponential correlation function C_{exp}. The correlation length along the warp tow

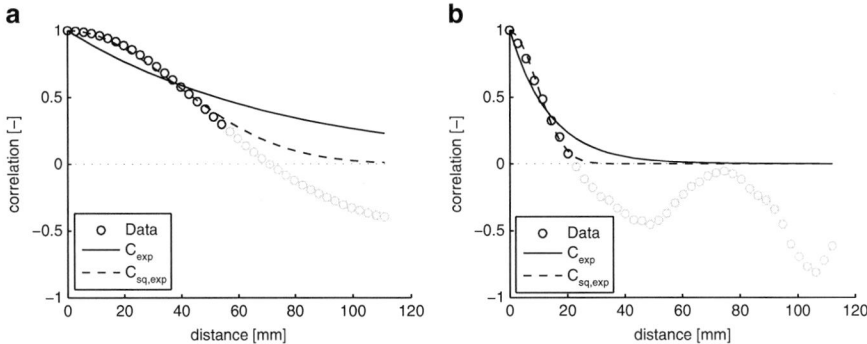

Fig. 9 Correlation graphs of the weft tows for the (**a**) auto-correlation and (**b**) cross-correlation. The data points in *lighter colour* are not considered for the fitting procedure

Table 5 Auto- and cross-correlation lengths obtained from exponential and squared exponential function fitting for the warp and weft tows using the combined data set

	ξ_{exp} (mm)	E_{res} (%)	$\xi_{sq,exp}$ (mm)	E_{res} (%)
Warp tows – C_{auto}	114.89	0.5	70.03	8.9
Warp tows – C_{cross}	4.49	0.3	4.55	6.8
Weft tows – C_{auto}	75.72	18.6	52.89	1.0
Weft tows – C_{cross}	13.89	6.7	13.16	0.6

is found to be very high and of similar size of the area of interest: around ten unit cells. This value reflects the straightness of the warp tows, which corresponds to the production process of a 2/2 twill fabric where the warp tows are kept straightened. The observation of the cross-correlation data shows that shifts of the in-plane warp centroid only affect near-neighbouring tows within one unit cell distance. The auto- and cross-correlation of the weft tows have a squared exponential correlation behaviour $C_{sq,exp}$. In-plane deviations along the weft direction seem to persist between four and five unit cells. This corresponds to only half the warp auto-correlation length, caused by the larger variability in the weft tow path. This is already reflected in the high standard deviation of this tow type (Table 4) and the higher COV in tow width and spacing (Table 2). The cross-correlation length of the weft tows shows an influence exceeding the unit cell size. Positions of the weft tows are less restricted due to the lack of tensioning during production. This affects near- and further-neighbouring tows, causing the band behaviour to appear in the composite tow paths as mentioned in Sect. 3 and shown by the gaps distribution in Fig. 3.

A different dependency structure is present than for the tow orientations of a five harness satin weave (Skordos and Sutcliffe 2008). There, a higher auto-correlation is observed for the weft tows, while auto- and cross-correlation of the warp tow orientations are negligible. These dissimilarities can be attributed to the differences in the manufacturing process of the weave.

5   Towards Virtual Modelling of Realistic Multiple Unit Cell Structures

A full characterisation of the short and long range deviations of the different tow path parameters enables to construct realistic descriptions of the tow geometry. All statistical information is given as input to a stochastic multi-scale modelling strategy which has the objective to generate random reinforcements that possess the same statistical information as quantified by the experiments.

Virtual specimens are build by combining the systematic and handling trends with zero-mean deviations. The systematic trends at the short-range and handling effect at the long-range can be taken directly from the experimental data. The zero-mean stochastic deviations need to be generated such that they represent the sample standard deviation and correlation lengths of each tow parameter.

Different generator techniques are used to represent the full randomness of the tow reinforcement at the meso- and macro-scale. Tow path parameters which vary within the unit cell size (out-of-plane centroid, tow area and tow aspect ratio) are generated using a Monte Carlo Markov Chain algorithm for textile structures, recently proposed by Blacklock et al. (2012) and already successfully applied in the generation of random unit cell structures of the 2/2 twill woven composite (Vanaerschot et al. 2013b). Generation of the long-range in-plane centroid deviations requires a different approach due to the occurrence of cross-correlation. For this purpose, a methodology described by Vořechovský (2008) is applied where series expansion methods based on Karhunen-Loève decomposition produce cross-correlated Gaussian random fields. The in-plane position of each tow is represented by a single random field, sharing an identical auto-correlation structure for all tows, which is cross-correlated at the same time with neighbouring tows of the same type.

More details and results of this generation of multiple unit cell structures is ongoing work and will be addressed in future publications. The realistic representations of internal geometry can be applied to (i) improve the understanding of damage progression in textile composites and (ii) obtain a quantitative measure of the spatial variation of the mechanical properties over the extend of composite components caused by variation in the reinforcement structure.

6   Conclusions

The long range statistical behaviour of the in-plane centroid of a one ply 2/2 twill woven carbon epoxy composite is investigated over a region of ten unit cells by ten unit cells. The in-plane position is decomposed in a mean trend and stochastic zero-mean deviations. The mean trend represents the handling effect of the fabric before it is impregnated with resin and is distinct for each individual sample. No periodic systematic pattern is present for the in-plane centroid. The scatter around the mean trend shows significant differences in warp and weft direction. The weft tows are subjected to a much larger variability, quantified by the standard deviation which is six times higher compared to the warp direction. Also the spatial dependency of the deviations significantly differs. Correlation of the in-plane centroid along the tow path (auto-correlation) is found to be twice as high for the warp tows, spanning ten unit cells. The correlation of deviations between different tows (cross-correlation) demonstrates that the neighbouring warp tows are only affected within the unit cell dimension, while this effect exceeds the unit cell size for the weft tows. This results in bundling behaviour which is mainly present for the weft tows. A possible explanation for the difference in scatter can be attributed to the manufacturing process of the weave, where the warp tows are kept in tension, while the weft tows are inserted by handles. The stochastic information is prepared such that it can be used as input for a stochastic multiple unit cell modelling procedure.

Acknowledgements This study is supported by the Flemish Government through the Agency for Innovation by Science and Technology in Flanders (IWT) and FWO-Vlaanderen.

References

Bale H, Blacklock M, Begley M, Marshall D, Cox B, Ritchi R (2012) Characterizing three-dimensional textile ceramic composites using synchrotron X-ray micro-computed-tomography. J Am Ceram Soc 95:392–402

Blacklock M, Bale H, Begley M, Cox B (2012) Generating virtual textile composite specimens using statistical data from micro-computed tomography: 1D tow representations for the Binary Model. J Mech Phys Solids 60:451–470

Charmpis DC, Schuëller GI, Pellisetti MF (2007) The need for linking micromechanics of materials with stochastic finite elements: a challenge for materials science. Comput Mater Sci 41:27–37

Cox B, Yang Q (2006) In quest of virtual tests for structural composites. Science 314:1102–1107

Desplentere F, Lomov SV, Woerdeman DL, Verpoest I, Wevers M, Bogdanovich A (2005) Micro-CT characterization of variability in 3D textile architecture. Compos Sci Technol 65:1920–1930

Endruweit A, McGregor P, Long AC, Johnson MS (2006) Influence of the fabric architecture on the variations in experimentally determined in plane permeability values. Compos Sci Technol 66:1778–1792

Gan JM, Bickerton S, Battley M (2012) Quantifying variability within glass fibre reinforcements using an automated optical method. Compos Part A-Appl S 43:1169–1176

HexForce G0986 SB 1200 (2014), Product data hexcel, edition May 2014. http://hexply.com/hexforce/database/web/front/main/index.php

Mehrez L, Doostan A, Moens D, Vandepitte D (2012) Stochastic identification of composite material properties from limited experimental databases, part 1: experimental database construction. Mech Syst Signal Pr 27:471–483

Rousseau C, Engelstad S, Owens S (2012) Industry perspectives on composite structural certification and design. In: Proceedings of the 53rd AIAA/ASME/ASCE/AHS/ASC conference. Honolulu, Hawai, pp 1–10

Skordos AA, Sutcliffe MPF (2008) Stochastic simulation of woven composites forming. Compos Sci Technol 68:283–296

Vanaerschot A, Cox BN, Lomov SV, Vandepitte D (2013a) Stochastic framework for quantifying the geometrical variability of laminated textile composites using micro-computed tomography. Compos Part A-Appl S, 44:122–131

Vanaerschot A, Cox BN, Lomov SV, Vandepitte D (2013b) Stochastic multi-scale modelling of textile composites based on internal geometry variability. Comput Struct 122:55–64

Vandepitte D, Moens D (2009) Quantification of uncertain and variable model parameters in non-deterministic analysis. In: Proceedings of the IUTAM symposium on the vibration analysis of structures with uncertainties, St. Petersburg, Russia, pp 15–28

Vořechovský M (2008) Simulation of simply cross correlated random fields by series expansion methods. Struct Saf 30:337–363

Zhu TL (1993) A reliability-based safety factor for aircraft composite structures. Comput Struct 48:745–748

A Variability Response-Based Adaptive SSFEM

Dimitris G. Giovanis and Vissarion Papadopoulos

Abstract The present work sets up a methodology that allows the estimation of the spatial distribution of the second-order error of the response, as a function of the number of terms used in the truncated Karhunen-Loève (KL) series representation of the random field involved in the problem. For this purpose, the concept of the variability response function (VRF) is adopted, as it is well recognized that VRF depends only on deterministic parameters of the problem as well as on the standard deviation of the random parameter. The criterion for selecting the number of KL terms at different parts of the structure is the uniformity of the spatial distribution of the second-order error. This way a significantly reduced number of polynomial chaos (PC) coefficients, with respect to classical PC expansion, is required in order to reach a target second-order error.

Keywords Karhune-Loève expansion • Variability response function • Spectral stochastic finite element • PC expansion

1 Introduction

Over the last two decades, the importance of stochastic approach to engineering problems has been recognized by the scientific community and has received significant attention. The majority of research work has focused on developing stochastic finite element methodologies which, with the aid of powerful computing resources and technologies can be applicable to realistic engineering systems. In the

D.G. Giovanis (✉) • V. Papadopoulos
Institute of Structural Analysis and Antiseismic Research, National Technical
University of Athens, Iroon Polytechniou 9, Zografou Campus, Athens 15780, Greece
e-mail: dgiov16@yahoo.gr; vpapado@central.ntua.gr

M. Papadrakakis and G. Stefanou (eds.), *Multiscale Modeling and Uncertainty Quantification of Materials and Structures*, DOI 10.1007/978-3-319-06331-7_13,
© Springer International Publishing Switzerland 2014

context of stochastic finite element analysis (SFEM), the spectral stochastic finite element method (SSFEM), introduced by Ghanem and Spanos is a powerful tool for treating systems with uncertain input parameters. In this approach, the classical finite element discretization is combined with the Karhunen-Loève (KL) decomposition of the input random fields. Then, the system's response is represented by a set of polynomials of the basic random variables, namely the polynomial chaos expansion (PC). The coefficients of the PC are obtained using a Galerkin scheme that leads to a system of coupled deterministic equations. However, solving this coupled system is the main computational burden of the method which, becomes more challenging as the size of the physical system and the level of uncertainty grows.

2   Variability Response Function

The major difficulties in quantifying uncertainty are a lack of available data and the inability to accurately simulate complex random fields. In light of the aforementioned limitations, the variability response function (VRF) was introduced by Shinozuka in the late 1980s. The VRF is a deterministic function dependent on the structure, its boundary conditions, and loading. It is independent of the distributional and spectral characteristics of the uncertain system parameters. It identifies the influence of the correlation structure of the uncertain parameters on the variability of the response. Different aspects and applications of the VRF were introduced in Wall and Deodatis (1994), Deodatis et al. (2003), Papadopoulos et al. (2005), Papadopoulos and Deodatis (2006), Miranda and Deodatis (2010), and Arwade and Deodatis (2010). A development of this approach was presented in Papadopoulos and Deodatis (2006), where the existence of closed-form integral expressions for the variance of the response displacement of the form

$$Var[u] = \int\limits_{-\infty}^{\infty} VRF(\mathbf{x}, \omega, \sigma_{ff}) S_{ff}(\omega) d\omega \tag{1}$$

was demonstrated for linear stochastic finite element systems under static loads using a flexibility-based formulation. In these works it was demonstrated that the VRF depends on deterministic parameters of the problem as well as on the standard deviation σ_{ff} of the random parameter but appears to be independent of the functional form of the spectral density function (SDF) S_{ff} modeling the random property. For statically determinate structures, VRF is independent of σ_{ff} as well (Miranda and Deodatis 2010). The variability response function can be estimated numerically using a fast Monte Carlo simulation (FMCS) approach.

2.1 Fast Monte Carlo Simulation

The basic steps of the fast Monte Carlo simulation (FMCS) are described next

1. Generate N sample functions of a random sinusoid with standard deviation σ_{ff} and wave number ω modeling a stochastic field $f(\mathbf{x})$ that describes the elastic modulus:

$$f_j(\mathbf{x}) = \sqrt{2}\sigma_{ff}\cos(\omega\mathbf{x} + \phi_j) \quad j = 1, 2, \ldots, N \tag{2}$$

where ϕ_j are random phase angles uniformly distributed in the range $[0, 2\pi]$. Rather than picking up the ϕ_j's randomly in $[0, 2\pi]$, they can be selected at N equal intervals in $[0, 2\pi]$ for significant computational savings.
2. Using these N generated sample functions $f_j(\mathbf{x})$, it is straightforward to compute the corresponding N displacement responses either analytically or numerically. Then, the mean value of the response $\varepsilon[u(\mathbf{x})]_\omega$ and its variance $Var[u(\mathbf{x})]_\omega$ can be easily determined for the specific value of ω considered by ensemble averaging the N computed responses.
3. The value of the variability response function (VRF) at wave number ω and for standard deviation σ_{ff} is computed from

$$VRF(\mathbf{x}, \omega, \sigma_{ff}) = \frac{Var[u(\mathbf{x})]_\omega}{\sigma_{ff}^2} \tag{3}$$

Steps 1–3 are repeated for different values of the wave number ω of the random sinusoid. Consequently $VRF(\mathbf{x}, \omega, \sigma_{ff})$ are computed over a wide range of wave numbers, wave number by wave number. The entire procedure can be eventually repeated for different values of the standard deviation σ_{ff} and for different locations over the domain (if necessary).

3 Karhunen-Loève Series Representation

The Karhunen-Loève (KL) expansion (Loève 1977; Huang et al. 2001; Grigoriu 2006) of a multi-dimensional random field $f(\mathbf{x}, \theta)$ is written as:

$$f(x, \theta) = \mu(\mathbf{x}) + \sum_{i=1}^{\infty} \sqrt{\lambda_i}\xi_i(\theta)\phi_i(\mathbf{x}) \tag{4}$$

$\mathbf{x}$ being a position vector and θ the random event. In Eq. (4) $\mu(\mathbf{x})$ is the mean value of the random field, λ_i and $\phi_i(\mathbf{x})$ are the eigenvalues and eigenfunctions of its covariance function $C(\mathbf{x}_1, \mathbf{x}_2)$, which may be calculated, in the domain D of the random field $f(\mathbf{x}, \theta)$, from the solution of the homogeneous Fredholm integral equation of the second kind given by:

$$\int_D C(\mathbf{x}_1, \mathbf{x}_2) \phi_i(\mathbf{x}_1) = \lambda_i \phi_i(\mathbf{x}_2) \tag{5}$$

and $\xi_i(\theta)$ is a set of uncorrelated Gaussian random variables with mean and covariance given by:

$$E[\xi_i(\theta)] = 0$$
$$E[\xi_i(\theta)\xi_j(\theta)] = \delta_{ij} \tag{6}$$

For all practical purposes, the KL series expansion of Eq. (4) is approximated by a finite number of M terms, giving

$$\hat{f}(\mathbf{x}, \vartheta) = \mu(\mathbf{x}) + \sum_{i=1}^{M} \sqrt{\lambda_i} \xi_i(\theta) \phi_i(\mathbf{x}) \tag{7}$$

4   Spectral Stochastic Finite Element Method

The spectral stochastic finite element method – SSFEM has been introduced by Ghanem and Spanos (1990) as an extension of the deterministic finite element method for the solution of boundary value problems with random material properties. The Young's modulus of a structure is considered to vary randomly over space with a Gaussian distribution function. The Karhunen-Loève decomposition of the random field reads (see Sect. 3)

$$f(\mathbf{x}, \theta) = \mu(\mathbf{x}) + \sum_{i=1}^{\infty} \sqrt{\lambda_i} \phi_i(\mathbf{x}) \xi_i(\theta) \tag{8}$$

In this context, the stochastic matrix of a finite element (e) has the following form:

$$\mathbf{k}^{(e)}(\theta) = \mathbf{k}_0^{(e)} + \sum_{i=1}^{\infty} \mathbf{k}_i^{(e)} \xi_i(\theta) \tag{9}$$

where $\mathbf{k}_0^{(e)}$ is the mean value of $\mathbf{k}^{(e)}(\theta)$, $\mathbf{k}_i^{(e)}$ are deterministic matrices given by

$$\mathbf{k}_i^{(e)} = \sqrt{\lambda_i} \int_{\Omega_e} \phi_i(\mathbf{x}) \mathbf{B}^T \mathbf{D}_0 \mathbf{B} d\Omega_e \tag{10}$$

B is the strain-displacement matrix and $\mathbf{D}_0$ is the mean value of the constitutive matrix. Assuming deterministic loading, the finite element equilibrium equation has the form:

$$\left(\sum_{i=0}^{\infty} \mathbf{K}_i \xi_i(\theta) \right) \cdot \mathbf{U}(\theta) = \mathbf{F} \tag{11}$$

In the context of SSFEM, the vector $\mathbf{U}(\theta)$ is expanded in a series of random Hermite polynomials $\{\Psi_j(\xi)\}_{j=0}^{\infty} = \{\Psi_j((\xi_1(\theta), \dots, \xi_M(\theta)))\}_{j=0}^{\infty}$ as follows

$$\mathbf{U}(\theta) = \sum_{j=0}^{\infty} \mathbf{U}_j \Psi_j(\theta) \tag{12}$$

and the final equilibrium equation reads:

$$\left(\sum_{i=0}^{\infty} \mathbf{K}_i \xi_i(\theta) \right) \cdot \left(\sum_{j=0}^{\infty} \mathbf{U}_j \Psi_j(\theta) \right) = \mathbf{F} \tag{13}$$

A finite number of terms is finally retained in both expansions for practical purposes (say M+1 terms in the KL expansion and P−1 terms in the polynomial chaos expansion – PCE), leading to a residual $\epsilon_{M,P}$ that has to be minimized in the mean square sense in order to obtain the optimal approximation of the exact solution $\mathbf{U}(\theta)$ in the space H_P spanned by the polynomials $\{\Psi_k\}_{k=0}^{P-1}$ (Galerkin approach)

$$\epsilon_{M,P} = \sum_{i=0}^{M+1} \sum_{j=0}^{P-1} \mathbf{K}_i \mathbf{U}_j \xi_i(\theta) \Psi_j(\theta) - \mathbf{F} \tag{14}$$

$$E[\epsilon_{M,P} \cdot \Psi_k] = 0, \quad k = 0, 1, \dots, P-1 \tag{15}$$

where, $P = \frac{(M+p)!}{M!p!}$ and p is the order of chaos polynomials. After some algebraic manipulations the following system of equations is obtained

$$\mathcal{K} \cdot \mathcal{U} = \mathcal{F} \tag{16}$$

where,

$$c_{ijk} = E[\xi_i \Psi_j \Psi_k] \tag{17}$$

$$\mathcal{K} = \begin{bmatrix} \sum_{i=0}^{M} c_{i,0,0}\mathbf{K}_i & \sum_{i=0}^{M} c_{i,1,0}\mathbf{K}_i & \cdots & \sum_{i=0}^{M} c_{i,P-1,0}\mathbf{K}_i \\ \sum_{i=0}^{M} c_{i,0,1}\mathbf{K}_i & \sum_{i=0}^{M} c_{i,1,1}\mathbf{K}_i & \cdots & \sum_{i=0}^{M} c_{i,P-1,1}\mathbf{K}_i \\ \vdots & \vdots & \ddots & \vdots \\ \sum_{i=0}^{M} c_{i,0,P-1}\mathbf{K}_i & \sum_{i=0}^{M} c_{i,1,P-1}\mathbf{K}_i & \cdots & \sum_{i=0}^{M} c_{i,P-1,P-1}\mathbf{K}_i \end{bmatrix}$$

$$\mathcal{U} = \begin{bmatrix} \mathbf{U}_0, \mathbf{U}_1, \cdots, \mathbf{U}_{P-1} \end{bmatrix}^T$$

$$\mathcal{F} = \begin{bmatrix} \mathbf{F}_0, \mathbf{F}_1, \cdots, \mathbf{F}_{P-1} \end{bmatrix}^T$$

After solving this system for $\mathcal{U} = \{\mathbf{U}_k, k = 0, \ldots, P-1\}$, $\mathbf{U}(\theta)$ is expressed as:

$$\mathbf{U}(\theta) = \sum_{j=0}^{P-1} \mathbf{U}_j \Psi_j(\theta) \tag{18}$$

5 Adaptive SSFEM based on VRF

The proposed work is conducted in two phases

Phase 1: Error estimation

1. Estimate the variability response function $VRF(\mathbf{x}, \omega)$, numerically with FMCS over a wide range of wave numbers.
2. For a target autocorrelation function with corresponding two sided power spectrum $S_{ff}(\omega)$ calculate an "exact" value for the response variance as

$$Var^T[u(\mathbf{x})] = 2 \cdot \int_0^{\infty} VRF(\mathbf{x}, \omega) \cdot S_{ff}(\omega)d\omega \tag{19}$$

3. Generate zero mean Gaussian sample functions of the stochastic field using Karhunen-Loève expansion for various number of terms M and estimate the corresponding SDF of the sample function in an ensemble average sense as follows:

$$S_{ff}^M(\omega) = \frac{1}{2\pi L} \left| \int_0^L f_M(\mathbf{x})e^{-i\omega\mathbf{x}}d\mathbf{x} \right|^2 \tag{20}$$

where L is the length of the sample functions of the random field.

4. Calculate the variance of the displacement from:

$$Var^M[u(\mathbf{x})] = 2 \cdot \int_0^\infty VRF(\mathbf{x}, \omega) \cdot S_{ff}^M(\omega)d\omega \qquad (21)$$

5. Estimate the error:

$$Error^M(\mathbf{x}_i) = \frac{\left|Var^T[u(\mathbf{x})] - Var^M[u(\mathbf{x})]\right|}{Var^T[u(\mathbf{x})]} \cdot 100(\%) \qquad (22)$$

Repeat steps 3–5 with increasing number of terms M until the error of Eq. (22) reaches a target value (e.g. <10 %). In general, a different number M_i will be required in different parts of the domain in order to have a uniform error distribution over all the domain.

Phase 2: Building a sparse PC coefficient matrix

1. For every element of the position vector $\mathbf{x}$ estimate the number of terms P required in the PC expansion of the corresponding displacement
2. Assemble the corresponding sparse PC coefficient matrix and solve the linear system.

6   Numerical Example

The proposed method is implemented in the 2-D plain stress plate of Fig. 1. The two-dimensional domain is a rectangle of length $L_x = 1$ m and width $L_y = 1$ m, with holes in the center of the domain of radius $R = 0.1$ m and two symmetric cut-offs at the middle, with dimensions $a = b = 0.33$ m. The domain is divided with a mesh of 20×20 quadrilateral elements. The model is subjected to a constant uniform load $p = 25$ KN/m along its boundary at its upper side. The values of $\sigma = 0.1$ and $b = 1$ are selected for this example.

The variability response function $VRF(x_i, \kappa)$ $i = 1, \ldots, 15$ is estimated numerically at 15 points on the model (Figs. 2 and 3). The VRFs are computed in the range $\omega \in [0, \omega_u]$ with a step of $d_k = 0.66$, ω_u being an upper cut-off frequency, taken equal to 20 rad/m. Plots of the VRFs calculated at various points x_i are depicted in Figs. 2 and 3.

Figure 4 presents the plots of S_{ff}^M for various values of M computed via Eq. (20), together with the target SDF. Inspection of this figure reveals once again that the error of the variance depends not only on the number M but also on the values of the VRF at the frequencies that are not well represented in the power spectral density due to the Karhunen-Loève truncation.

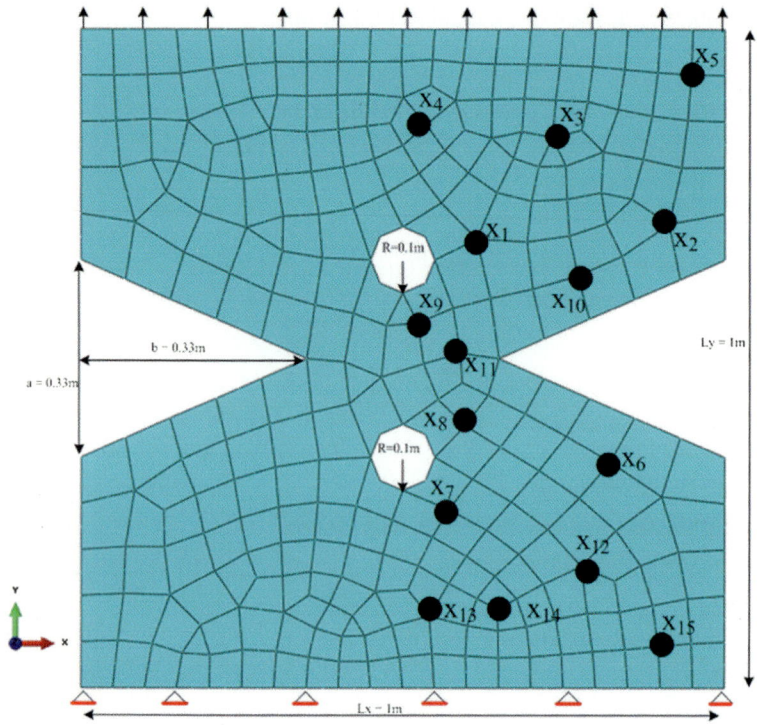

Fig. 1 VRF points x_i

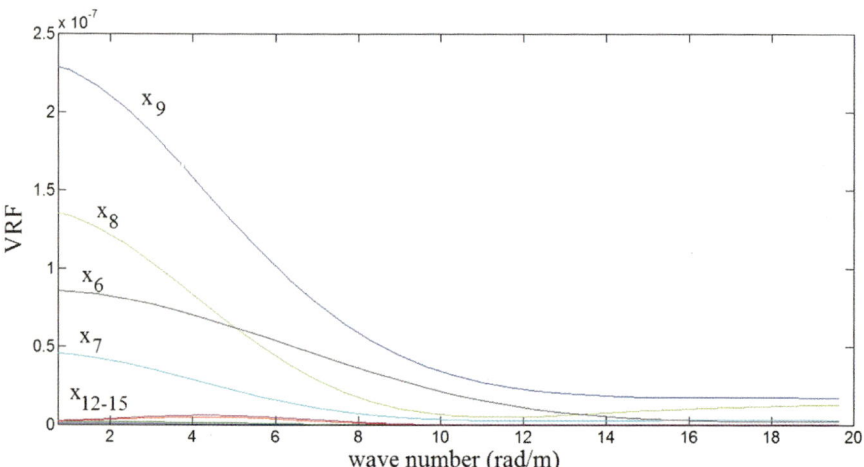

Fig. 2 VRF at points x_i $i = 6-9, 12-15$

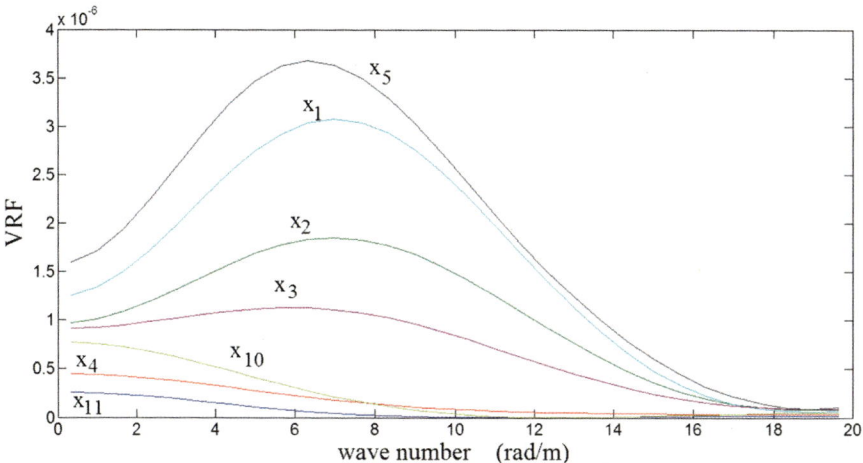

Fig. 3 VRF at points x_i $i = 1-5, 10, 11$

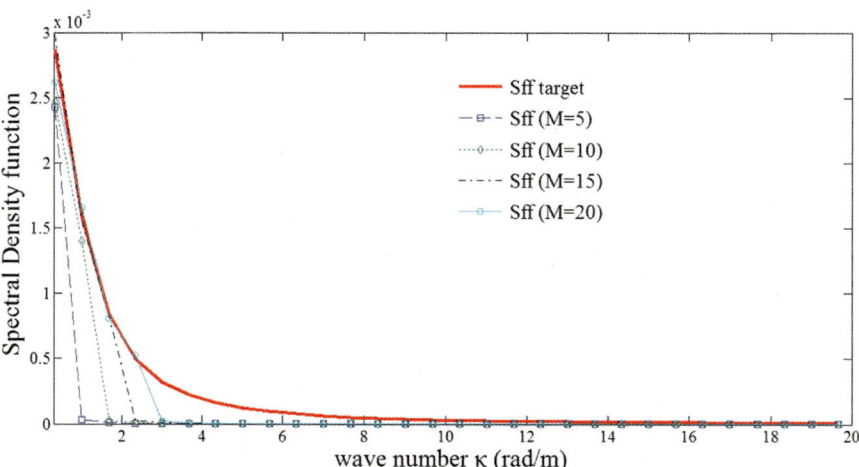

Fig. 4 SDFs obtained from KL expansion for different number of terms M and target SDF

This error behavior for this case is presented in Fig. 5 for three representative locations of the domain, namely the points x_5, x_9 and x_{13}, as a function of M. This figure plots the (%) error computed via Eq. (22). Figure 6 depicts the sub-domains with equal M values required to reach a target error of about 20 %. More specifically, for points x_5, x_1, x_2 and x_3 in sub-domain I the value of $M_I = 15$ is required, for points x_{10} and x_4 in sub-domain II we need $M_{II}=12$, while for all the other points in sub-domain III, $M_{III}=10$ is satisfying the requirement for a uniform error.

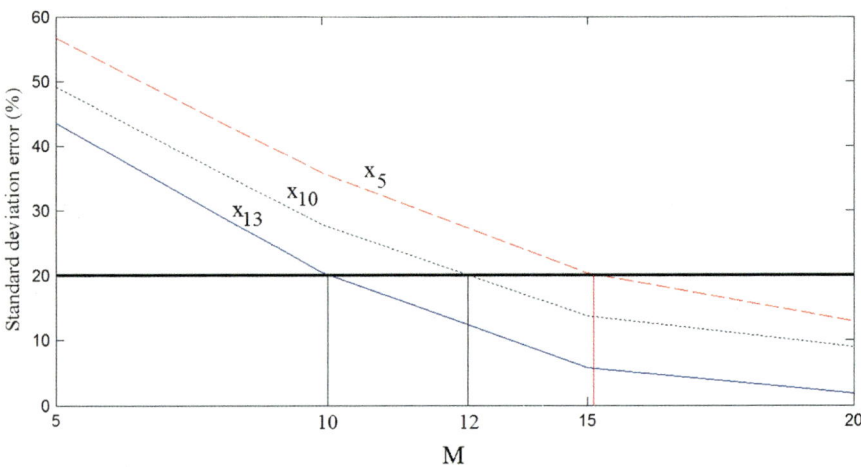

Fig. 5 Standard deviation error (%) as a function of M for three different points

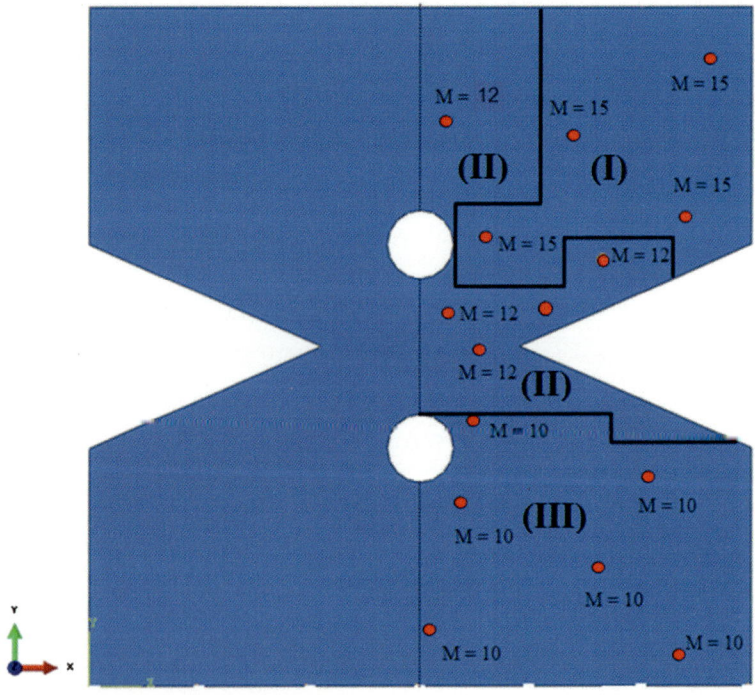

Fig. 6 Sub-domains with equal M values required to reach a target error of about 20 %

Table 1 Standard PC expansion

(M, p)	$U_y(\times 10^{-3})$	$Std(\times 10^{-4})$	$Error^{(M)}(\%)$
(15,2)	1.24839	1.2809	30.16
(12,2)	1.24812	1.2845	30.55
(10,2)	1.24795	1.3205	32.44

Table 2 Adaptive PC expansion

$(M_I, M_{II}, M_{III}, p)$	$U_y(\times 10^{-3})$	$Std(\times 10^{-4})$	$Error_{rel}(\%)$	Sparsity
(15,12,10,2)	1.24291	1.2884	0.58	+31.3

The problem is initially solved three times using the classical PC methodology with $M = 10$, $M = 12$ and $M = 15$ all over the domain. The displacement of the upper-right node along the y-axis is monitored. The results are gathered in Table 1.

The same calculation is now repeated using the proposed adaptive sparse PC methodology. For this calculation the different values of $M = 15$, 12 and 10, for $p = 2$ are used for sub-domains (I), (II) and (III) respectively (see Fig. 6). As we can see in Table 2 the results of the proposed methodology coincide with the ones obtained with the classic PC. Indeed, the (%) relative error, with respect to the classical PC method for $M = 15$ in the entire domain is 0.58 % while the sparsity increase is 31.3 %.

7 Conclusions

In this work a methodology is described to construct an adaptive sparse polynomial chaos (PC) expansion of the response of stochastic systems whose input parameters are modeled with random fields. The proposed methodology adopts the concept of variability response function (VRF), as it is well recognized that VRF depends only on deterministic parameters of the problem as well as on the standard deviation of the random parameter. This way an a priori inexpensive estimation of the spatial distribution of the second-order error of the response as a function of the number of terms used in the truncated Karhunen-Loève (KL) series representation of the random fields involved in the problems is achieved. As a result a spatial adaptation of the number of terms used for describing the random field is achieved in order to obtain a uniform error distribution, leading to a significant reduction of the number of PC coefficients and to an increase in the sparsity of the corresponding deterministic matrix. This sparsity increase is expected to improve significantly the computational performance of the SSFEM. The benefits of the aforementioned sparsity increase in the computational efficiency of PC are expected to be significant especially if a preconditioned conjugate gradient (PCG) solver is used with a block diagonal preconditioner (Papadrakakis and Babilis 1994; Chung et al. 2005; Ghanem and Kruger 1996; Stavroulakis and Papadrakakis 2009; Papadrakakis and Kotsopoulos 1999).

Acknowledgements This work has been supported by the European Research Council Advanced Grant MASTER – Mastering the computational challenges in numerical modeling and optimum design of CNT reinforced composites (ERC-2011-ADG-20110209).

References

Arwade SR, Deodatis G (2010) Variability response functions for effective material properties. Probab Eng Mech 26:174–181

Chung DB, Guttierez MA, Graham-Brady LL, Lingen FJ (2005) Efficient numerical strategies for spectral stochastic finite element models. Int J Numer Methods Eng 44(2):1334–1349

Deodatis G, Graham L, Micaletti R (2003) A hierarchy of upper bounds on the response of stochastic systems with large variation of their properties: random variable case. Probab Eng Mech 18(4):349–364

Ghanem R, Kruger RM (1996) Numerical solution of spectral stochastic finite element systems. Comput Methods Appl Mech Eng 129:289–303

Ghanem R, Spanos PD (1990) Polynomial chaos in stochastic finite elements. J Appl Mech ASME 57:197–202

Grigoriu M (2006) Evaluation of Karhunen-Loève, spectral and sampling representations for stochastic processes. J Eng Mech 132:179–189

Huang SP, Quek ST, Phoon KK (2001) Convergence study of the truncated Karhunen-Loève expansion for simulation of stochastic processes. Int J Numer Methods Eng 52:1029–1043

Loève M (1977) Probability theory I, 4th Edition, Springer-Verlag Inc. doi:10.1007/978-1-4684-9464-8

Miranda M, Deodatis G (2010) On the response variability of beams with large stochastic variations of system parameters. Safety, Reliability and Risk of Structures, Infrastructures and Engineering Systems, CRC Press

Papadrakakis M, Babilis G (1994) Solution techniques for the p-version of the adaptive finite element method. Int J Numer Methods Eng 37:1413–1431

Papadopoulos V, Deodatis G (2006) Response variability of stochastic frame structures using evolutionary field theory. Comput Methods Appl Mech Eng 195(9–12):1050–1074

Papadrakakis M, Kotsopoulos A (1999) Parallel solution methods for stochastic finite element analysis using Monte Carlo simulation. Comput Methods Appl Mech Eng 168:305–320

Papadopoulos V, Deodatis G, Papadrakakis M (2005) Flexibility-based upper bounds on the response variability of simple beams. Comput Methods Appl Mech Eng 194(12–16): 1385–1404

Stavroulakis MG, Papadrakakis M (2009) Advances on the domain decomposition solution of large-scale porous media problems. Comput Methods Appl Mech Eng 198:1935–1945

Wall FJ, Deodatis G (1994) Variability response functions of stochastic plane stress/strain problems. J Eng Mech 120(9):1963–1982

Monte Carlo Simulation vs. Polynomial Chaos in Structural Analysis: A Numerical Performance Study

George Stavroulakis, Dimitris G. Giovanis, Manolis Papadrakakis, and Vissarion Papadopoulos

Abstract The present work revisits the computational performance of non-intrusive Monte Carlo versus intrusive Galerkin methods for large-scale stochastic systems in the framework of high performance computing environments. The purpose of this work is to perform an assessment of the range of the relative superiority of these approaches with regard to a variety of stochastic parameters. In both approaches, the solution of the resulting algebraic equations is performed with a combination of primal and dual domain decomposition methods implementing specifically tailored preconditioners.

Keywords Spectral stochastic finite element method • Monte Carlo methods • High performance computing • Primal–dual domain decomposition • FETI method

1 Monte Carlo Simulation in High Performance Computing Environments

MC methods require the solution of problems of the form

$$\mathbf{K}_i \mathbf{u}_i = \mathbf{f} \quad (i = 1, \dots, n_{sim}) \tag{1}$$

where $\mathbf{K}_i$ is the stiffness matrix corresponding to the stochastic realization of the ith simulation, $\mathbf{u}_i$ is the corresponding vector of unknown nodal displacements, n_{sim} is the number of Monte Carlo simulations and $\mathbf{f}$ is the vector of nodal loads.

G. Stavroulakis (✉) • D.G. Giovanis • M. Papadrakakis • V. Papadopoulos
Institute of Structural Analysis & Antiseismic Research, National Technical University of Athens, Iroon Polytechniou 9, Zografou Campus, Athens 15780, Greece
e-mail: stavroulakis@nessos.gr; dgiov16@yahoo.gr; mpapadra@central.ntua.gr; vpapado@central.ntua.gr

M. Papadrakakis and G. Stefanou (eds.), *Multiscale Modeling and Uncertainty Quantification of Materials and Structures*, DOI 10.1007/978-3-319-06331-7_14, © Springer International Publishing Switzerland 2014

The size of the stiffness matrix and the corresponding vectors is equal to the size of the equivalent deterministic problem. Thus, if $\mathbf{K}_0$ is the stiffness matrix of the deterministic problem with dimensions $N \times N$, Eq. (1) can be written as

$$(\mathbf{K}_0 + \Delta \mathbf{K}_i)\mathbf{u}_i = \mathbf{f}, \quad i = 1, \ldots, n_{sim} \tag{2}$$

which specifies a set of near-by problems.

If the uncertainties in the input parameters are modeled by Gaussian random fields then the truncated KL expansion is defined as (Grigoriu 2006; Huang et al. 2001):

$$\hat{a}(\mathbf{x}, \theta) = a_0(\mathbf{x}) + \sum_{i=1}^{M} \sqrt{\lambda_i} \xi_i(\theta) \phi_i(\mathbf{x}) \tag{3}$$

where, $a_0(\mathbf{x})$ denotes the mean value of the random field, $\xi_i(\theta)$ is a set of uncorrelated zero mean Gaussian random variables, θ being the random event. λ_i and $\phi_i(\mathbf{x})$ are the eigenvalues and mutually orthogonal eigenfunctions of its covariance function $C(\mathbf{x}_1, \mathbf{x}_2)$ which may be calculated in the domain D of the random field $a(\mathbf{x}, \theta)$, from the solution of a homogeneous Fredholm integral equation of the second kind.

Thus, in the case of Gaussian random fields Eq. (2) can be written as

$$\left(\mathbf{K}_0 + \sum_{j=1}^{M} \mathbf{K}_j \xi_j(\theta)\right) \mathbf{u}(\theta) = \mathbf{f} \tag{4}$$

$\mathbf{K}_j$ are deterministic and are given by:

$$\mathbf{K}_j = \mathbf{K}(\sqrt{\lambda_j} \phi_j(\mathbf{x})) \quad j = 1, \ldots, M \tag{5}$$

These repeated solutions can be performed either with a standard direct method based on Cholesky factorization or with preconditioned iterative methods. In high performance computing environments iterative schemes are more advantageous since they manage to harness the computational power of such environments while being more easily custom tailored to the particular properties of the equilibrium equations arising in the context of the numerical simulation used. In this work, we have further improved two variants of the MC-PCG-Skyline method previously proposed in Papadrakakis and Kotsopulos (1999) and Charmpis and Papadrakakis (2005).

Table 1 The PCG algorithm

Solution estimate		$\mathbf{x}^k = \mathbf{x}^{k-1} + \eta^{k-1}\mathbf{p}^{k-1}$
Residual vector		$\mathbf{r}^k = \mathbf{r}^{k-1} - \eta^{k-1}\mathbf{q}^{k-1}$
Preconditioned residual vector		$\mathbf{z}^k = \tilde{\mathbf{A}}^{-1}\mathbf{r}^k$
Search vector	Using re-orthogonalization	$\mathbf{p}^k = \mathbf{z}^k - \sum_{i=0}^{k-1} \frac{\mathbf{z}^{k^T}\mathbf{q}^i}{\mathbf{p}^{i^T}\mathbf{q}^i}\mathbf{p}^i$
A matrix product vector		$\mathbf{q}^k = \mathbf{A}\mathbf{p}^k$
η estimation	Using re-orthogonalization	$\eta^k = \frac{\mathbf{p}^{k^T}\mathbf{r}^k}{\mathbf{p}^{k^T}\mathbf{q}^k}$

1.1   The MC-PCG-Skyline Method

The PCG algorithm, when solving a linear system of the form $\mathbf{Ax} = \mathbf{b}$ with a preconditioner $\tilde{\mathbf{A}}$, is depicted in Table 1 for iteration k.

- Initialization phase: $\mathbf{r}^0 = \mathbf{b} - \mathbf{Ax}^0$ $\mathbf{z}^0 = \tilde{\mathbf{A}}^{-1}\mathbf{r}^0$, $\mathbf{p}^0 = \mathbf{z}^0$, $\mathbf{q}^0 = \mathbf{Ap}^0$, $\eta^0 = \frac{\mathbf{p}^{0^T}\mathbf{r}^0}{\mathbf{p}^{0^T}\mathbf{q}^0}$,
- Repeat for $k = 1, 2 \ldots$ until convergence.

The PCG algorithm equipped with a preconditioner following the rationale of incomplete Cholesky preconditioning features an error matrix $\mathbf{E}_i$. This matrix is dependent on the discarded elements of the lower triangular matrix produced by the incomplete Cholesky factorization procedure, which do not satisfy a specified magnitude or position criterion (Papadrakakis 1993). Considering the near-by problems of the form (2), if matrix $\mathbf{E}_i$ is taken as $\Delta\mathbf{K}_i$, the preconditioning matrix becomes the initial matrix $\tilde{\mathbf{A}} = \mathbf{K}_0$. The PCG algorithm equipped with the latter preconditioner throughout the entire solution process constitutes the MC-PCG-Skyline method for the solution of the n_{sim} near-by problems of Eq. (2).

The original MC-PCG-Skyline algorithm proposed in Charmpis and Papadrakakis (2005) uses a Cholesky direct solver for performing the proconditioning step, where $\mathbf{K}_0$ is factorized to LL^T at the beginning of the Monte Carlo simulation procedure and each evaluation of the preconditioned residual vector is carried out by a forward substitution, a vector operation and a backward substitution. In the present work, each evaluation of the preconditioned residual vector is carried out using a PFETI solver (Fragakis and Papadrakakis 2003), optimized for multiple right-hand sides (Fragakis and Papadrakakis 2004), as described in the following section, adhering to the rationale of the PCG method where the preconditioning step is performed with the FETI method (Papadrakakis and Kotsopulos 1999).

1.2 Optimizing the Solution with Multiple Right-Hand Sides

When using the PCG algorithm to solve problems with multiple right-hand sides, convergence can be accelerated by utilizing appropriately information accumulated during the previous solutions. In particular, given a sequence of linear systems with a constant left-hand side matrix $\mathbf{A}$ and multiple right-hand side vectors of the form

$$\mathbf{A}\mathbf{x}_i = \mathbf{b}_i, \quad i = 1, \cdots, j, j + 1, \cdots, n_a \tag{6}$$

where n_a is the number of solutions required, the number of PCG iterations required for each linear system may be reduced using the Krylov subspaces generated from search vectors $\mathbf{p}$ during the previous solutions. For the solution of the linear system $j + 1$, the following first solution estimate is considered:

$$\mathbf{x}_{j+1}^0 = \mathbf{P}_{np}\mathbf{x}_p \tag{7}$$

with

$$\mathbf{P}_{np} = [\mathbf{p}_1 \cdots \mathbf{p}_{n0}]$$

$$\mathbf{x}_p = \left(\mathbf{Q}_{n_p}^T \mathbf{P}_{n_p}\right)^{-1} \mathbf{P}_{n_p}^T \mathbf{b}_{j+1} \mathbf{Q}_{n_p} \tag{8}$$

$$\mathbf{Q}_{n_p} = \mathbf{A}\mathbf{P}_{n_p} = [\mathbf{A}\mathbf{p}_1 \cdots \mathbf{A}\mathbf{p}_{n0}] = [\mathbf{q}_1 \cdots \mathbf{q}_{n0}]$$

Given that search vector $\mathbf{p}$ and matrix product vector $\mathbf{q}$ using a re-orhtogonalization procedure are ensured to be A-orthogonal, the evaluation of $\mathbf{x}_p$ is trivial since $\mathbf{Q}_{n_p}^T \mathbf{P}_{n_p}$ has values only in its diagonal. Moreover, the search vector evaluation step can be carried out using not necessarily all but a fraction of the vectors stored from all the accumulated solutions.

In this work, we have used the first 600 search vectors, achieving a reduction between 90 and 95 % of the required iterations for each PFETI solution when compared to solving the same problems without using the aforementioned technique.

2 SSFEM in High Performance Computing Environment

In the SSFEM approach, the system response is projected in a PC basis as follows

$$\mathbf{u}(\theta) = \sum_{j=0}^{Q-1} \mathbf{u}_j \Psi_j(\xi) \tag{9}$$

where $\{\Psi_j(\xi)\}_{j=0}^{Q-1} = \{\Psi_j((\xi_1(\theta), \ldots, \xi_M(\theta)))\}_{j=0}^{Q-1}$ is the PC basis, consisting of the M−dimensional zero mean and orthogonal Hermite polynomials of order p.

The value of Q in Eq. (9) is determined by the following formula

$$Q = \frac{(M + p)!}{M! p!} \tag{10}$$

If the stochastic field is Gaussian Eq. (4) can be written

$$\left(\sum_{i=0}^{M} \mathbf{K}_i \xi_i(\theta) \right) \cdot \left(\sum_{j=0}^{Q-1} \mathbf{u}_j \Psi_j(\theta) \right) = \mathbf{f} \tag{11}$$

while for a lognormal stochastic field

$$\left(\sum_{i=0}^{Q-1} \mathbf{K}_i \Psi_i(\theta) \right) \cdot \left(\sum_{j=0}^{Q-1} \mathbf{u}_j \Psi_j(\theta) \right) = \mathbf{f} \tag{12}$$

After solving the augmented system for $\mathbf{u} = \{\mathbf{u}_k, k = 0, \ldots, Q - 1\}$, the required $\mathbf{u}(\theta)$ is computed from:

$$\mathbf{u}(\theta) = \sum_{j=0}^{Q-1} \mathbf{u}_j \Psi_j(\theta) \tag{13}$$

Once the coefficients $\mathbf{u}_j$ of the expansion are computed, approximate statistics of the solution can be derived by MC simulations. In this case however, the MC simulation computational effort is trivial since it is applied directly to the polynomial representation of Eq. (13) without the need of solving a system of equations at each simulation.

2.1 Solution of the Augmented Systems

The augmented systems that are generated when using SSFEM are suitable candidates for iterative solvers since they are flexible enough to be custom tailored to the particular architecture of the augmented systems while they are amenable to be efficiently implemented in high performance computing environments. Solution techniques are based on either Gauss-Jacobi (Anders and Hori 2001; Chung et al. 2005; Ghanem and Spanos 1990; Li et al. 2006) or PCG (Chung et al. 2005; Desceliers et al. 2005; Fraunfelder et al. 2005; Ghanem and Kruger 1996; Ghosh et al. 2008; Keese and Matthies 2005; Matthies and Keese 2005; Panayirci 2010; Pellissetti and Ghanem 2000) iterative solvers for addressing this problem. In this work two specialized preconditioners that take advantage of the properties of the augmented SSFEM linear systems are proposed which were found to be effective for both Gaussian and log-normal distributions.

Consider the preconditioning matrix for the case of Gaussian distribution of the form

$$
\tilde{\mathbf{A}} =
\begin{bmatrix}
a_1 \mathbf{K}_0 & 0 & \cdots & 0 \\
0 & a_2 \mathbf{K}_0 & \cdots & 0 \\
\vdots & \vdots & \ddots & \vdots \\
0 & 0 & \cdots & a_n \mathbf{K}_0
\end{bmatrix}
\tag{14}
$$

where a_n are the coefficients as calculated from the PC bases. For each evaluation of the preconditioned residual vector, the same $\mathbf{K}_0$ matrix needs to be "inverted" n times, as in the case of the MC-PCG-Skyline method. This matrix "inversion" is implemented as the solution of n linear systems. Since matrix $\tilde{\mathbf{A}}$ is block diagonal, the solution process can be pipelined as the successive solution of n linear systems with multiple right-hand sides. The PCG algorithm equipped with preconditioning matrix $\tilde{A}$ and utilizing the FETI method for solving the successive linear systems is proposed in Ghosh et al. (2008, 2009). A variant of this approach is tested in this work by employing PFETI for the solution of the repeated linear systems involved in the preconditioning steps of PCG. This algorithm is abbreviate as SSFEM-PCG-B for the solution of the augmented linear system that occurs from SSFEM.

The second preconditioner is based on the SSOR-type preconditioning matrix. In particular, the augmented matrix $\mathbf{K}$ is decomposed into the diagonal component $\mathbf{D}$ as it appears in Eq. (14), and a strictly lower triangular component $\mathbf{L}$ of the form:

$$
\mathbf{L} =
\begin{bmatrix}
0 & 0 & 0 & 0 \\
\mathbf{K}_{21} & 0 & \cdots & 0 \\
\vdots & \mathbf{K}_{m2} & \ddots & \vdots \\
\mathbf{K}_{n1} & \mathbf{K}_{n2} & \cdots & 0
\end{bmatrix}
\tag{15}
$$

Using this decomposition, the SSOR-type preconditioner is of the form:

$$
\tilde{\mathbf{A}} = (\mathbf{D} - \mathbf{L})\mathbf{D}^{-1}(\mathbf{D} - \mathbf{L}^T) \Leftrightarrow \tilde{\mathbf{A}}^{-1} = (\mathbf{D} - \mathbf{L}^T)^{-1}\mathbf{D}(\mathbf{D} - \mathbf{L})^{-1}
\tag{16}
$$

The evaluation of the preconditioned residual vector of the PCG algorithm is implemented as follows:

1. Solve:

$$
(\mathbf{D} - \mathbf{L})\mathbf{z}_1^k = \mathbf{r}^k
\tag{17}
$$

2. Evaluate:

$$
\mathbf{z}_2^k = \mathbf{D}\mathbf{z}_1
\tag{18}
$$

3. Solve:

$$(\mathbf{D} - \mathbf{L}^T)\mathbf{z}^k = \mathbf{z}_2^k \tag{19}$$

The PCG algorithm equipped with the above block-SSOR preconditioner and utilizing the PFETI method for solving the linear systems occurring at each step of the occurring forward and backward substitutions, constitutes the SSFEM-PCG-S method for the solution of the augmented linear system that occurs from SSFEM.

2.2 A Full Block Preconditioning Scheme

The existence of a number of linear combinations of the deterministic matrix with stochastic ones at the block diagonal part of the augmented stiffness matrix, for the log-normal case, can really deteriorate the convergence rate of both the preconditioner of the SSFEM-PCG-B method and the preconditioner of the SSFEM-PCG-S method, especially at large input covariances where the magnitude of the stochastic matrices is comparable to the magnitude of the deterministic one.

In order to overcome this deficiency, a MC-PCG-PFETI solver is used instead of a regular PFETI solver, in order to evaluate the preconditioned residual of each iteration of the SSFEM-PCG-B solver. Thus the MC-PCG-PFETI solver takes into account the full linear combination of the block diagonal, enhancing the convergence rate of the SSFEM-PCG-B and SSFEM-PCG-S solvers, instead of taking into account only $\mathbf{K}_0$ matrix which gives an approximation to the preconditioned residual.

As in the case of the Monte Carlo simulations, the repeated solutions required for the preconditioning step of the MC-PCG-PFETI algorithm can be treated as problems with multiple right-hand sides, since the entries in the residual vector are updated at each PCG iteration k of each block diagonal part of the coefficient matrix.

The SSFEM-PCG-B algorithm equipped with this preconditioner and utilizing the MC-PCG-PFETI method constitutes the SSFEM-PCG-BF variant for the solution of the augmented linear system that occurs from the SSFEM.

In the same fashion, the PCG algorithm equipped with the block SSOR preconditioner of the SSFEM-PCG-S method and utilizing the MC-PCG-Skyline method for solving the linear systems occurring at the preconditioned residual vector evaluation, constitutes the SSFEM-PCG-SF variant for the solution of the augmented linear system that occurs from the SSFEM.

3 Numerical Test

Numerical tests are performed for stochastic finite element analysis implementing the proposed versions of SSFEM and MC. The computer platform used is an Intel Core i7 X980 with 6 physical cores at 3.33 GHz with 24 GB of RAM.

a   b

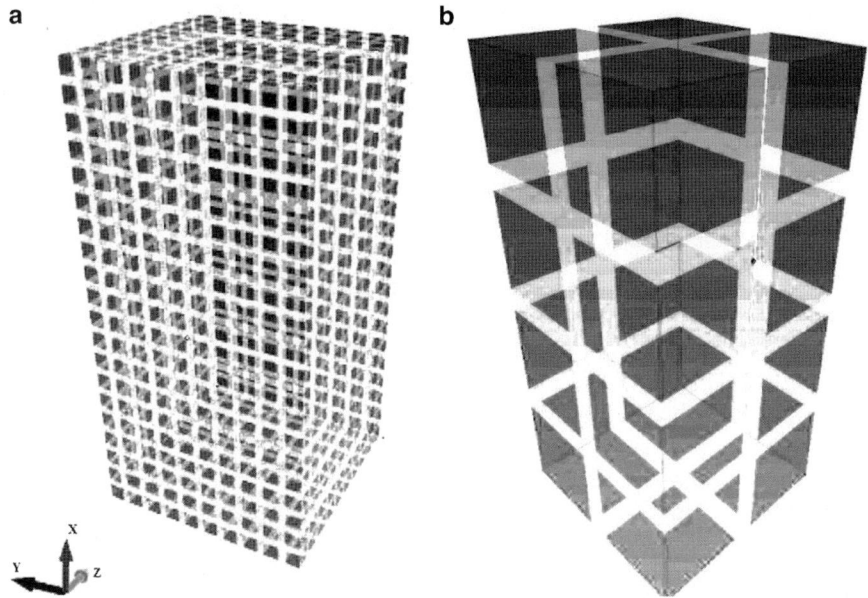

Fig. 1 Domain decomposition of a quarter of the deterministic soil problem with 10k dof.
(**a**) Element mesh (**b**) Subdomain mesh

In order to assess the computational efficiency of the MC and SSFEM methods
for the analysis of systems with uncertain properties, a soil cube of $10 \times 10 \times 20$
meters with uncertain Young modulus E under load in the center of its upper surface
due to a large footing was considered, resulting to a finite element mesh of 10k dof
approximately (Fig. 1).

Three test cases regarding coefficients σ_E are examined: (a) $\sigma_E = 15\%$
(Gaussian), (b) $\sigma_E = 30\%$ (log-normal) and (c) $\sigma_E = 80\%$ (log-normal).
Moreover, four correlation length values are assumed: (a) b=0.1a, (b) b=1a,
(c) b–10a and (d) b=100a, with a being the height of the cube. Setting $a = 20$ m,
the correlation lengths that were examined for this example were 2, 20, 200
and 2,000 m.

3.1 Solver Assessment Procedure

In order to set an objective basis for assessing the computational performance of
the numerical algorithms discussed, a parametric study was conducted, regarding
different values for standard deviation σ_E and correlation length b. For the
computation of the second moments of the response field, the following procedure
was followed:

1. A series of Monte Carlo analyses of 100k simulations was carried out, using
   $M = 1$ as the order of the KL expansion, in order to estimate the necessary

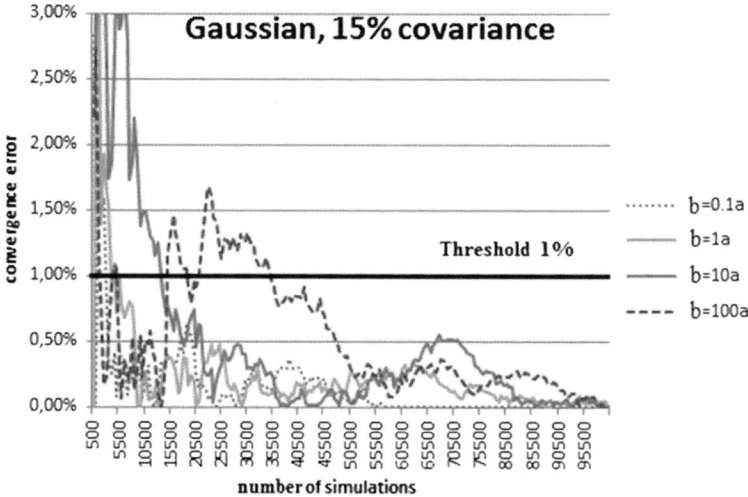

Fig. 2 Step 1: $COV(\%)$ convergence error of MC for the Gaussian field with $\sigma_E = 15\%$

number of simulations for a convergence error of less than 1 % for each value of σ_E and b examined. This error is computed as the normalized difference of the COV (%) at each simulation with respect to the COV (%) computed at the end of the $100k$ simulations.

2. Assuming that the convergence behavior of the previous step remains invariant for increasing M, another series of Monte Carlo analyses was carried out, in the range of $M = 2$ to $M = 12$, in order to estimate the appropriate order of the KL expansion for a convergence error of less than 1 %. In this case an "exact" solution was assumed at $M = 12$ in order to compute the relative error (%) for different M.

   Using the results of the previous step, the same procedure was carried out performing SSFEM analyses, in order to estimate the appropriate order of the PC expansion required for convergence to the corresponding MC results.

3.2 Computation of the Second Moments of the Response Field

Figures 2–4 show the convergence error for each field as per step 1 of the assessment procedure. Based on these figures, the number of simulations necessary for evaluating the second moments of the response field are shown in Table 2.

Following step 2, Figs. 5–7 show the convergence error for each field as per step 3 of the assessment procedure for the selection of PC expansion order (p) required for the SSFEM to converge at an error less than 1 % using the KL expansion orders M shown in Table 3. This relative error is computed with respect to the corresponding MC simulations with the same parameter M.

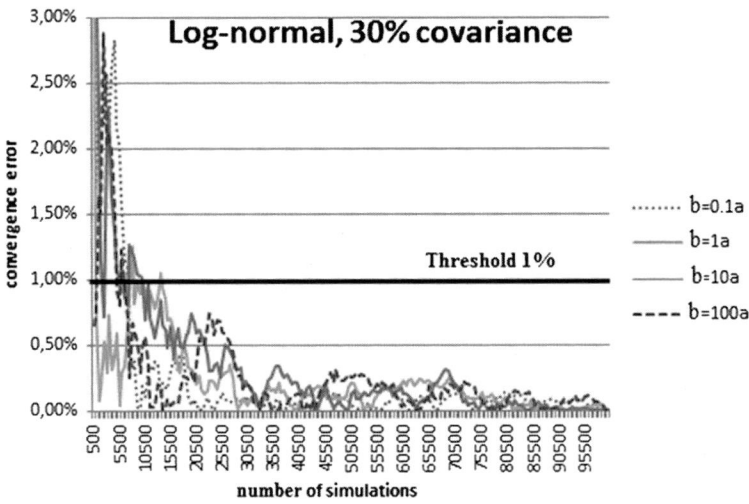

Fig. 3 Step 1: $COV(\%)$ convergence error of MC for the log-normal field with $\sigma_E = 30\%$

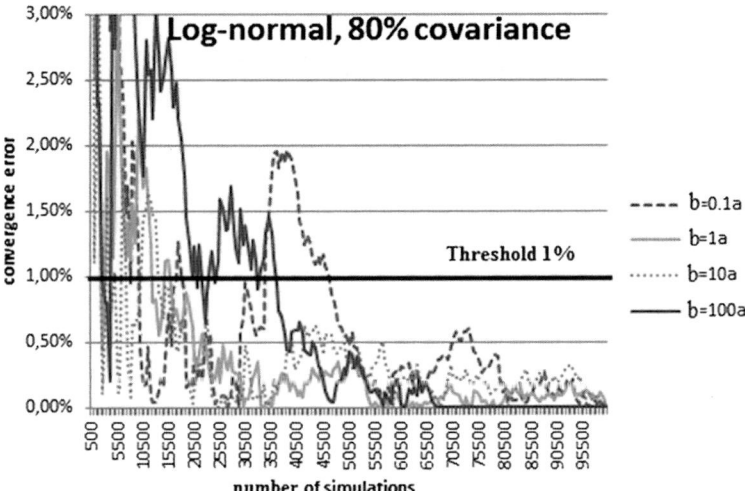

Fig. 4 Step 1: $COV(\%)$ convergence error of MC for the log-normal field with $\sigma_E = 80\%$

Table 2 Required number of MC simulations for achieving a COV error less than 1 %

Correlation length b	$\sigma_E = 15\%$	$\sigma_E = 30\%$	$\sigma_E = 80\%$
0,1a	20,000	10,000	53,000
1a	25,000	18,000	28,000
10a	23,000	23,000	34,000
100a	50,000	43,000	45,000

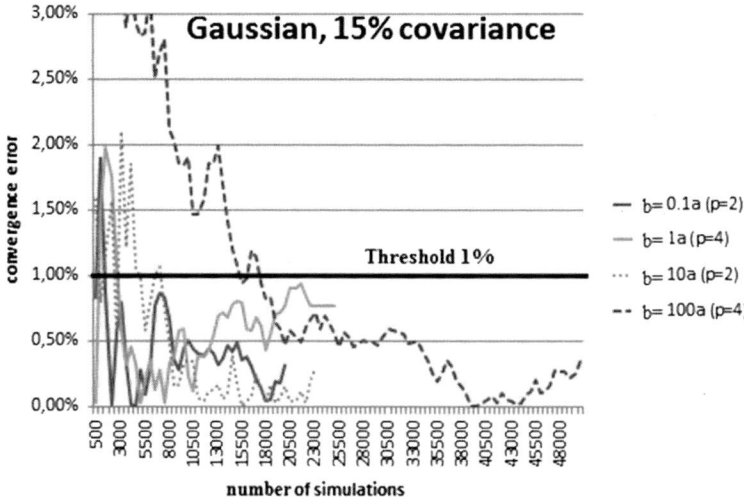

Fig. 5 Step 3: $COV(\%)$ convergence error of the SSFEM for the Gaussian field with $\sigma_E = 15\%$ and $p = 2, 4$

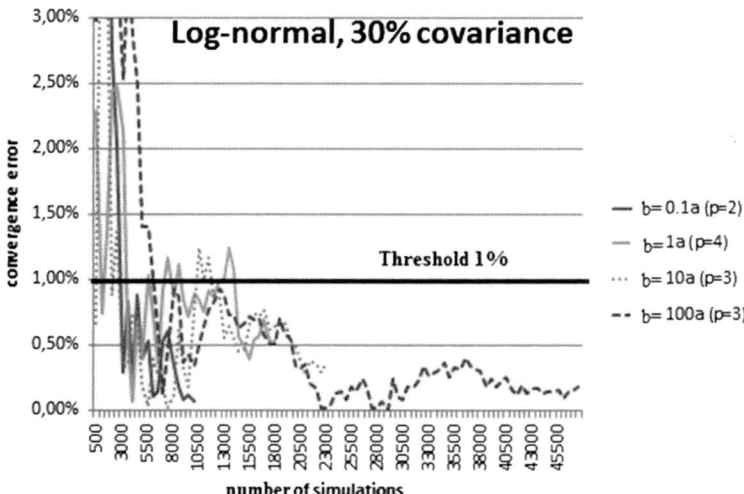

Fig. 6 Step 3: $COV(\%)$ convergence error of the SSFEM for the log-normal field with $\sigma_E = 30\%$ and $p = 2, 3, 4$

Table 4 summarizes the convergence of the SSFEM (relative error %) with respect to MC, for all cases considered.

It is worth noting that for the case of $b = 0.1a$, the SSFEM failed to provide a solution within the acceptable error margin when compared to the MC solution. While increasing the p-order of the PC expansion, the SFFEM method was asymptotically converging to a solution which exhibited a 30 % error when compared to the corresponding Monte Carlo solution.

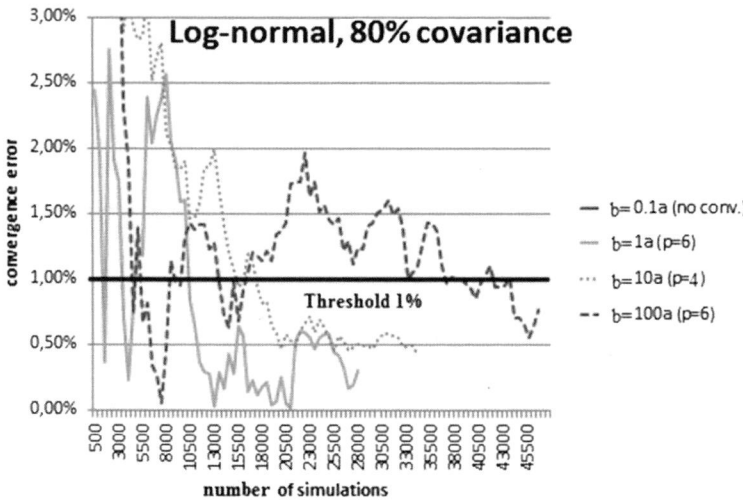

Fig. 7 Step 3 : $COV(\%)$ convergence error of the SSFEM for the log-normal field with $\sigma_E = 80\%$ and $p = 2, 3, 4$

Table 3 Step 2: $COV(\%)$ convergence errors for the various KL expansion orders

Correlation length b	$\sigma_E = 15\%$		$\sigma_E = 30\%$		$\sigma_E = 80\%$	
	M	Error (%)	M	Error (%)	M	Error (%)
0.1a	12	"exact"	10	0.43	4	0.75
1a	6	0.93	4	0.75	4	0.57
10a	2	0.36	2	0.85	4	0.26
100a	2	0.48	2	0.53	4	0.96

Table 4 Convergence errors for the SSFEM

Correlation length b	$\sigma_E = 15\%$		$\sigma_E = 30\%$		$\sigma_E = 80\%$	
	p	Error (%)	p	Error (%)	p	Error (%)
0.1a	2	0.23	2	0.07	6	30.00
1a	4	0.09	4	0.69	6	0.52
10a	2	0.03	3	0.36	4	0.68
100a	4	0.36	3	0.74	6	0.88

3.3 Performance of the Proposed Solution Procedures

Using all previous numerical data (number of simulations, KL expansion order and PC expansion order), a series of numerical tests were performed in order to assess the performance of the various solution techniques discussed and proposed in this work. For all cases considered the normalized solution accuracy was set to 10^{-7} while for the computation of the preconditioned residual vector, the required accuracy was set to 10^{-3}.

Table 5 Performance of the various MC-PCG-Skyline variants for the MC for evaluating the second moments of the response field for $\sigma_E = 30\%$ in sequential and parallel implementation

$\sigma_E = 30\%$	**Correlation length b**	0.1a	1a	10a	100a
	MC simulations	10,000	18,000	23,000	43,000
	PCG iterations	110,100	221,531	114,541	153,825
MC-PCG-Skyline	Time (s)-sequential	18,922	105,391	161,203	241,235
	Time (s)-parallel	2,783	15,499	23,706	35,476
MC-PCG-FETI	FETI iterations	314,475	107,198	23,442	37,087
		(1,761,600)	(3,544,496)	(1,832,656)	(2,461,200)
	Time (s)-sequential	33,053	32,437	22,189	38,625
	Time (s)-parallel	4,861	4,770	3,263	5,680
MC-PCG-PFETI	PFETI iterations	294,235	99,478	21,777	34,375
		(1,761,600)	(3,544,496)	(1,832,656)	(2,461,200)
	Time (s)-sequential	25,337	24,956	17,393	30,080
	Time (s)-parallel	3,726	3,670	2,558	4,423

Table 6 Performance of the various MC-PCG-Skyline variants for the MC for evaluating the second moments of the response field for $\sigma_E = 80\%$ in sequential and parallel implementation

$\sigma_E = 80\%$	**Correlation length**	0.1a	1a	10a	100a
	MC simulations	53,000	28,000	34,000	45,000
	PCG iterations	1,193,825	695,100	272,340	253,350
MC-PCG-Skyline	Time (s)-sequential	205,082	68,030	370,320	400,378
	Time (s)-parallel	30,159	10,004	54,459	58,879
MC-PCG-FETI	FETI iterations	3,413,444	1,624,400	72,507	56,799
		(19,101,200)	(11,121,600)	(4,357,440)	(4,053,600)
	Time (s)-sequential	358,806	97,472	65,108	60,159
	Time (s)-parallel	52,765	14,334	9,575	8,847
MC-PCG-PFETI	PFETI iterations	3,265,860	1,530,760	67,320	52,650
		(19,101,200)	(11,121,600)	(4,357,440)	(4,053,600)
	Time (s)-sequential	281,182	75,286	50,653	46,932
	Time (s) -parallel	41,350	11,071	7,449	6,902

Tables 5 and 6 show the performance of proposed MC-PCG-PFETI solver for the evaluation of the second order moments of the response field using the MC method, in comparison to MC-PCG-Skyline and MC-PCG-FETI. The PFETI and FETI iterations correspond to the sum of the PFETI and FETI iterations needed for all the MC simulations using the A-orthogonalization technique, while in parentheses the corresponding PFETI and FETI iterations without A-orthogonalization are given. These numbers show a drastic decrease of iterations ranging from one to two orders of magnitude as a result of the A-orthogonalization procedure. Moreover, from these tables, it is evident that the PFETI variant outperforms the FETI one in all tests, showing a 1.25× speedup. This performance increase occurs for two reasons: (i) PFETI needs ∼10% less iterations when compared to FETI. (ii) The cost for each reorthogonalization of the PFETI method is about 35% less when compared to the FETI method. This stems from the fact that the interface problem of the

Table 7 Performance metrics for the log-normal case ($\sigma_E = 80\%$ covariance)

Log-normal 80%					
Correlation length b		$0.1a$	$1a$	$10a$	$100a$
SSFEM-PCG-B	MC simulations	53,000	28,000	34,000	45,000
	PCG iterations	48	89	33	58
	PFETI iterations	685	1,010	523	584
	Total time (s)-sequential	266,702	272,836	14,171	321,161
	Total time (s)-parallel	46,799	49,175	2,447	55,380
	Total time cached (s)-sequential	112,142	114,721	9,443	134,402
	Total time cached (s)-parallel	1988	2079	131	2393
SSFEM-PCG-BF	PCG iterations	47	117	47	82
	PFETI iterations	2,827	12,393	423	467
	Total time (s)-sequential	273,786	280,083	19,072	452,675
	Total time (s)-parallel	48,047	50,475	3,291	78,050
	Total time cached (s)-sequential	12,246	12,563	1,237	1,885
	Total time cached (s)-parallel	21,495	22,580	2,130	32,538
SSFEM-PCG-S	PCG iterations	16	186	36	59
	PFETI iterations	528	1,167	414	338
	Total time (s)-sequential	177,970	368,937	28,790	651,255
	Total time (s)-parallel	31,236	66,484	4,968	112,289
	Total time cached (s)-sequential	74,930	154,271	18,473	276,205
	Total time cached (s)-parallel	13,180	27,807	3,189	47,624
SSFEM-PCG-SF	PCG iterations	12	25	13	20
	PFETI iterations	461	449	273	228
	Total time (s)-sequential	133,533	276,818	10,529	220,894
	Total time (s)-parallel	23,430	49,887	1,818	38,096
	Total time cached (s)-sequential	56,253	115,818	6,803	92,094
	Total time cached (s)-parallel	9,879	2,087	1,175	15,888

PFETI method is based on the boundary dof of each subdomain while the interface problem of the FETI method is based on the lagrange multipliers which, due to the existence of a considerable number of subdomains crosspoints, are significantly larger in quantity than the boundary dof.

Table 7 presents the performance metrics for the log-normal case with 80% covariance. As previously, the SSFEM-PCG-S and SSFEM-PCG-SF variants

Table 8 Monte Carlo vs. SSFEM for the Gaussian case (15 % covariance)

Gaussian 15 %					
Correlation length b		0.1a	1a	10a	100a
MC	PCG iterations	184,398	196,875	114,715	144,592
	PFETI iterations	425,281	124,133	20,157	35,761
	Time (s)-sequential	36,771	28,076	16,443	30,590
	Time (s)-parallel	5,407	4,129	2,418	4,498
SSFEM	PCG iterations	3	4	3	4
	PFETI iterations	377	269	74	74
	Time (s)-sequential	1,026	1,178	77	100
	Time (s)-parallel	179	204	13	17

Table 9 Monte Carlo vs. SSFEM for the log-normal case (30 % covariance)

Log-normal 30 %					
Correlation length b		0.1a	1a	10a	100a
MC	PCG iterations	110,100	221,531	114,541	153,825
	PFETI iterations	294,235	99,478	21,777	34,375
	Time (s)-sequential	25,337	24,956	17,393	30,080
	Time (s)-parallel	3,726	3,670	2,558	4,423
SSFEM	PCG iterations	4	5	4	4
	PFETI iterations	403	260	100	79
	Time (s)-sequential	1,568	2,884	162	132
	Time (s)-parallel	277	498	28	23

Table 10 Monte Carlo vs. SSFEM for the log-normal case (80 % covariance)

Log-normal 80 %					
Correlation length b		0.1a	1a	10a	100a
MC	PCG iterations	1,194,328	695,100	272,340	253,350
	PFETI iterations	3,265,860	1,530,760	67,320	52,650
	Time (s)-sequential	281,182	75,286	50,653	46,932
	Time (s)-parallel	41,350	11,071	7,449	6,902
SSFEM	PCG iterations	–	89	13	20
	PFETI iterations	–	1,010	273	228
	Time (s)-sequential	–	114,721	6,803	92,094
	Time (s)-parallel	–	20,679	1,175	15,888

outperform the SSFEM-PCG-B and SSFEM-PCG-BF methods, showing a speedup up to 2.8×. For this covariance of the log-normal case, the proposed caching scheme proves to be quite efficient, providing up to 3× speedup when compared to the corresponding uncached method.

Tables 8–10 compare the performance of the MC and SSFEM when using the most computationally efficient solution method for evaluating the second order moments of the response field. It can be seen that for the Gaussian input field the SSFEM outperforms Monte Carlo method. The same conclusion can be reached for the log-normal case with 30 % covariance. For the log-normal case with 80 % covariance MC method outperforms SSFEM in all cases except for the $b = 10a$ correlation length. This is due to the small order of $p = 4$ required by the PC expansion, compared to the other cases which required an expansion of order $p = 6$. For the $b = 0.1a$ case, SSFEM fails to converge.

4 Conclusions

An investigation of high performance solution techniques amenable to parallelization has been made in the context of stochastic problems solved with the finite element method. In order to address the stochastic part of the problem, both Monte Carlo and spectral finite element methods have been explored. The numerical performance of their solution methods has been demonstrated utilizing primal and dual domain decomposition methods with enhanced preconditioners, custom tailored to the specific numerical properties of the corresponding formulation of the stochastic problem with no loss of parallel scalability.

When comparing the novel solution techniques proposed for the MC procedure, a speedup of $1.25\times$ was exhibited while for the SSFEM, a speedup of $3\times$ was exhibited when utilizing the block-SSOR preconditioning combined with caching techniques with respect to the diagonally block preconditioning, making the SSFEM even more attractive for solving large scale stochastic problems in high performance computing environments.

Acknowledgements This work has been supported by the European Research Council Advanced Grant MASTER – Mastering the computational challenges in numerical modeling and optimum design of CNT reinforced composites (ERC-2011-ADG-20110209).

References

Anders M, Hori M (2001) Three-dimensional stochastic finite element method for elasto-plastic bodies. Int J Numer Methods Eng 51:449–478
Charmpis DC, Papadrakakis M (2005) Improving the computational efficiency in finite element analysis of shells with uncertain properties. Comput Methods Appl Mech Eng 194:1447–1478
Chung DB, Guttierez MA, Graham-Brady LL, Lingen F-J (2005) Efficient numerical strategies for spectral stochastic finite element models. Int J Numer Meth Eng 64:1334–1349
Desceliers C, Ghanem R, Soize C (2005) Polynomial chaos representation of a stochastic preconditioner. Int J Numer Methods Eng 64(5):618–634
Fragakis Y, Papadrakakis M (2003) The mosaic of high performance domain decomposition methods for structural mechanics: formulation, interrelation and numerical efficiency of primal and dual methods. Comput Methods Appl Mech Eng 192:35–36
Fragakis Y, Papadrakakis M (2004) The mosaic of high performance domain decomposition methods for structural mechanics-part II: formulation enhancements, multiple right-hand sides and implicit dynamics. Comput Methods Appl Mech Eng 193:4611–4662
Fraunfelder P, Schwab C, Todor RA (2005) Finite elements for elliptic problems with stochastic coefficients. Comput Methods Appl Mech Eng 194:205–228

Ghanem R, Kruger RM (1996) Numerical solution of spectral stochastic finite element systems. Comput Methods Appl Mech Eng 129:289–303

Ghanem R, Spanos PD (1990) Polynomial chaos in stochastic finite elements. J Appl Mech ASME 57:197–202

Ghosh D, Avery P, Farhat C (2008) A method to solve spectral stochastic finite element problems for large-scale systems. Int J Numer Meth Eng 00:1–6

Ghosh D, Avery P, Farhat C (2009) A FETI-preconditioned conjugate gradient method for large-scale stochastic finite element problems. Int J Numer Meth Eng 80:914–931

Grigoriu M (2006) Evaluation of Karhunen-Loève, spectral and sampling representations for stochastic processes. J Eng Mech 132:179–189

Huang SP, Quek ST, Phoon KK (2001) Convergence study of the truncated Karhunen-Loève expansion for simulation of stochastic processes. Int J Numer Methods Eng 52:1029–1043

Keese A, Matthies HG (2005) Hierarchical parallelisation for the solution of stochastic finite element equations. Comput Struct 83:1033–1047

Li CF, Feng YT, Owen DRJ (2006) Explicit solution to the stochastic system of linear algebraic equations $(\acute{a}_1 a_1 + \acute{a}_2 a_2 + \cdots + \acute{a}_m a_m)x = b$. Comput Methods Appl Mech Eng 195(44–47):6560–6576

Matthies HG, Keese A (2005) Galerkin methods for linear and nonlinear elliptic stochastic partial differential equations. Comput Methods Appl Mech Eng 194:1295–1331

Panayirci HM (2010) Efficient solution of Galerkin-based polynomial chaos expansion systems. Adv Eng Softw 41:1277–1286

Papadrakakis M (1993) Solving large-scale linear problems in solid and structural mechanics. In: Solving large-scale problems in mechanics, Wiley, pp 1–37

Papadrakakis M, Kotsopulos A (1999) Parallel solution methods for stochastic finite element analysis using Monte Carlo simulation. Comput Methods Appl Mech Eng 168:305–320

Pellissetti MF, Ghanem R (2000) Iterative solution of systems of linear equations arising in the context of stochastic finite elements. Adv Eng Softw 31:607–616

Effects of POD-Based Components of Turbulent Wind on the Aeroelastic Stability of Long Span Bridges

Vincenzo Sepe and Marcello Vasta

Abstract In this paper the effects of low-frequency wind speed fluctuations on the aeroelastic stability of long span bridges are analyzed and discussed. The spatial-temporal field of wind turbulence is described by means of orthogonal loading components representing the eigenfunctions of the cross power spectral density function (cpsdf) of the wind process. The stability condition, derived through stochastic averaging technique and moment stability method, are finally applied to a bridge designed for crossing the Messina Strait (Italy).

Keywords Long span bridges • Aeroelastic stability • Stochastic averaging • POD wind representation

1 Introduction

The effects on the aeroelastic stability of long span bridges of low-frequency wind speed fluctuations, i.e. those components of the atmospheric turbulence characterized by frequencies below 0.2 Hz and by a strong along-span coherence, has been dealt with in previous papers of the writers, under simplifying assumptions on the turbulence model (Sepe and Vasta 2005a, b; Sepe and D'Asdia 2003).

Namely, in Sepe and D'Asdia (2003), it was shown that the effects of low-frequency wind-speed fluctuations can be modeled as a perturbation of the critical state of the 1-dof system (flutter mode) representing the 3D oscillations (vertical, transversal and torsional) of the bridge under a non-turbulent critical wind-speed.

From a mathematical point of view, the contribution of the low-frequency turbulence introduces parametric excitations, both on the stiffness and on the

---

V. Sepe • M. Vasta (✉)
Department of Engineering and Geology INGEO, University "G. D'Annunzio" of Chieti-Pescara, V.le Pindaro 42, I-65127 Pescara, Italy
e-mail: v.sepe@unich.it; mvasta@unich.it

M. Papadrakakis and G. Stefanou (eds.), *Multiscale Modeling and Uncertainty Quantification of Materials and Structures*, DOI 10.1007/978-3-319-06331-7_15,
© Springer International Publishing Switzerland 2014

damping terms of the equation describing the time-evolution of the flutter mode. As a consequence, if frequency and coherence of the fluctuating part of the wind speed are unfavorably tuned with the bridge characteristics, they can make the critical state unstable.

To evaluate in a closed form the worst effects that can ever be expected as a consequence of this kind of turbulence, in Sepe and Vasta (2005a) only low-frequency wind speed fluctuations strongly correlated along the span were taken into account, according to Sepe and D'Asdia (2003); namely, the longitudinal shapes of the components significant from the dynamic point of view were assumed as deterministic sine waves along the bridge span, whose wavelength increases for a decreasing time-frequency of the turbulence component; in such a way, the beneficial effects due to loss of coherence along the span were neglected (or at least underestimated).

Developing the approach described in Sepe and D'Asdia (2003), where the stability conditions were derived taking into account only the effects of the parametric excitation related to the angular speed (i.e. assuming the aeroelastic behaviour depending only on the "total equivalent damping", as is typical for flutter under non turbulent wind), in Sepe and Vasta (2005b) the effects of both the damping and stiffness terms were considered and the stability condition was derived through stochastic averaging technique and moment stability method; this requires an appropriate application of the stochastic averaging technique, that extends the results known for stochastically independent forcing components to the case, of interest for the present problem, of strongly correlated forcing terms.

However, the procedure developed in Sepe and Vasta (2005a, b) and Sepe and D'Asdia (2003) still assumes a deterministic longitudinal shape of the turbulence components, with both frequency and coherence unfavorably tuned (and therefore worse than any realistic wind field).

In this paper, such simplifying hypothesis on the spatial correlation of turbulence components is removed. Indeed, the spatial-temporal field of wind turbulence is described by means of orthogonal loading components representing the eigenfunctions of the cross power spectral density function (cpsdf) of the process, according to a POD-based representation (e.g. Carassale 2005; Carassale et al. 2007; Fiore and Monaco 2009; Di Paola and Gullo 2001), so obtaining a more realistic wind turbulence field.

The stability conditions derived in the paper are finally analyzed and discussed with reference to a bridge designed for crossing the Messina Strait (Italy).

2 Forced Oscillations Near the Critical Conditions

Let $\alpha(z, t), h(z, t)$ denote the rotation and the vertical displacement of the deck at the time t and abscissa z (Fig. 1). If a constant wind speed is assumed along the bridge span (i.e. wind turbulence neglected), as typical for aeroelastic instability analyses, a critical value can be found $U(z, t) \equiv U_c$ that makes the structure oscillate with a fixed

Fig. 1 Displacements and
aeroelastic forces on the deck

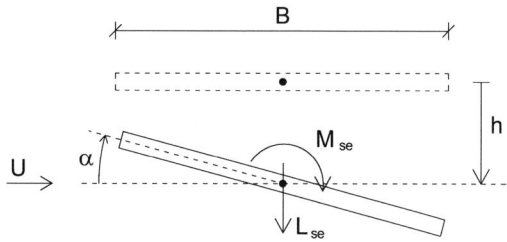

shape ("flutter mode"), with prevalent vertical and torsional components, behaving as an undamped 1-dof system with critical angular frequency ω_c (Simiu and Scanlan 1996).

In this case, denoting as $f_\alpha(z), f_h(z)$ the components of the flutter mode and by φ the phase-lag among them, it results (Sepe and D'Asdia 2003)

$$\alpha(z,t) = f_\alpha(z)A(t) = f_\alpha(z)A_0 \cos \omega_c t$$
$$h(z,t) = f_h(z)H(t) = f_h(z)H_0 \cos(\omega_c t + \varphi) \tag{1}$$

with $H_0 = \beta\, BA_0$ where B denotes the deck width, β a non-dimensional coefficient, $A(t)$ the torsional amplitude and $H(t)$ the flexural amplitude, related each other as

$$\dot{H}(t) = \beta B \cos \varphi\, \dot{A}(t) - \beta B \omega_c \sin \varphi\, A(t) \tag{2}$$

Equation (2) maintains its validity if the oscillation amplitude varies slowly in time; therefore Eq. (2) will be used also for a wind speed slightly different from U_c

$$U(z,t) = U_c\,[1 + \psi_0 + \psi(z,t)] \qquad \psi_0 << 1 \quad \psi(z,t) << 1 \tag{3}$$

where a variation $\psi_0 U_c$ of the mean speed is separated from the fluctuations $\psi(z,t)\, U_c$. For this fluctuating wind speed, therefore, the motion is still described by a single scalar function, with a constant spatial shape assumed as equal to the flutter mode.

Under these conditions the aeroelastic moment and lift acting on the deck (Simiu and Scanlan 1996) can be expressed as (D'Asdia and Sepe 1998; Sepe et al. 2000)

$$M_{se} = \frac{\rho U_c^2 B^2}{2}\left[K_c A_1^*(K)\frac{\dot{h}}{U_c} + K_c A_2^*(K)\frac{B\dot{\alpha}}{U_c} + K_c^2 A_3^*(K)\alpha \right]$$

$$L_{se} = \frac{\rho U_c^2 B}{2}\left[K_c H_1^*(K)\frac{\dot{h}}{U_c} + K_c H_2^*(K)\frac{B\dot{\alpha}}{U_c} + K_c^2 H_3^*(K)\alpha \right] \tag{4}$$

respectively, where A_i^*, H_j^* are the aeroelastic derivatives evaluated by means of wind-tunnel tests or CFD (computational fluid dynamics) simulations (Simiu and

Scanlan 1996; Lin 1996), ρ the air density and K, K_c the reduced frequencies

$$K(z,t) = \frac{B\omega_c}{U(z,t)} \quad , \quad K_c = \frac{B\omega_c}{U_c} \tag{5}$$

In particular, it turns out

$$M_{se} = M_{se}{}^{(0)} + \Delta M_{se}; \quad L_{se} = L_{se}{}^{(0)} + \Delta L_{se} \tag{6}$$

where $M_{se}{}^{(0)}, L_{se}{}^{(0)}$ and $\Delta M_{se}, \Delta L_{se}$ denote the aeroelastic actions corresponding to the critical wind speed U_c and to the difference $U_c [\psi_0 + \psi(z,t)]$, respectively.

Under these assumptions and linearising the aeroelastic derivatives A_i^*, H_j^* with respect to the reduced velocity $v = 2\pi/K$, around the critical value $v_c = 2\pi U_c/(B\omega_c)$, the equation governing the motion becomes (Sepe and D'Asdia 2003)

$$I\ddot{A} + [-\mu_1 \beta B \cos\varphi (\Psi_{\alpha h0} + \Psi_{\alpha h}(t)) - \mu_2 (\Psi_{\alpha\alpha 0} + \Psi_{\alpha\alpha}(t))] \dot{A}$$
$$+ [I\omega_c{}^2 + \mu_1 \beta B\omega_c \sin\varphi (\Psi_{\alpha h0} + \Psi_{\alpha h}(t)) - \mu_3 (\Psi_{\alpha\alpha 0} + \Psi_{\alpha\alpha}(t))] A = 0 \tag{7}$$

where I is the generalised inertia of the bridge and with the positions

$$\Psi_{\alpha h0} = \psi_0 \int_L f_\alpha(z) f_h(z) dz = \psi_0 F_{\alpha h} \qquad \Psi_{\alpha\alpha 0} = \psi_0 \int_L f_\alpha{}^2(z) dz = \psi_0 F_{\alpha\alpha}$$

$$\Psi_{\alpha h}(t) = \int_L \psi(z,t) f_\alpha(z) f_h(z) dz \qquad \Psi_{\alpha\alpha}(t) = \int_L \psi(z,t) f_\alpha{}^2(z) dz \tag{8}$$

The constants μ_1, μ_2, μ_3 depends on the geometry and the aerodynamic characteristic of the bridge

$$\mu_1 = \pi\rho U_c B^2 a_1^*, \qquad \mu_2 = \pi\rho U_c B^3 a_2^*, \qquad \mu_3 = \pi\rho U_c^2 B^2 K_c a_3^* \tag{9}$$

where

$$a_i^* = \left(\frac{d}{dv} A_i^*\right)_{v=v_c}, \quad i = 1, 2, 3 \tag{10}$$

Equation (7) can be written as

$$\ddot{A} + \left[2\tilde{\zeta}\omega_c + \eta(t)\right] \dot{A} + \omega_c^2 \left[1 + \varepsilon + \tilde{\xi}(t)\right] A = 0 \tag{11}$$

with

$$\tilde{\zeta} = -\frac{1}{2I\omega_c} [\mu_1 \beta B \cos \varphi \ \Psi_{\alpha h0} + \mu_2 \Psi_{\alpha\alpha0}] \qquad (12)$$

$$\eta(t) = -\frac{1}{I} [\mu_1 \beta B \cos \varphi \ \Psi_{\alpha h}(t) + \mu_2 \Psi_{\alpha\alpha}(t)] \qquad (13)$$

$$\tilde{\xi}(t) = \frac{\mu_1 \beta B \omega_c \sin \varphi \ \Psi_{\alpha h}(t) - \mu_3 \Psi_{\alpha\alpha}(t)}{I\omega_c^2} \qquad (14)$$

$$\varepsilon = \frac{1}{I\omega_c^2} [\mu_1 \beta B \omega_c \sin \varphi \ \Psi_{\alpha h0} - \mu_3 \Psi_{\alpha\alpha0}] \qquad (15)$$

Notwithstanding the presence of ε, Eq. (11) can be recast in standard form by setting

$$\overline{\omega}_c = \sqrt{1+\varepsilon} \ \omega_c, \quad \xi(t) = \frac{\tilde{\xi}(t)}{1+\varepsilon}, \quad \zeta = \frac{\tilde{\zeta}}{\sqrt{1+\varepsilon}} \qquad (16)$$

to obtain (Sepe and Vasta 2005b)

$$\ddot{A} + [2\zeta\overline{\omega}_c + \eta(t)]\dot{A} + \overline{\omega}_c^2 [1 + \xi(t)]A = 0 \qquad (17)$$

The constant ε in Eq. (15) depends only on the variation of the mean speed ψ_0 and can be iteratively evaluated, as in the sample case of Sect. 6, to estimate the correction on the frequency ω_c due to ψ_0; as a simplifying approach, in Sepe and Vasta (2005a, b) and Sepe and D'Asdia (2003) the correction on the frequency was neglected, resulting $\varepsilon << 1$ ($\varepsilon = 0.05$ for the example in Sect. 6), referring to Eq. (11) with $\zeta = \tilde{\zeta}$, $\xi(t) = \tilde{\xi}(t)$ and $\varepsilon = 0$ (i.e. $\overline{\omega}_c = \omega_c$).

Equation (17) describes a problem of parametric excitation, with both coefficients of $\dot{A}$ and A depending on time.

In (Sepe and Vasta 2005b) the stability condition was derived through stochastic averaging technique and moment stability method (Lin and Cai 1995). Denoting with $S_{\xi\xi}(\omega)$ and $S_{\eta\eta}(\omega)$ the power spectral densities of the parametric excitations ξ and η, respectively, and by $S_{\xi\eta}(\omega)$ their cross power spectral density, the stability of the second order amplitude moment requires that

$$\zeta > \frac{\pi\overline{\omega}_c}{2} S_{\xi\xi}(2\overline{\omega}_c) + \frac{\pi}{8\overline{\omega}_c} [2S_{\eta\eta}(2\overline{\omega}_c) + 4S_{\eta\eta}(0) + 7\overline{\omega}_c \text{Im}(S_{\xi\eta}(2\overline{\omega}_c))] \qquad (18)$$

Equation (18) will be used in the numerical example to derive the stability conditions of a designed bridge undergoing critical conditions.

3 Modeling of the Wind Field

The wind velocity field can be described as the sum of a mean value, function of the position, and of zero mean stationary fluctuations, function of the position and of the time (Simiu and Scanlan 1996).

For the sake of simplicity, the wind actions are considered applied only to the deck and orthogonal to the bridge axis, neglecting the across-wind components of turbulence

$$U(z,t) = \overline{U} + u(z,t) \tag{19}$$

with the mean value $\overline{U}(h)$ depending on the height h on the ground, and therefore approximately constant in the case here considered (the bridge deck is almost horizontal).

According to Simiu and Scanlan (1996), the turbulence along wind in a given point is described by means of the Kaimal spectrum

$$S_u(\omega) = u_*^2 \frac{200f}{n(1+50f)^{5/3}}; \quad f = \frac{\omega h}{2\pi \overline{U}} \tag{20}$$

where n is the frequency, ω is the circular frequency, u_* the friction velocity (Simiu and Scanlan 1996) and h is the height (constant) of the bridge deck.

The cross power spectral density of the wind turbulence in different points, assumed as real according to Carassale and Solari (2002), is described by means of a "coherence function" that introduces an exponential decay factor

$$S_u(z,z',\omega) = \overline{S}_u(\omega) \, Coh_u(z,z',\omega) \tag{21}$$

$$Coh_u(z,z',\omega) = e^{-\frac{C_{uz}|z-z'|}{2\pi \overline{U}}|\omega|} \tag{22}$$

Denoting by $\theta_k(z,\omega)$ and $\gamma_k(\omega)$ the eigenfunctions and eigenvalues of the cpsdf $S_u(z,z',\omega)$, respectively, the spectral proper transformation SPT allows the following representation of the wind velocity fluctuations $u(z,t)$ (Carassale 2005; Carassale et al. 2007; Fiore and Monaco 2009; Di Paola and Gullo 2001)

$$S_u(z,z',\omega) = \sum_{k=1}^{N_s} \theta_k(z,\omega)\,\theta_k^*(z',\omega)\,\gamma_k(\omega) \tag{23}$$

where N_s denotes the number of significant terms.

According to Carassale and Solari (2002), the eigenfunctions and the eigenvalues of the cpsdf can be represented in the following approximate closed form

$$\theta_k (z, \omega) = \overline{A}_k (\alpha) \sin \left[\frac{k\pi - 2\varepsilon_k (\alpha)}{L} z + \varepsilon_k (\alpha) \right] \tag{24}$$

$$\gamma_k (\omega) = \overline{S}_u (\omega) \frac{2}{\alpha} \cos^2 \left[\varepsilon_k (\alpha) \right] \tag{25}$$

with L the overall bridge length and

$$\alpha = \frac{C_{uz} L}{2\pi \overline{U}} |\omega| ; \quad \varepsilon_k (\alpha) = \arctan \left[\frac{\mu_k (\alpha)}{\alpha} \right] ; \quad 2 \cot (\mu_k) = \frac{\mu_k}{\alpha} - \frac{\alpha}{\mu_k} \tag{26}$$

$$\overline{A}_k (\alpha) = \sqrt{\frac{2 (k\pi - 2\varepsilon_k)}{\sin (2\varepsilon_k) + k\pi - 2\varepsilon_k}} \tag{27}$$

4   Simplified Flutter Mode Shape

For a large span bridge the flutter mode can be approximated (Sepe and D'Asdia 2003) by sinusoidal functions with n half-wavelength, here normalised so to have a unitary generalised torsional inertia in the main span (with length L)

$$f_\alpha(z) = f_h(z) = \sqrt{\frac{2}{\overline{I} L}} \sin \frac{n\pi z}{L} \tag{28}$$

where $\overline{I}$ is the torsional inertia per unit length.

With these assumptions it results

$$\Psi_{\alpha h}(t) = \Psi_{\alpha\alpha}(t) = \int_L \psi (z, t) f_\alpha{}^2(z) dz; \quad \Psi_{\alpha h0} = \Psi_{\alpha\alpha 0} \tag{29}$$

In the equation of motion (cf. Eq. 17)

$$\ddot{A} + [2\zeta\omega_c + \eta(t)] \dot{A} + \omega_c^2 [1 + \xi(t)] A = 0 \tag{30}$$

a constant "damping" coefficient can be observed, defined as (see Eqs. 12, 13, 14, 15 and 16, with $I = 1$)

$$\zeta = -\frac{1}{2\overline{\omega}_c} (\mu_1 \beta B \cos \varphi + \mu_2) \Psi_{\alpha\alpha 0} \tag{31}$$

and therefore positive if the mean wind speed is reduced ($\psi_0 < 0$) with respect to U_c (cf. Eq. 3), and two parametric forcing terms

$$\eta(t) = -k_\eta \Psi_{\alpha\alpha}(t), \quad \xi(t) = \frac{k_\xi}{\omega_c^2} \Psi_{\alpha\alpha}(t) \tag{32}$$

with $k_\eta = \mu_1 \beta B \cos\varphi + \mu_2$, $k_\xi = \mu_1 \beta B \omega_c \sin\varphi - \mu_3$ and ε, defined in Sect. 2, iteratively evaluated as in the sample case in Sect. 6 (or neglected, as in Sepe and Vasta (2005b), being $\varepsilon << 1$). The parametric forcing terms depend therefore both on the forcing function $\Psi_{\alpha\alpha}(t)$, with

$$\Psi_{\alpha\alpha}(t) = \int_L \psi(z,t) f_\alpha^{\,2}(z) dz, \quad \Psi_{\alpha\alpha}(\omega) = \int_L \psi(z,\omega) f_\alpha^{\,2}(z) dz \tag{33}$$

Around the critical wind speed $\overline{U} = U_c$, the power spectral density of the forcing function can be therefore rewritten as

$$S_{\Psi_{\alpha\alpha}\Psi_{\alpha\alpha}}(\omega) = \Psi_{\alpha\alpha}(\omega) \Psi_{\alpha\alpha}^*(\omega) = \iint_{L\,L} \psi(z,\omega) \psi^*(z',\omega) f_\alpha^{\,2}(z) f_\alpha^{\,2}(z') \, dz dz'$$

$$= \iint_{L\,L} S_{\psi\psi}(z,z',\omega) f_\alpha^{\,2}(z) f_\alpha^{\,2}(z') \, dz dz' \tag{34}$$

where $S_{\psi\psi}(z,z',\omega)$ represents the cpsdf of the normalized turbulence $u(z,z',t)/U_C$ (see Eq. 19), whose SPT representation is as follows (see Eq. 23)

$$S_{\psi\psi}(z,z',\omega) = \frac{1}{U_C^2} \sum_{k=1}^{Ns} \theta_k(z,\omega) \theta_k^*(z',\omega) \gamma_k(\omega) \tag{35}$$

The psd of the forcing term acting on both the terms of the governing equation can be therefore written as follows

$$S_{\Psi_{\alpha\alpha}\Psi_{\alpha\alpha}}(\omega) = \frac{1}{U_C^2} \sum_{k-1}^{Ns} \gamma_k(\omega) \int_L \theta_k(z,\omega) f_\alpha^2(z) dz \int_L \theta_k^*(z',\omega) f_\alpha^2(z') dz'$$

$$= \sum_{k=1}^{Ns} \frac{\gamma_k(\omega)}{U_C^2} \Theta_k(\omega) \Theta_k^*(\omega) \tag{36}$$

where the contribution of the wind-mode θ_k on the flutter mode f_α is represented by mixed wind-structure coefficients $\Theta_k(\omega)$ defined as

$$\Theta_k(\omega) = \int_L \theta_k(z,\omega) f_\alpha^2(z) dz \tag{37}$$

The frequency content of $\Psi_{\alpha\alpha}(t)$, and so the frequency content of parametric excitations $\xi(t)$ and $\eta(t)$, depend in fact both on the frequency content of the turbulence $\psi(z,t)$ and on the shape $f_\alpha(z)$ of the flutter mode.

The psd of the forcing term $S_{\Psi_{\alpha\alpha}\Psi_{\alpha\alpha}}(\omega)$ can conveniently be expressed as

$$S_{\Psi_{\alpha\alpha}\Psi_{\alpha\alpha}}(\omega) = \frac{c(\omega)}{\overline{I}^2} S_{\psi\psi}(\omega) \tag{38}$$

by introducing a multiplying factor $c(\omega)$ denoted in the following as "shape coefficient", that for each frequency component ω of the turbulence describes the "similarity" between along-span variation of that turbulence component and the flutter mode.

With these notation, the psd of the parametric forcing terms can be rewritten as

$$S_{\eta\eta}(\omega) = \frac{k_\eta^2}{\overline{I}^2} c(\omega) S_{\psi\psi}(\omega), \quad S_{\xi\xi}(\omega) = \frac{k_\xi^2}{\overline{I}^2 \omega_c^4} c(\omega) S_{\psi\psi}(\omega),$$

$$S_{\xi\eta}(\omega) = -\frac{k_\xi k_\eta}{\overline{I}^2 \omega_c^2} c(\omega) S_{\psi\psi}(\omega) \tag{39}$$

The parametric excitations $\xi(t)$ and $\eta(t)$ are proportional to each other and then $\mathrm{Im}(S_{\xi\eta}) = 0$; the stability equation becomes therefore

$$-\psi_0 > \frac{\pi}{\overline{I}} \left(\frac{k_\xi^2}{k_\eta \omega_c^2} + \frac{k_\eta}{2} \right) c(2\overline{\omega}_c) S_{\psi\psi}(2\overline{\omega}_c) + \frac{\pi}{\overline{I}} k_\eta S_{\psi\psi}(0) \tag{40}$$

$S_{\psi\psi}(2\overline{\omega}_c)$ represents the psd of the turbulence evaluated for a frequency $\omega = 2\overline{\omega}_c$, twice than the critical one (with the correction factor $\sqrt{1+\varepsilon}$ due to the variation of the mean wind speed ψ_0, see Sect. 2).

For $S_{\psi\psi}(0)$ in the Kaimal model, the following expression is assumed (see Simiu and Scanlan (1996), with the normalisation factor U_c^2 of Eq. (3))

$$S_{\psi\psi}(0) = \frac{1}{U_c^2} \frac{4\overline{u}^2 L_u^x}{\overline{U}} \tag{41}$$

where $\overline{U} = U_c(1 + \psi_0)$ is the average speed, $\overline{u}^2 = 6u_*^2$ is the mean square value of the wind-speed fluctuations, and L_u^x the integral length of longitudinal turbulence.

When turbulence has a non negligible low-frequency content $S_{\psi\psi}(2\overline{\omega}_c) > 0$, therefore, aeroelastic stability can be assured only if the mean wind speed is lowered with respect to critical value, i.e. $\psi_0 < 0$ (that introduces a positive damping), meaning that the aeroelastic stability requires a reduction of the average wind speed with respect to the critical value U_c defined in Sect. 2, as shown in previous papers for a different turbulence (Sepe and Vasta 2005a, b).

Fig. 2 Multi-box deck of the
Messina bridge

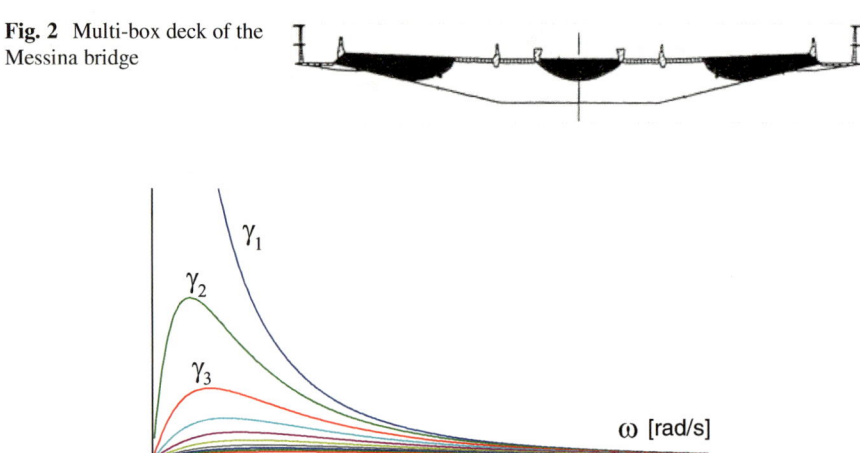

Fig. 3 Wind modes: eigenvalues

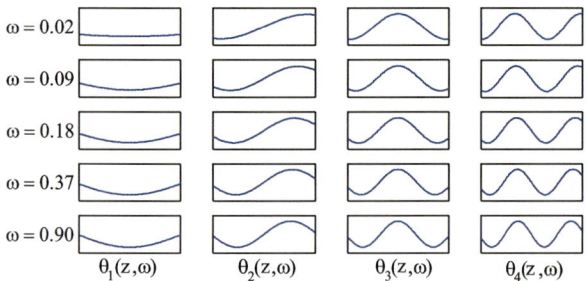

Fig. 4 Wind modes: eigenvectors # 1-2-3-4 (ω [rad/s])

5 Example

As an example, it is considered a case (Fig. 2) already studied in previous papers
(Sepe and D'Asdia 2003; D'Asdia and Sepe 1998; Sepe et al. 2000). The example
considered is the bridge designed for crossing the Messina Strait (Italy), with a
main span of L = 3,300 m and a multi-box deck optimized through wind tunnel
tests (Fig. 2).

It results also (Sects. 2 and 3)

$L = 3,300$ m, $h = 100$ m, $B = 60.4$ m, $\overline{I} = 2.8 \cdot 10^7$ kgm^2/m, $u_* = 3.3$ m/s,
$L_u = 200$ m, $C_{uz} = 10$,

$U_c = 94$ m/s, $\omega_c = 0.418$ rad/s, $n = 2$, $\beta = 2.33$, $\varphi = 59.9°$, $\mu_1 = 3 \cdot 10^5$ kg s^{-1},
$\mu_2 = -1.5 \cdot 10^7$ kg m s^{-1}, $\mu_3 = 2.1 \cdot 10^7$ kg m s^{-2}

Eigenvalues and eigenvectors of the wind turbulence, described according to
Sect. 3, are reported in Figs. 3 and 4, respectively. The "shape coefficient" $c(\omega)$ in
Fig. 5, obtained with the contribution of ten wind modes, is indeed "dominated" for

Fig. 5 Shape coefficient $c(\omega)$: 10 wind modes; the relevant contribution at frequency $2\omega_c \cong 0.84$ rad/s is $c(\omega) = 0.07$

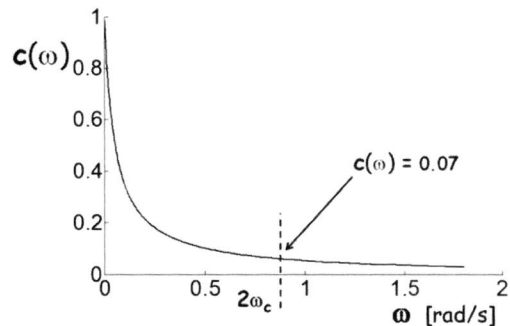

the sample case here considered by the first wind mode and only marginally affected by the 3rd and 5th ones (with even modes irrelevant for the assumed flutter mode shape). As shown in Fig. 5, the value of $c(\omega)$ at the relevant frequency $\omega = 2\omega_c$ (see Sect. 4) is about 0.07; according to the stability equation (Eq. 40), aeroelastic stability can be assured only if the mean wind speed is lowered with respect to critical value, (that introduces a positive damping); in the sample case discussed here, a reduction $\psi_0 = -0.042$ of mean wind speed is required with respect to "ideal" value U_C.

After few iterations, the coefficient ε due to the variation of the mean wind speed ψ_0 and affecting the actual frequency $\overline{\omega}_c = \omega_c \sqrt{1 + \varepsilon}$ (Eqs. 15 and 16 of Sect. 2) has been evaluated as $\varepsilon = 0.05$.

6   Concluding Remarks

The paper describes the effects on the aeroelastic stability of long span bridges due to low-frequency wind speed fluctuations. The spatial-temporal field of wind turbulence is described by means of orthogonal loading components representing the eigenfunctions of the cross power spectral density function (cpsdf) of the process, according to a POD-based representation. It is shown that for turbulent wind the aeroelastic stability can be assured only if the infinite-time average of the wind speed is appropriately lowered. In a sample case (Messina bridge), about 4 % reduction of mean wind speed is required with respect to the "ideal" value U_C, but reduction is expected to be more relevant for bridges with worse aeroelastic behavior

References

Carassale L (2005) POD-based filters for the representation of random loads on structures. Prob Eng Mech 20:263–280

Carassale L, Solari G (2002) Wind modes for structural dynamics: a continuous approach. Prob Eng Mech 17:157–166

Carassale L, Solari G, Tubino F (2007) Proper orthogonal decomposition in wind engineering. Part 2: Theoretical aspects and some applications. Wind Struct 10(2):177–208

D'Asdia P, Sepe V (1998) Aeroelastic instability of long span suspended bridges: a multi-mode approach. J Wind Eng Ind Aerodyn 74–76:849–857

Di Paola M, Gullo I (2001) Digital generation of multivariate wind field processes. Prob Eng Mech 16:1–10

Fiore A, Monaco PI (2009) POD-based representation of the alongwind equivalent static force for long-span bridges. Wind Struct 12(3):239–257

Lin YK (1996) Stochastic stability of wind-excited long-span bridges. Prob Eng Mech 11:257–261

Lin YK, Cai GQ (1995) Probabilistic structural dynamics. McGraw-Hill, New York

Sepe V, D'Asdia P (2003) Influence of low-frequency wind speed fluctuations on the aeroelastic stability of suspension bridges. J Wind Eng Ind Aerodyn 91: 1285–1297. ISSN: 0167–6105

Sepe V, Vasta M (2005a) Turbulence effects on the aeroelastic stability of long span bridges. In: Soize C, Schueller GI (eds) Sixth European conference on structural dynamics EURODYN 2005, Paris, September 2005, vol 1, Millpress, Rotterdam, pp 421–426. ISBN: 978-90-5966-033-5

Sepe V, Vasta M (2005b) Aeroelastic stability of long span bridges under turbulent wind. In: Proceedings of the 16th Italian conference on theoretical and applied mechanics, AIMETA'05

Sepe V, Caracoglia L, D'Asdia P (2000) Aeroelastic instability of long-span bridges: contributions to the analysis in frequency and time domains. Wind Struct 3(1):41–58

Simiu E, Scanlan RH (1996) Wind effects on structures, 3rd edn. Wiley, New York

Part V
Stochastic Dynamics

PDEM-Based Response Analysis of Nonlinear Systems with Double Uncertainties

Jian-Bing Chen, Pei-Hui Lin, and Jie Li

Abstract Large degree of uncertainties may exist simultaneously in system parameters and external excitations of engineering structures. To capture the performance of such nonlinear multi-degree-of-freedom structures is still a great challenge in stochastic dynamics. In the present paper, the probability density evolution method is adopted and extended to reduce the dimension of parametric FPK equation of an uncertain-parameter structure subjected to additively white noise process. Numerical examples validate the proposed algorithm. Problems to be further studied are discussed.

Keywords Nonlinear systems • Dimension reduction • FPK equation • Probability density evolution method (PDEM)

1 Introduction

Engineering structures in service will almost unavoidably be attacked by random disastrous dynamic actions such as earthquakes, strong wind etc. Simultaneously, the system parameters (e.g., the strength parameters and stiffness parameters) usually could not be captured exactly. Therefore, to grasp the performance of an engineering structure we have to deal with the randomness involved in both external loadings and system parameters (Li 1996; Li and Chen 2009). To stress this we call it a structure with double uncertainties. Moreover, generally the system will exhibit nonlinear behaviors under such extreme disastrous actions (Roberts and Spanos 1990). The analysis of such systems involving coupled double randomness and nonlinearity is still a challenging problem (Wen 2004; Goller et al. 2013).

J.-B. Chen (✉) • P.-H. Lin • J. Li
School of Civil Engineering and State Key Laboratory of Disaster Reduction in Civil
Engineering, Tongji University, Shanghai 200092, China
e-mail: chenjb@tongji.edu.cn

M. Papadrakakis and G. Stefanou (eds.), *Multiscale Modeling and Uncertainty*
Quantification of Materials and Structures, DOI 10.1007/978-3-319-06331-7_16,
© Springer International Publishing Switzerland 2014

Traditionally, the randomness is dealt with separately in stochastic dynamics, resulting in the random vibration theory and the stochastic structural analysis theory (or stochastic finite element method as in some literature) (Li 1996; Lutes and Sarkani 2004). In random vibration, the methods for analysis of linear systems were well developed. Actually, there are elegant transfer relationships between the moments and/or power spectral density (PSD) function of input to the counterparts of output (Li and Chen 2009). However, in nonlinear problems the so-called closure problem exists (Lutes and Sarkani 2004). In the level of probability density, the FPK equation was derived nearly one century ago but the solution is available still only for some special simple systems, mainly approximate, stationary, and for only single-degree-of-freedom systems (Caughey and Ma 1982). The solution for multi-degree-of-freedom (MDOF) nonlinear systems is still a great challenge, although some important progress has been made in the past two decades (Soize 1994; Naess and Moe 2000; von Wagner and Wedig 2000; Zhu 2006). The situation for stochastic structural analysis where the uncertainty in system parameters is dealt with is similar: the methods for linear systems were well developed but there are almost no effective approaches for the problems involving nonlinearity in MDOF systems encountered in engineering practice (Schuëller 1997). Actually the difficulty arising in the above two branches is essentially of the same origin, i.e., the coupling of nonlinearity and randomness in MDOF systems could not be broken in the traditional framework of stochastic dynamics when a lot of state variables are involved simultaneously. When the randomness is involved in both external loadings and system parameters, the investigations were extremely seldom (Li 1996). It seems that only the Monte Carlo simulation is available in these cases if the prohibitively large computational efforts are not cared about (Shinozuka 1972; Au and Beck 2001).

Under such background, the development of the probability density evolution method (PDEM) in the past decade paved a new path for nonlinear stochastic dynamics (Li and Chen 2009; Li et al. 2012a). This method deals with the randomness on a unified basis by invoking the random event description of the principle of preservation of probability and the embedded physical mechanism. By doing so, a completely decoupled partial differential equation, i.e., the generalized density evolution equation was derived and the numerical methods were extensively studied. Most recently, it was extended to reduce the dimension of FPK equations (Chen and Yuan 2014; Chen and Lin 2014). In the present paper, a further step is made by constructing the equivalent flux of probability so that when the randomness is involved in both system parameters and external loadings a one-dimensional partial differential equation still exists.

2 Probability Density Evolution Method (PDEM)

As mentioned, in engineering practice the uncertainties will be exhibited both in system parameters and in external excitations. To be specific, in the present paper we only consider the structure subjected to earthquake excitations. In this case, the equation of motion of a generic MDOF system is

$$\mathbf{M\ddot{X}} + \mathbf{C\dot{X}} + \mathbf{f}\,(\mathbf{\Theta}_k, \mathbf{X}) = -\mathbf{MI}\ddot{X}_g(t) \tag{1}$$

where $\mathbf{M}$ and $\mathbf{C}$ are the n by n mass and damping matrices, respectively, $\mathbf{f}$ is the n by 1 internal force vector, $\mathbf{\Theta}_k = (\Theta_{1,k}, \cdots, \Theta_{s_k,k})$ is the basic random vector characterizing the randomness involved in the restoring forces, $\mathbf{I}$ is a column vector with all the components being 1, $\ddot{X}_g(t)$ is the stochastic ground motion acceleration.

The response information of the system (1) could be captured by PDEM (Li and Chen 2009) where a generalized density evolution equation (GDEE) is derived and solved. To this end, firstly the stochastic ground acceleration $\ddot{X}_g(t)$ should be represented by an explicit function of the basic random variables rather than a function of the abstract sample points. This could be implemented by the physical stochastic model for ground motions (Wang and Li 2011; Li et al. 2012b), or could also be represented by the summation of stochastic harmonic functions (Chen et al. 2013) if the PSD, say the Kanai-Tajimi spectrum, is given. Finally, we have

$$\ddot{X}_g(t) = F\left(\mathbf{\Theta}_g, t\right) \tag{2}$$

where $\mathbf{\Theta}_g = \left(\Theta_{1,g}, \cdots, \Theta_{s_g,g}\right)$ is the basic random vector characterizing the randomness involved in the stochastic ground acceleration, $F(\cdot, t)$ is an explicit known function of $\mathbf{\Theta}_g$ and t, determined by either the physical stochastic model or by the stochastic harmonic functions.

By doing so, Eq. (1) becomes

$$\mathbf{M\ddot{X}} + \mathbf{C\dot{X}} + \mathbf{f}\,(\mathbf{\Theta}_k, \mathbf{X}) = -\mathbf{MI}F\left(\mathbf{\Theta}_g, t\right) \tag{3}$$

For notational convenience, we denote $\mathbf{\Theta} = (\mathbf{\Theta}_k, \mathbf{\Theta}_g) = (\Theta_1, \cdots, \Theta_s)$ as the basic random vector with known joint probability density function (PDF) $p_{\mathbf{\Theta}}(\boldsymbol{\theta})$, where $\boldsymbol{\theta} = (\theta_1, \cdots, \theta_s)$, and $s = s_k + s_g$ is the total number of the involved random variables.

If some physical quantities related to the system (2), denoted by $\mathbf{Z} = (Z_1, \cdots, Z_m)$, are of concern, then these physical quantities could be captured by solving Eq. (3) and then adopting the connection between them and the displacement and velocity vector, e.g. the geometric matrix and constitutive law via the finite element assembling. Thus, finally the solution of $\mathbf{Z}$ could be expressed as the function of the basic random vector $\mathbf{\Theta}$ and t, i.e.

$$\mathbf{Z} = \mathbf{H}\,(\mathbf{\Theta}, t) \tag{4}$$

where $\mathbf{H} = (H_1, \cdots, H_m)$. Note that all the randomness involved in the system (4) comes from $\mathbf{\Theta}$. The augmented system $(\mathbf{Z}(t), \mathbf{\Theta})$ is consequently probability preserved and thus the joint PDF of $(\mathbf{Z}(t), \mathbf{\Theta})$, denoted by $p_{\mathbf{Z\Theta}}(\mathbf{z}, \boldsymbol{\theta}, t)$, should satisfy

$$\frac{d}{dt} \int_{\Omega_t \times \Omega_\theta} p_{\mathbf{Z\Theta}}\,(\mathbf{z}, \boldsymbol{\theta}, t)\, d\mathbf{z} d\boldsymbol{\theta} = 0 \tag{5}$$

for any arbitrary $\Omega \times \Omega_\theta \in \Omega_Z \times \Omega_\Theta$, where Ω_Z is the value domain of $\mathbf{Z}(t)$ and Ω_Θ is the distribution domain of Θ (e.g., the support of $p_\Theta(\theta)$). By a series of manipulations on Eq. (5) we are led to (Li et al. 2012a)

$$\int_{\Omega_\theta} \left(\frac{\partial p_{Z\Theta}(\mathbf{z}, \theta, t)}{\partial t} + \sum_{j=1}^{m} \dot{Z}_j(\theta, t) \frac{\partial p_{Z\Theta}(\mathbf{z}, \theta, t)}{\partial z_j} \right) d\theta = 0 \qquad (6)$$

which could finally be reduced to the following generalized density evolution equation (GDEE) (Li and Chen 2008, 2009)

$$\frac{\partial p_{Z\Theta}(\mathbf{z}, \theta, t)}{\partial t} + \sum_{j=1}^{m} \dot{Z}_j(\theta, t) \frac{\partial p_{Z\Theta}(\mathbf{z}, \theta, t)}{\partial z_j} = 0 \qquad (7)$$

because of the arbitrariness of Ω_θ.

If we are only interested in one single physical quantity, i.e., $m = 1$, Eq. (7) becomes a one-dimensional partial differential equation

$$\frac{\partial p_{Z\Theta}(z, \theta, t)}{\partial t} + \dot{Z}(\theta, t) \frac{\partial p_{Z\Theta}(z, \theta, t)}{\partial z} = 0 \qquad (8)$$

Equations (7) and (8) are quite different from the traditional equations such as the FPK equation because the dimension of GDEE is completely untied from the original dynamical system (1). This is because the source of randomness was involved together with the physical quantities of concern. In other words, the embedded physical mechanism is incorporated in the partial differential equation.

After solving Eq. (8) the PDF of $Z(t)$ could be given by

$$p_Z(z, t) = \int_{\Omega_\Theta} p_{Z\Theta}(z, \theta, t) d\theta \qquad (9)$$

The initial condition of Eq. (8) is usually specified by $p_{Z\Theta}(z, \theta, t)|_{t=0} = \delta(z - z_0)$ $p_\Theta(\theta)$ if z_0 is the deterministic initial value of $Z(t)$. Here $\delta(\cdot)$ is Dirac's delta function. For instance, if $Z(t)$ is the displacement, then z_0 could take zero in most cases, whereas if $Z(t)$ is the stress at a crucial point, then z_0 is the initial stress mainly due to the weight of structure before the earthquake action is exerted.

3   FPK-Like Equation and Its Dimension Reduction

If we introduce $\mathbf{Y} = \left(\dot{\mathbf{X}}^T, \mathbf{X}^T \right)^T = (Y_1, \cdots, Y_{2n})^T$, then the system (1) could be transformed into a state equation

$$\dot{\mathbf{Y}} = \mathbf{A}(\Theta_k, \mathbf{Y}, t) + \mathbf{B}\ddot{X}_g(t) \qquad (10)$$

where $\mathbf{A} = (A_1, \cdots, A_{2n})^T$ and $\mathbf{B} = (B_1, \cdots, B_{2n})^T$ could be obtained according to the transform. If the earthquake ground motion $\ddot{X}_g(t)$ is idealized as a white noise process with the mean $E\left[\ddot{X}_g(t)\right] = 0$ and correlation function $E\left[\ddot{X}_g(t)\ddot{X}_g(t+\tau)\right] = D\delta(\tau)$, then Eq. (10) could be understood as an Itô stochastic differential equation for given $\boldsymbol{\Theta}_k$. In this case, the joint PDF of $(\mathbf{Y}, \boldsymbol{\Theta}_k)$, denoted by $p_{\mathbf{Y}\boldsymbol{\Theta}_k}(\mathbf{y}, \boldsymbol{\theta}_k, t)$, follows the FPK equation

$$
\frac{\partial p_{\mathbf{Y}\boldsymbol{\Theta}_k}(\mathbf{y}, \boldsymbol{\theta}_k, t)}{\partial t} = -\sum_{j=1}^{2n} \frac{\partial}{\partial y_j}\left[p_{\mathbf{Y}\boldsymbol{\Theta}_k}(\mathbf{y}, \boldsymbol{\theta}_k, t) A_j(\boldsymbol{\theta}_k, \mathbf{y}, t) \right]
$$

$$
+ \frac{1}{2}\sum_{i=1}^{2n}\sum_{j=1}^{2n} \sigma_{ij} \frac{\partial p_{\mathbf{Y}\boldsymbol{\Theta}_k}(\mathbf{y}, \boldsymbol{\theta}_k, t)}{\partial y_i \partial y_j} \tag{11}
$$

where σ_{ij} is the component of $\boldsymbol{\sigma} = D\mathbf{B}\mathbf{B}^T$.

Equation (11) could of course be understood as a parametric FPK equation for a given $\boldsymbol{\theta}_k$. If we marginalize Eq. (11) in terms of $\mathbf{y}, \boldsymbol{\theta}_k$ excludes y_ℓ, i.e., let

$$
p_{Y_\ell}(y_\ell, t) = \int_{-\infty}^{\infty} \cdots \int_{-\infty}^{\infty} \cdots \int_{\Omega_{\boldsymbol{\Theta}_k}} p_{\mathbf{Y}\boldsymbol{\Theta}_k}(\mathbf{y}, \boldsymbol{\theta}_k, t)\, dy_1 \cdots dy_{\ell-1} dy_{\ell+1} \cdots dy_{2n} d\boldsymbol{\theta}_k
$$

$$
\tag{12}
$$

then integrating on both sides of Eq. (11) in terms of $y_1, \cdots, y_{\ell-1}, y_{\ell+1}, \cdots, y_{2n}, \boldsymbol{\theta}_k$ yields

$$
\frac{\partial p_{Y_\ell}(y_\ell, t)}{\partial t} = -\frac{\partial J(y_\ell, t)}{\partial y_\ell} + \frac{1}{2}\sigma_{\ell\ell}\frac{\partial p_{Y_\ell}(y_\ell, t)}{\partial y_\ell^2} \tag{13}
$$

where the flux of probability related to drift effect

$$
J(y_\ell, t) = \int_{-\infty}^{\infty} \cdots \int_{-\infty}^{\infty} \int_{\Omega_{\boldsymbol{\Theta}_k}} p_{\mathbf{Y}\boldsymbol{\Theta}_k}(\mathbf{y}, \boldsymbol{\theta}_k, t) A_\ell(\boldsymbol{\theta}_k, \mathbf{y}, t)
$$

$$
\times dy_1 \cdots dy_{\ell-1} dy_{\ell+1} \cdots dy_{2n} d\boldsymbol{\theta}_k \tag{14}
$$

Equation (13) is in the form a one-dimensional partial differential equation. However, the flux of probability related to drift in Eqs. (13) and (14) involves a high-dimensional integral related to the original high-dimensional joint PDF $p_{\mathbf{Y}\boldsymbol{\Theta}_k}(\mathbf{y}, \boldsymbol{\theta}_k, t)$, which is unknown and is what we want to circumvent due to its high dimension. This forms a loop that could not be broken in the traditional theoretical frame. Some researchers have made efforts to obtain the needed information by involving a linearized solution of the marginalized part state vector for white noise excited systems without parametric uncertainty (Er 2011).

For the FPK equation without θ_k, an equivalent flux of probability was proposed recently (Chen and Yuan 2014; Chen and Lin 2014; Yuan et al. 2012) by invoking the solution of GDEE to construct the flux due to drift. This is established on the basis of the equivalence between the state space description and the random event description of the principle of preservation of probability. For the parametric FPK equation, similar ideas could be adopted. In this case, Eq. (14) is replaced by

$$J\left(y_\ell, t\right) = \int_{\Omega_{\Theta_e}} \int_{\Omega_{\Theta_k}} p_{Y_\ell \Theta_k \Theta_e}\left(y_\ell, \theta_k, \theta_e, t\right) \widehat{A}_\ell\left(\theta_k, \theta_e, t\right) d\theta_k d\theta_e$$

$$= \int_{\Omega_\Theta} p_{Y_\ell \Theta}\left(y_\ell, \theta, t\right) \widehat{A}_\ell\left(\theta, t\right) d\theta \qquad (15)$$

where $p_{Y_\ell \Theta}\left(y_\ell, \theta, t\right)$ is the solution of GDEE (8) when Z is replaced by Y_ℓ, and $\widehat{A}_\ell\left(\theta, t\right)$ is given by $\widehat{A}_\ell\left(\theta, t\right) = \dot{Y}_\ell\left(\theta, t\right) - B_\ell F\left(\theta_g, t\right)$ as the component of Eq. (10), in which $F(\theta_g, t)$ is specified by Eq. (2) as $\Theta_g = \theta_g$.

Once the information of the flux of probability related to drift is known from Eq. (15), the one-dimensional partial differential equation (Eq. (13)) could be solved at least easily by numerical methods.

4   Numerical Examples

To verify and validate the proposed method for the evaluation of stochastic systems with double uncertainties, two numerical examples with both stochastic excitation and random system parameters are investigated.

Example 1: First-Order Nonlinear System   Consider a first-order stochastic differential equation

$$\dot{X} = \frac{1}{2}\left(\gamma X - X^3 - \alpha X^5\right) + b\xi(t) \qquad (16)$$

where γ is a random parameter uniformly distributed over $[\gamma_1, \gamma_2]$ and $\xi(t)$ is a Gaussian white noise with zero mean, unit variance and correlation function $E[\xi(t)\xi(t+\tau)] = \delta(\tau)$; α is a deterministic nonlinear factor and b is the diffusion coefficient. The exact stationary PDF of $X(t)$ is known if γ is deterministic (Er 2000) and thus the stationary PDF of the stochastic process $X(t)$ is given by

$$p_X(x) = \frac{1}{\gamma_2 - \gamma_1} \int_{\gamma_1}^{\gamma_2} C\left(\gamma\right) \exp\left[\frac{1}{2b^2}\left(\gamma x^2 - \frac{x^4}{2} - \alpha \frac{x^6}{3}\right)\right] d\gamma \qquad (17)$$

where $C\left(\gamma\right) = \left\{\int_{-\infty}^{\infty} \exp\left[\frac{1}{2b^2}\left(\gamma x^2 - \frac{x^4}{2} - \alpha \frac{x^6}{3}\right)\right] dx\right\}^{-1}$ is a normalization factor.

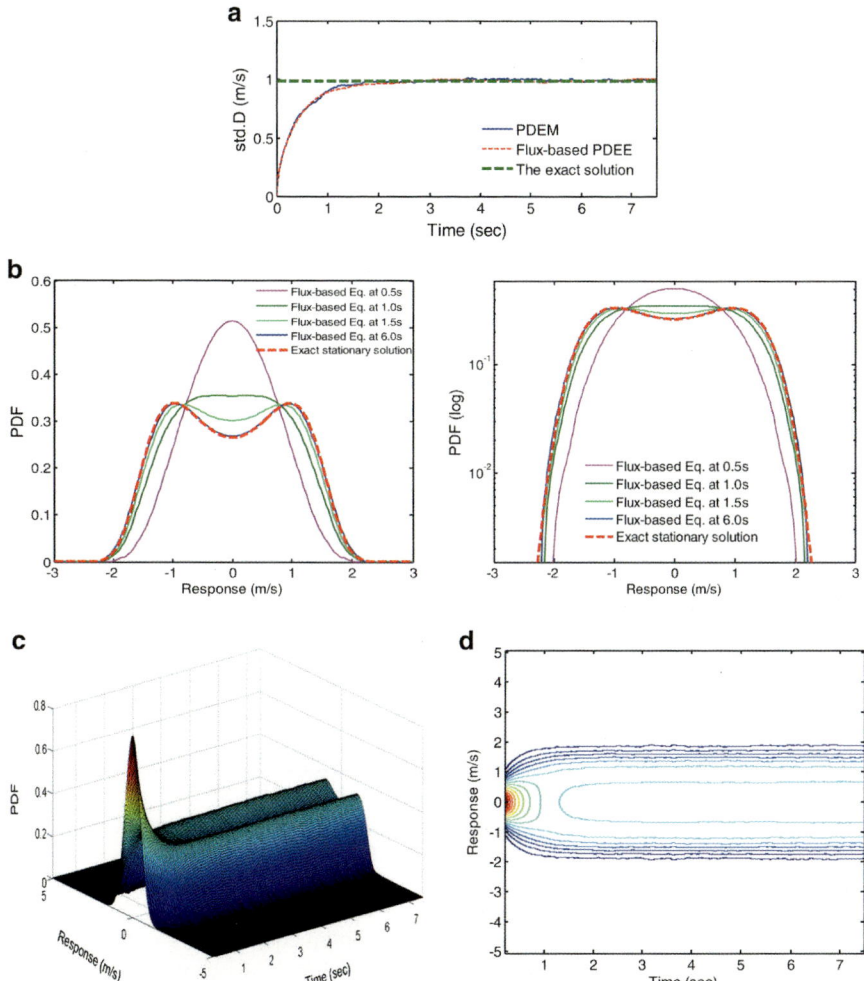

Fig. 1 Probability information of the first-order nonlinear system with double uncertainties. (**a**) Standard deviation of stochastic process. (**b**) Typical PDF of stochastic process (*left*: in ordinary coordinate system; *right*: in logarithmic coordinate system). (**c**) PDF evolution surface. (**d**) PDF contour

In the present example, we take $\gamma_1 = 0.9$, $\gamma_2 = 1.1$, $\alpha = 1/20$ and $b = 1$. Through the proposed approach, one can get the evolution of PDF and other statistical information. In the numerical solving process, the spectral representation method is employed and the Sobol' set is chosen as the representative point set in order to generate representative time histories of Gaussian white noise. The time step takes $\Delta t = 0.001$ sec. A total number of 1,000 representative time histories are employed. The results by the proposed approach and the exact stationary solution of standard deviation computed by Eq. 17 are plotted in Fig. 1. The results include: (a) standard deviation of $X(t)$; (b) typical PDFs of $X(t)$ in ordinary coordinate and logarithmic coordinate systems; (c) PDF surface evolving with time; and (d) the contour of PDF surface. Clearly, it is seen that the proposed approach is of fair accuracy.

Example 2: A 9-Story Shear Frame Consider a linear 9-story shear frame subjected to Gaussian white noise excitation. The lateral inter-story stiffness from top to bottom are defined as a uniform distributed random variable $k \sim U(2.664, 3.256)$ ($\times 10^7$ N/m) and the lumped masses of each story are deterministic with the same value of 9.78×10^4 kg. Rayleigh damping is used, i.e. $\mathbf{C} = a\mathbf{M} + b\mathbf{K}$, where $a = 0.2150$ and $b = 0.0088$. Similarly, the spectral representation method is employed and the Sobol' set is adopted as the representative point set to generate representative time histories. The standard deviation of the seismic excitation is taken as 0.1 g. The time step takes $\Delta t = 0.002$ s. A total number of 2,000 representative time histories are employed here. Through the proposed approach, the probability information of the top floor velocity is obtained. Again, the corresponding exact solution is computed for comparison. The standard deviation, typical PDFs, the PDF evolution surface and the corresponding contour are shown in Fig. 2. Again, it is seen that the results by the proposed approach accord quite well with the exact solutions.

5 Concluding Remarks

In the present paper, the structures with double randomness, i.e., uncertain structures subjected to stochastic excitations, are studied. The probability density evolution method is adopted to construct the equivalent flux of probability in the marginalized parametric FPK equation. The major results include:

1. For additively white-noise excited system, the parametric FPK equation could be reduced to a one-dimensional partial differential equation and then solved by combining the solution of the generalized probability density evolution equation and the reduced flux-equivalent equation;
2. Two numerical examples, including a one-dimensional system and a MDOF system, are studied. The results show the feasibility of the proposed method.

There are problems to be further studied: (1) Extension of the proposed method to multiplicatively excited systems; (2) More robust numerical algorithm for one- and higher-dimensional flux-equivalent probability density evolution equation; (3) Extension of the method from the macro-scale structural systems to multi-scale structural systems.

Acknowledgments Financial supports from the National Natural Science Foundation of China (NSFC Grant Nos. 11172210 and 51261120374), the Shuguang Program of Shanghai (Grant No.11SG21), the National Key Technology R&D Program (Grant No. 2011BAJ09B03-02) and the fundamental funding for Central Universities of China are gratefully appreciated.

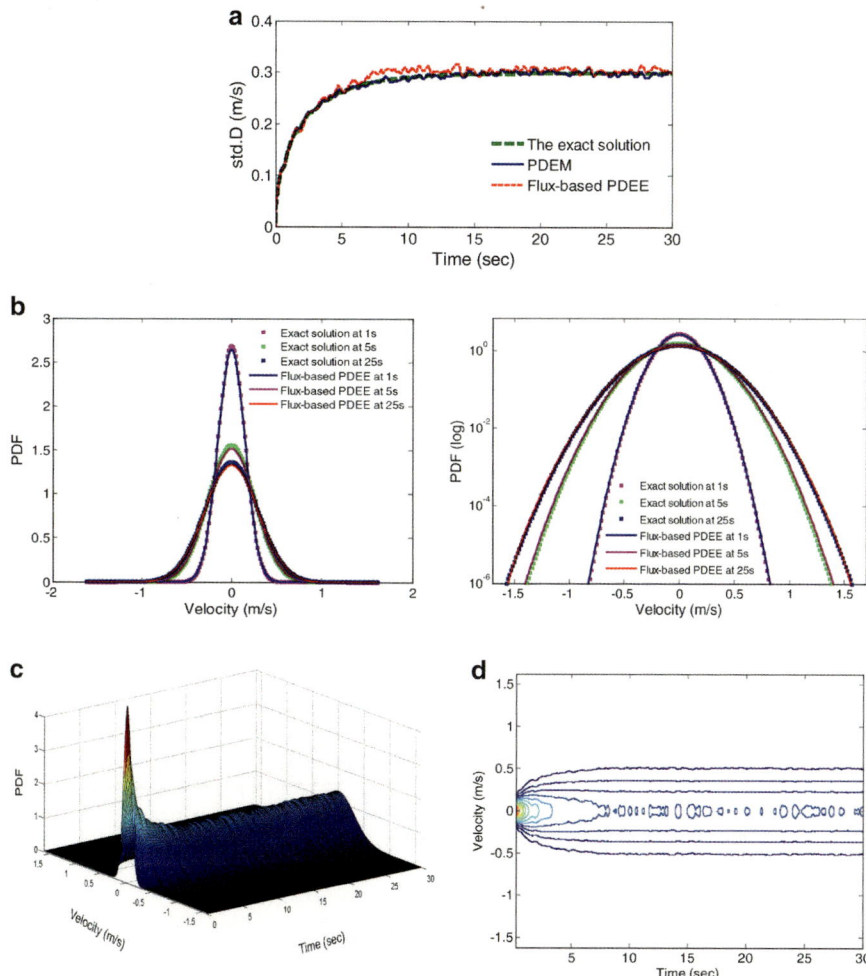

Fig. 2 Probability information of the 9-story shear frame with double uncertainties. (**a**) Standard deviation of the top floor velocity. (**b**) Typical PDF at three different time instants (*left*: in ordinary coordinate system; *right*: in logarithmic coordinate system). (**c**) PDF evolution surface. (**d**) Contour of the PDF surface

References

Au SK, Beck JL (2001) Estimation of small failure probabilities in high dimensions by subset simulation. Prob Eng Mech 16(4):263–277

Caughey TK, Ma F (1982) The exact steady-state solution of a class of non-linear stochastic systems. J Non-Linear Mech 17(3):137–142

Chen JB, Lin PH (2014) Dimension-reduction of FPK equation via equivalent drift coefficient. Theor Appl Mech Lett 4(1): No. 013002

Chen JB, Yuan SR (2014) Dimension reduction of FPK equation via equivalence of probability flux for additively excited systems. J Eng Mech. doi:10.1061/(ASCE)EM.1943-7889.0000804

Chen JB, Sun WL, Li J, Xu J (2013) Stochastic harmonic function representation of stochastic processes. J Appl Mech 80(1) No.011001

Er GK (2000) Exponential closure method for some randomly excited non-linear systems. Int J Non-Linear Mech 35:69–78

Er GK (2011) Methodology for the solutions of some reduced Fokker-Planck equations in high dimensions. Ann Phys 523(3):247–258

Goller B, Pradlwarter HJ, Schuëller GI (2013) Reliability assessment in structural dynamics. J Sound Vib 332(10):2488–2499

Li J (1996) Stochastic structural systems: analysis and modelling. Science Press, Beijing (in Chinese)

Li J, Chen JB (2008) The principle of preservation of probability and the generalized density evolution equation. Struct Saf 30:65–77

Li J, Chen JB (2009) Stochastic dynamics of structures. Wiley, Singapore, pp 191–230

Li J, Chen JB, Sun WL, Peng YB (2012a) Advances of probability density evolution method for nonlinear stochastic systems. Prob Eng Mech 28:132–142

Li J, Yan Q, Chen JB (2012b) Stochastic modeling of engineering dynamic excitations for stochastic dynamics of structures. Prob Eng Mech 27(1):19–28

Lutes LD, Sarkani S (2004) Random vibrations: analysis of structural and mechanical systems. Elsevier, Amsterdam

Naess A, Moe V (2000) Efficient path integration methods for nonlinear dynamic systems. Prob Eng Mech 15(2):221–231

Roberts JB, Spanos PD (1990) Random vibration and statistical linearization. Wiley, Chichester

Schuëller GI (1997) A state-of-the-art report on computational stochastic mechanics. Prob Eng Mech 12(4):197–321

Shinozuka M (1972) Monte Carlo simulation of structural dynamics. Comput Struct 2:855–874

Soize C (1994) The Fokker-Planck equation for stochastic dynamical systems and its explicit steady state solutions. World Scientific, Singapore

von Wagner U, Wedig WV (2000) On the calculation of stationary solutions of multi-dimensional Fokker–Planck equations by orthogonal functions. Nonlinear Dyn 21(3):289–306

Wang D, Li J (2011) Physical random function model of ground motions for engineering purposes. Sci China Technol Sci 54(1):175–182

Wen YK (2004) Chapter 7: Probabilistic aspects of earthquake engineering. In: Bozorgnia Y, Bertero VV (eds) Earthquake engineering: from engineering seismology to performance-based engineering. CRC Press, Boca Raton, pp 7–1–7–45

Yuan SR, Chen JB, Li J (2012) Dimension reduction of FPK equation and its applications in seismic response of structures. In: Dai JG, Zhu SY (eds) Proceedings of the fourth Asia-Pacific young researchers & graduates symposium, The Hong Kong Polytechnic University, Hong Kong, China, pp 46–52

Zhu WQ (2006) Nonlinear stochastic dynamics and control in Hamiltonian formulation. Appl Mech Rev 59:230–248

The Probabilistic Solutions of the Cantilever Excited by Lateral and Axial Excitations Being Gaussian White Noise

G.K. Er and V.P. Iu

Abstract The multi-degree-of-freedom system with both external and parametric excitations is formulated with Galerkin's method from the typical problem of the cantilever excited by both lateral excitation and axial excitation being correlated Gaussian white noises. The probabilistic solution of this multi-degree-of-freedom stochastic dynamical system is obtained by the state-space-split method and exponential polynomial closure method. The way for selecting the sub-state vector in the dimension reduction procedure with the state-space-split method is given for the analyzed cantilever. The solution procedure with the state-space-split method is presented for the system excited by both external excitation and parametric excitation being correlated Gaussian white noises. Numerical results are presented. The results obtained with the state-space-split method and exponential polynomial closure method are compared with those obtained by Monte Carlo simulation and Gaussian closure method to verify the effectiveness and efficiency of the state-space-split method and exponential polynomial closure method in analyzing the probabilistic solutions of the multi-degree-of-freedom stochastic dynamical systems with both external excitation and parametric excitation similar to that formulated from the cantilever excited by both lateral excitation and axial excitation being correlated Gaussian white noises.

Keywords Probabilistic solution • State-space-split method • EPC method • Cantilever • External and parametric excitations

G.K. Er (✉) • V.P. Iu
University of Macau, Macau SAR, China
e-mail: gker@umac.mo; vaipaniu@umac.mo

M. Papadrakakis and G. Stefanou (eds.), *Multiscale Modeling and Uncertainty Quantification of Materials and Structures*, DOI 10.1007/978-3-319-06331-7_17,
© Springer International Publishing Switzerland 2014

1   Introduction

The random vibrations of the cantilever excited by external Gaussian white noise and parametric Gaussian white noise can be found its application in modeling the tall buildings excited by both horizontal ground motion and vertical ground motion. The horizontal ground motion can cause lateral excitation on the structure and the vertical ground motion can cause axial excitation on the structure. There are both external or additive excitation due to lateral excitation and parametric or multiplicative excitation due to axial excitation in the dynamical system describing the motion of the cantilever. When both the external excitation and the parametric excitation exist, it was a challenging problem in obtaining the analytical probabilistic solution of the system when the system is modeled as multi-degree-of-freedom (MDOF) system. In this paper, the probabilistic solution of this structure is analyzed. It is known that the equation of motion of the cantilever is a partial differential equation in time and space. With Galerkin's method, the partial differential equation is reduced to a MDOF stochastic dynamical system. Many other real problems can also be modeled with the similar MDOF stochastic dynamical systems. It is known that no exact solutions of this type of MDOF stochastic dynamical systems are obtainable. Therefore, two methods can be used for the approximate solutions of the MDOF stochastic dynamical systems with both external excitation and parametric excitation. One of them is by using moment equations which can be solved to obtain the moments of the responses of the linear oscillator with both external excitation and parametric excitation being Gaussian white noise (Soong 1973; Iyengar and Dash 1978; Sun and Hsu 1987; Baratta and Zuccaro 1994). Another method applicable for analyzing this kind of stochastic dynamical systems is Monte Carlo simulation (MCS) method which is for the numerical solution of stochastic differential equations (Harris 1979; Kloeden and Platen 1995). The moment equation method only works well for single-degree-of-freedom system since huge number of equations in terms of moments needs to be formulated and solved when high order moments of the system responses are needed for MDOF systems. The number of moment equations increases exponentially as the number of degrees of freedom and the moment order increase. MCS can be used for analyzing a lot of MDOF stochastic dynamical systems, but the computational effort needed by MCS is huge when the system is large or the small probability of system responses is needed. Identifying the conditions about the numerical convergence, stability, round-off error, and the requirement for large sample size are also challenges for MCS method in analyzing MDOF stochastic dynamical systems. It is known that directly solving the Fokker-Planck-Kolmogorov (FPK) equation governing the probabilistic solution of the system in high dimensions is a challenge (Risken 1989; Gardiner 2009). Recently, a new method named state-space-split (SSS) method was proposed for obtaining the probabilistic solutions of MDOF stochastic dynamical systems or solving the FPK equations in high dimensions (Er 2011; Er and Iu 2011). The SSS method makes the problem of solving the FPK equation in high dimensions reduced to the problem of solving the FPK equations in low dimensions. Then, the FPK equations in low dimensions are solved with the exponential polynomial closure (EPC) method (Er 1998). The SSS-EPC method was extended to analyze the probabilistic solutions of stretched Euler-Bernoulli beam excited by lateral Poissonian white noise (Er and Iu 2012) or filtered Gaussian white noise (Er 2013). In this paper, the SSS-EPC method is further extended to analyze the probabilistic solutions of the vertical cantilever excited by Gaussian white noises both horizontally and vertically. The system formulated in this case is the MDOF system excited by fully correlated external excitation and parametric excitation being Gaussian white noise. The results obtained with the SSS-EPC method are also compared with those from MCS and Gaussian closure to show the effectiveness of the SSS-EPC method in this case. It can be concluded that the results obtained with the SSS-EPC method are close to MCS when the system excitation is under the reasonable level of real situation.

2 Dynamical System of the Cantilever Excited by Correlated External and Parametric Gaussian White Noises

Consider the cantilever excited by inclined ground motion as shown in Fig. 1. Under the action of the inclined ground acceleration, the motion of the vertical cantilever is governed by the following equation.

$$\rho A \ddot{Y}(x,t) + c\dot{Y}(x,t) + EI\frac{\partial^4 Y(x,t)}{\partial x^4} + \rho A\left[\frac{\partial^2 Y(x,t)}{\partial x^2}(L-x) - \frac{\partial Y(x,t)}{\partial x}\right]$$

$$[c_v W(t) + g] = \rho A c_h W(t) \tag{1}$$

where $Y(x,t)$ is the deflection of the cantilever; L is the length of the cantilever; E is Young's modulus; I is the moment of inertia of the cross section of the cantilever; A is the area of the cross section of the cantilever; $\rho(\mathrm{kg/m^3})$ is the mass density; c is the damping constant; $W(t)$ is Gaussian white noise representing the ground acceleration; g is gravitational acceleration; $c_h = \cos\theta$ and $c_v = \sin\theta$ where θ is the angle between horizontal line and the direction of ground acceleration. The boundary condition of the cantilever is

$$Y(0,t) = \frac{\partial Y(0,t)}{\partial x} = \frac{\partial^2 Y(L,t)}{\partial x^2} = \frac{\partial^3 Y(L,t)}{\partial x^3} \tag{2}$$

Approximately express the solution of the cantilever by

$$Y(x,t) = \sum_{i=0}^{m} A_i(t)\phi_i(x) \tag{3}$$

Fig. 1 Cantilever excited by
inclined ground acceleration

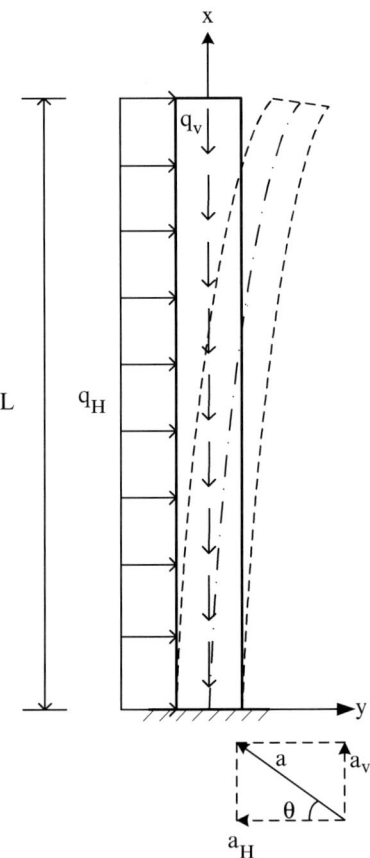

where $\phi_i(x)$ is the ith linear mode function of the cantilever which is given by

$$\phi_i(x) = \cos \alpha_i x - \cosh \alpha_i x - \frac{\cos \alpha_i L + \cosh \alpha_i L}{\sin \alpha_i L + \sinh \alpha_i L}(\sin \alpha_i x - \sinh \alpha_i x) \quad (4)$$

in which $\alpha_1 = 1.875/L$, $\alpha_2 = 4.694/L$, $\alpha_3 = 7.855/L$, and $\alpha_i = (i - 0.5)\pi/L$
for $i \geq 4$.

With Galerkin's method, the following stochastic dynamical system is obtained
for the deflection $Y_0(x_0, t) = \sum_{i=1}^{m} A_i(t)\phi_i(x_0)$ at $x = x_0$ if the damping ratio is
the same for each mode.

$$\ddot{Y}_0 + 2\xi\omega_1\dot{Y}_0 + \frac{EI}{\rho A}\sum_{k=1}^{m}\alpha_k^4\phi_k(x_0)A_k + \sum_{j=1}^{m}\sum_{k=1}^{m}e_{kj}\phi_k(x_0)A_j[c_v W(t) + g]$$

$$= W(t)\sum_{k=1}^{m}f_k\phi_k(x_0) \qquad (5)$$

$$\ddot{A}_i + 2\xi\omega_i\dot{A}_i + \omega_i^2 A_i + \sum_{j=1}^{m} e_{ij} A_j [c_v W(t) + g] = f_i W(t) \qquad i = 2, 3, \ldots, m \quad (6)$$

where $\omega_i = \alpha_i^2 \sqrt{EI/(\rho A)}$, ξ is the damping ratio of the system, $e_{ij} = (Lc_{ij} - b_{ij} - d_{ij})/a_{ii}$, $f_i = d_i c_h/a_{ii}$, $d_i = \int_0^L \phi_i(x)dx$, $a_{ii} = \int_0^L \phi_i^2(x)dx$, $b_{ij} = \int_0^L \phi_i(x)\phi_j'(x)dx$, $c_{ij} = \int_0^L \phi_i(x)\phi_j''(x)dx$, $D_{ij} = \int_0^L x\phi_i(x)\phi_j''(x)dx$, and $A_1 = \phi_1^{-1}(x_0)[Y_0 - \sum_{k=2}^{m} A_k \phi_k(x_0)]$.

3   Dimension Reduction with State-Space-Split Method

Equations 5 and 6 formulate a coupled m-degrees-of-freedom system with both external excitation and parametric excitation being Gaussian white noise. No exact solution is available for this system. Many systems similar to Eqs. 5 and 6 can be formulated from real problems in science and engineering. In order to solve the problem, a single-degree-of-freedom stochastic dynamical system is formulated with Eqs. 5 and 6 by setting $m = 1$. Another multi-degree-of-freedom stochastic dynamical system is formulated with Eqs. 5 and 6 by setting $m = 16$ since further increasing the number of shape functions cannot make the solution further changed obviously. The solution corresponding to $m = 16$ can be considered as converged solution. In the case of $m = 16$, the original FPK equation is in 32 dimensions. The joint probability density function (PDF) of the deflection and the velocity on the top of the cantilever at $x = L$ is analyzed with the SSS method and EPC method based on Eqs. 5 and 6. The state variables in the first sub vector is taken to be $\{Y_0(L,t), \dot{Y}_0(L,t)\} \in \mathfrak{R}^2$ in formulating the two-dimensional approximate FPK equation with the dimension reduction procedure of SSS method (Er 2011; Er and Iu 2011). The MCS is also conducted to verify the effectiveness of the SSS-EPC method in solving the FPK equations in high dimensions or analyzing the PDF solution of the multi-degree-of-freedom systems excited by the external excitation and parametric excitation being Gaussian white noise. The results obtained with Gaussian closure are also given for comparison in the following numerical analysis.

4   Numerical Analysis

In the numerical analysis, the cantilever which models a tower and is excited by correlated horizontal and vertical ground acceleration is analyzed. The material of the cantilever is reinforced concrete and the cross section of the cantilever is a ring with outer diameter 7 m and inner diameter 6 m. The parameters in the formulated system are given as $L = 300$ m, $E = 2.55 \times 10^{10}$ N/m^2, $A = 10.20$ m^2, $I = 54.21$ m^4, $\rho = 2.3 \times 10^3$ kg/m^3, $\xi = 0.01$, $S = 10$ m^2/s^5, and $\theta = 85°$. In solving

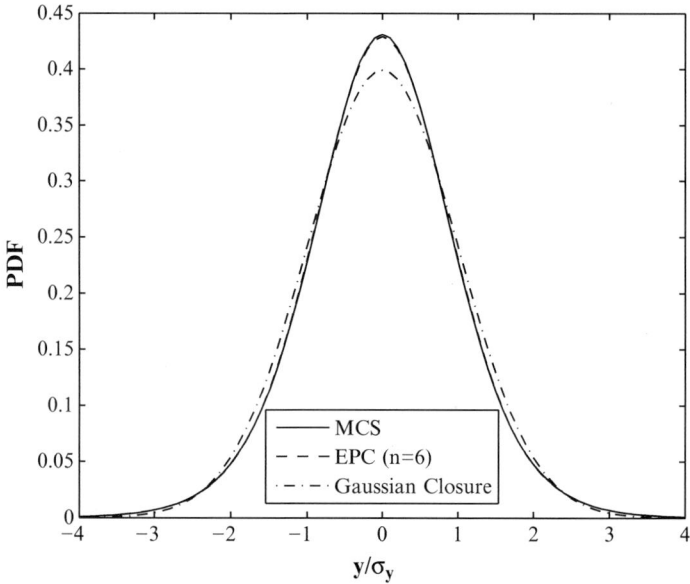

Fig. 2 PDFs of top deflection in Case 1

the two-dimensional FPK equations obtained with SSS method, the sixth degree polynomial is used in the EPC solution procedure (Er 1998). The sample size in the procedure of MCS is 10^8.

4.1   Case 1. SDOF System

When $m = 1$, the system is a SDOF system. The relevant FPK equation is in two dimensions. In the case of $\theta = 85°$, the acceleration is almost vertical. $c_h = \cos 85° = 0.087$ and $c_v = \sin 85° = 0.996$. The vertical component of the ground acceleration is much greater than the horizontal component in order to show the influence of the parametric excitation on the system response. It is known that the vertical acceleration causes axial excitation or parametric excitation in the system. Obtained with SSS-EPC, MCS, and Gaussian closure, the PDFs of the deflection and velocity on the top of the cantilever are shown in Figs. 2 and 4. In view that the tails of the PDF plays an important role in the system reliability analysis, the logarithms of the PDFs are also shown and compared in Figs. 3 and 5. It is seen from Figs. 2 to 5 that both the top deflection and velocity are far from being Gaussian in this case, particularly in the tails of the PDFs, though the system is linear and the excitation is Gaussian. The non-Gaussian behaviour of the deflection and velocity is caused by the parametric excitation or the axial excitation on the cantilever. By

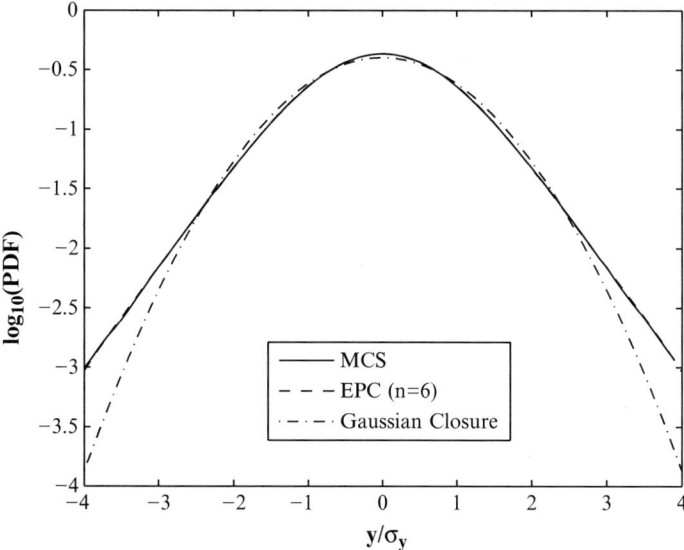

Fig. 3 Logarithm of PDFs of top deflection in Case 1

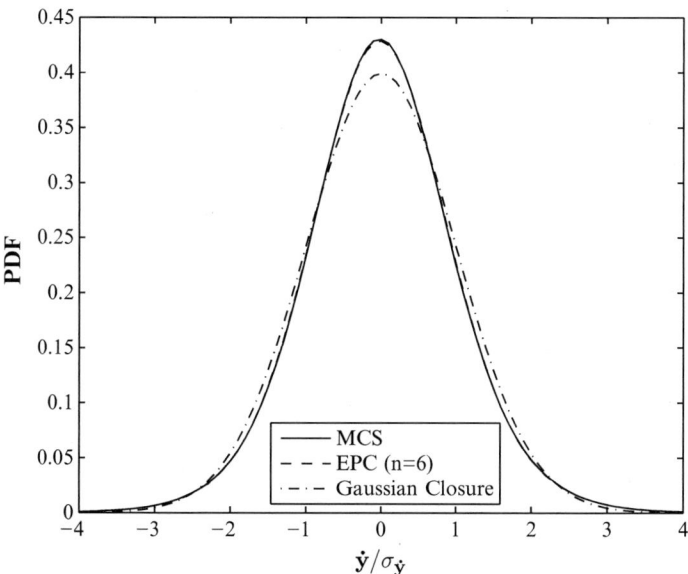

Fig. 4 PDFs of top velocity in Case 1

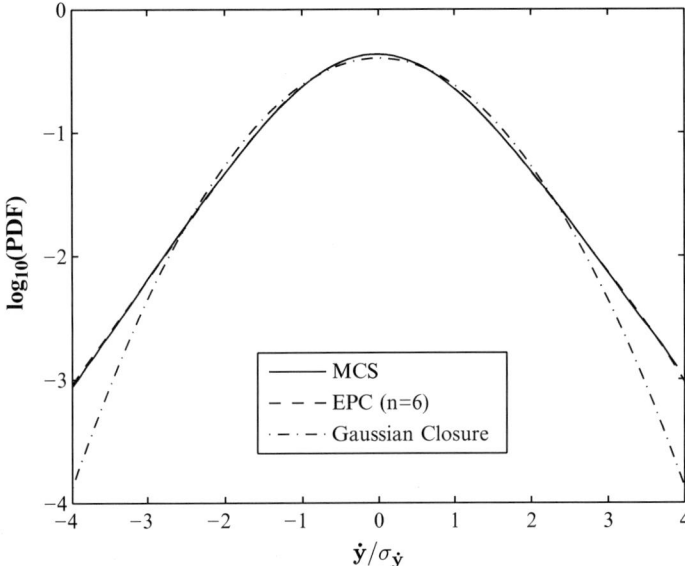

Fig. 5 Logarithm of PDFs of top velocity in Case 1

comparison it is observed that the results obtained with SSS-EPC are close to those obtained with MCS while the results from Gaussian closure deviate a lot from MCS.

The mean up-crossing rate (MCR) ν^+ is a quantity that is frequently used in system reliability analysis. It is defined as

$$\nu^+(y_0) = \int_0^{+\infty} \dot{y}_0 p(y_0, \dot{y}_0) d\dot{y}_0 \tag{7}$$

where $p(y_0, \dot{y}_0)$ is the joint PDF of Y_0 and $\dot{Y}_0$. The MCRs and logarithmic MCRs of the top deflection are also shown and compared in Figs. 6 and 7. It is seen that the MCR obtained with SSS-EPC is also in good agreement with that obtained with MCS.

4.2 Case 2. 16-DOF System

As the number of shape functions used in formulating the stochastic dynamic systems increases to 16, i.e., $m = 16$, the formulated system is a 16-DOF system. The relevant FPK equation is in 32 dimensions. Further increasing the number of shape functions cannot make the PDFs of top deflection and velocity obviously changed, so the solution corresponding to $m = 16$ is considered as converged solution. Still consider the case of $\theta = 85°$. In this case, there are both external

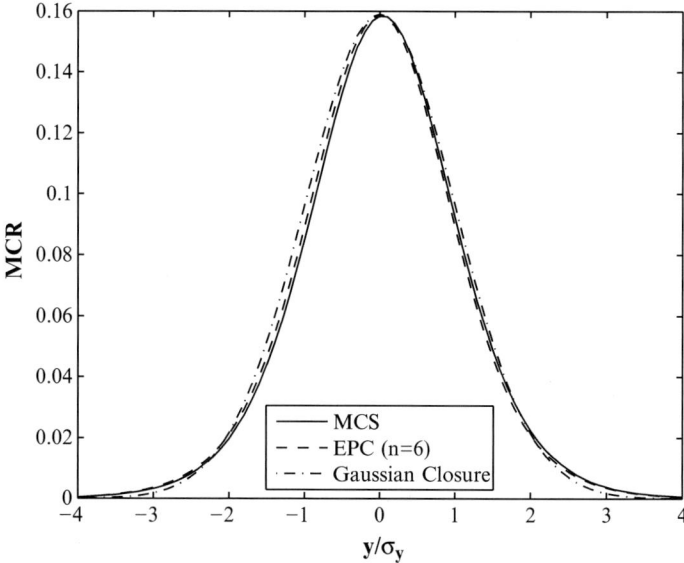

Fig. 6 MCRs of top deflection in Case 1

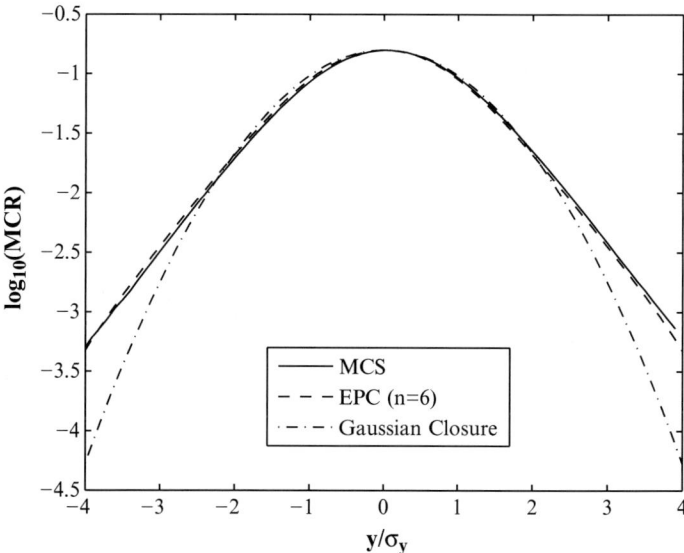

Fig. 7 Logarithm of MCRs of top deflection in Case 1

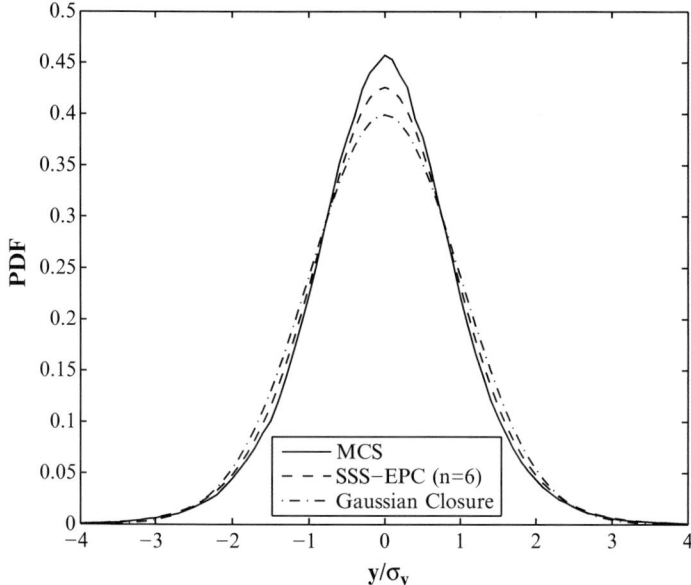

Fig. 8 PDFs of top deflection in Case 2

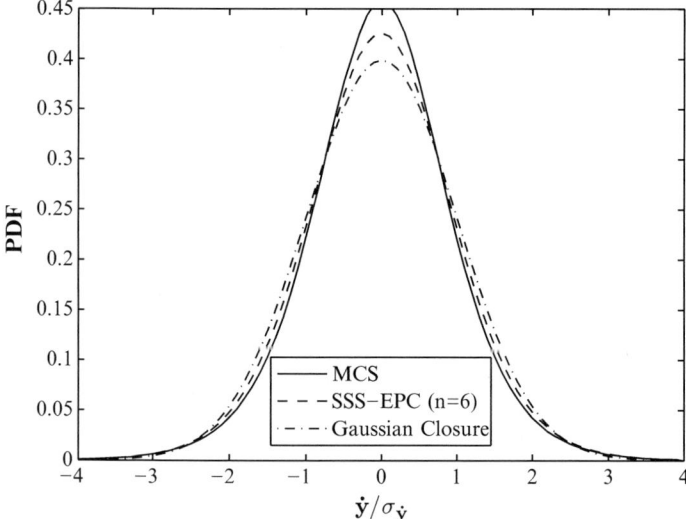

Fig. 9 PDFs of top velocity in Case 2

and parametric excitations in the formulated system. Obtained with SSS-EPC and polynomial degree being six in the EPC procedure, MCS, and Gaussian closure, the PDFs of the deflection and velocity on the top of the cantilever are shown in Figs. 8 and 9. The MCRs of the top deflection is shown in Fig. 10. The logarithms

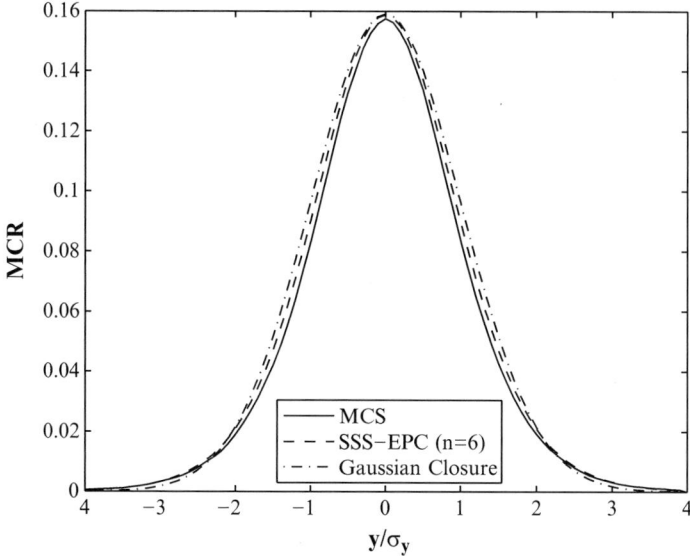

Fig. 10 MCRs of top deflection in Case 2

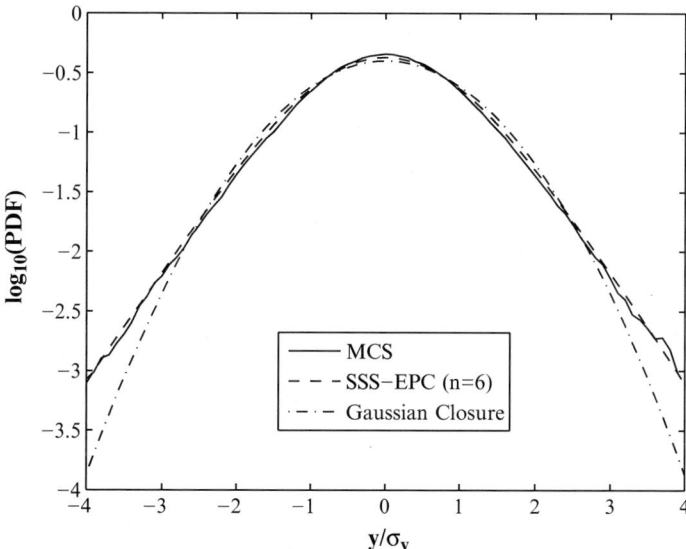

Fig. 11 Logarithm of PDFs of top deflection in Case 2

of the PDFs and MCRs are also shown and compared in Figs. 11–13. It is seen from these figures that both the top deflection and velocity are far from being Gaussian in this case though the system is linear and the excitation is Gaussian. The non-Gaussian behavior of the deflection and velocity is caused by the parametric

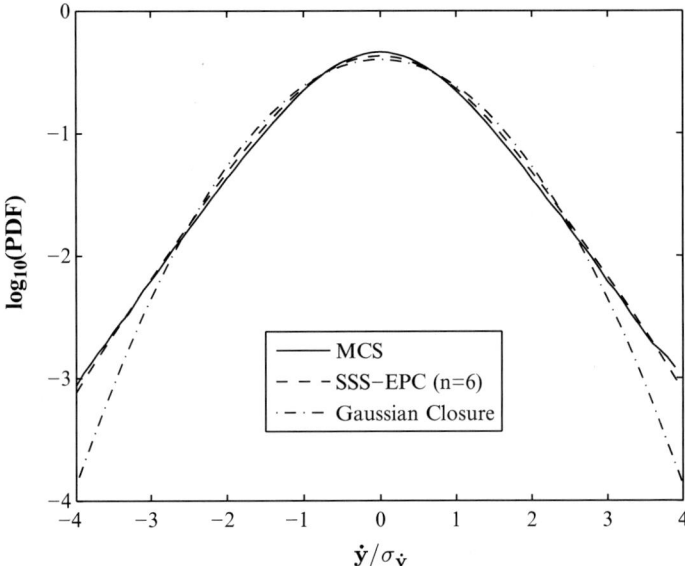

Fig. 12 Logarithm of PDFs of top velocity in Case 2

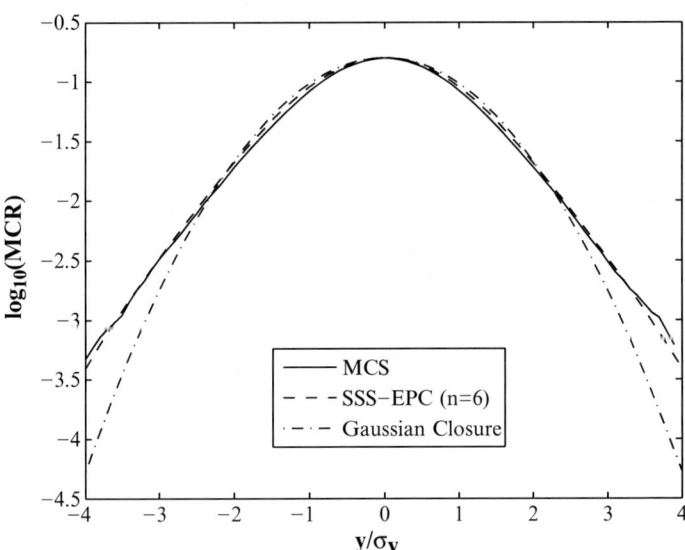

Fig. 13 Logarithm of MCRs of top deflection in Case 2

excitation or the axial excitation on the cantilever. It is observed that the results obtained with SSS-EPC are close to those obtained with MCS, but the computational time needed with SSS-EPC is about 200 s mainly spent on solving the moment equations with 2 s spent on the SSS-EPC procedure and the computational time needed with MCS is 1,060 s for this 16-DOF linear system in the same computer and running environment. It is seen that the PDF solutions obtained with the 16-DOF system in Case 2 is close to those obtained with the SDOF system in Case 1. It means that the solution from the SDOF system in Case 1 is accurate enough for this cantilever.

5 Conclusions

Under the action of correlated lateral and axial excitations being fully correlated Gaussian white noises, the equations of motion of the cantilever is given to be a partial differential equation in time and space. With Galerkin's method, this partial differential equation is reduced to be a SDOF or MDOF system with both external and parametric excitations being Gaussian white noise. The FPK equation governing the PDFs of the responses of the MDOF system with both external and parametric excitations can be reduced to two-dimensional FPK equations by the SSS method for obtaining the PDFs of the displacement and its corresponding velocity. The formulated FPK equation in two-dimensions can then be solved with the EPC method when the polynomial degree in the EPC procedure equals six. The procedure for the probabilistic solution of the cantilever is presented. Similar problems can also be observed from many real problems in science and engineering. Numerical analysis is conducted with the SSS-EPC method, MCS method, and Gaussian closure method for comparison and testing the effectiveness of the SSS-EPC procedure in investigating the high-dimensional cantilever systems with both external and parametric excitations. Two cases are considered. One is about a SDOF system formulated with one linear mode function with Galerkin's method and the other is about a 16-DOF system formulated with 16 linear mode functions with Galerkin's method. In each case, the results obtained by the SSS-EPC method are close to those obtained by the MCS when the polynomial degree equals six in the EPC procedure. From the PDFs of displacement and velocity it is also observed that both the displacement and velocity of the cantilever are far from being Gaussian, which is caused by the parametric excitation. The mean up-crossing rate obtained by SSS-EPC with polynomial degree being six are also close to those obtained by the MCS even if both the displacement and velocity deviate a lot from Gaussian. Hence the SSS-EPC method is effective in analyzing the systems similar to the cantilever excited by correlated lateral and axial Gaussian white noises. The solution from the 16-DOF system can be considered as converged solution since further increasing the number of mode functions cannot make the solution obviously changed. It is also found that the solution of the SDOF system is very close to the solution from the 16-DOF system, which means that it is accurate enough to model the motion of the cantilever as a SDOF system with the correlated lateral and axial excitations being Gaussian white noises. It is also noted that the number of linear modes needs to be even with SSS-EPC procedure in order to fit the results from MCS if the cantilever is modeled as MDOF system and there are fully correlated external and parametric excitations being Gaussian white noise. The reason behind this phenomenon needs to be further investigated.

Acknowledgements This research is supported by the Research Committee of the University of Macau (Grant No. MYRG138-FST11-EGK).

References

Baratta A, Zuccaro G (1994) Analysis of nonlinear oscillators under stochastic excitation by the Fokker-Planck-Kolmogorov equation. Nonlinear Dyn 5:255–271

Er GK (1998) An improved closure method for analysis of nonlinear stochastic systems. Nonlinear Dyn 17(3):285–297

Er GK (2011) Methodology for the solutions of some reduced Fokker-Planck equations in high dimensions. Ann Phys (Berlin) 523(3):247–258

Er GK (2013) The probabilistic solutions of some nonlinear stretched beams excited by filtered white noise. In: Proppe C (ed) IUTAM symposium on multiscale problems in stochastic mechanics. Karlsruhe, Germany, 25–29 June 2012. Procedia IUTAM 6. Elsevier, pp 141–150

Er GK, Iu VP (2011) A new method for the probabilistic solutions of large-scale nonlinear stochastic dynamical systems. In: Zhu WQ, Lin YK, Cai GQ (eds) IUTAM Symposium on nonlinear stochastic dynamics and control. Hangzhou, China, 10–14 May 2010. IUTAM book series, vol 29. Springer, Dordrecht, pp 25–34

Er GK, Iu VP (2012) State-space-split method for some generalized Fokker-Planck-Kolmogorov equations in high dimensions. Phys Rev E 85:067701

Gardiner CW (2009) Stochastic methods: a handbook for the natural and social sciences. Springer, Berlin

Harris CJ (1979) Simulation of multivariate nonlinear stochastic system. Int J Numer Methods Eng 14:37–50

Iyengar RN, Dash PK (1978) Study of the random vibration of nonlinear systems by the Gaussian closure technique. ASME J Appl Mech 45:393–399

Kloeden PE, Platen E (1995) Numerical solution of stochastic differential equations. Springer, Berlin

Risken H (1989) The Fokker-Planck equation, methods of solution and applications. Springer, Berlin

Soong TT (1973) Random differential equations in science and engineering. Academic, New York

Sun JQ, Hsu CS (1987) Cumulant-neglect closuremethod for nonlinear systems under random excitations. ASME J Appl Mech 54:649–655

Dynamic Response Variability of General FE-Systems

Vissarion Papadopoulos and Odysseas Kokkinos

Abstract In this paper a general formulation is proposed for the dynamic analysis of stochastic structures with uncertain material properties. A straightforward generalization of the mean and variability response function concept is introduced leading to closed form integral expressions for the dynamic mean and variability response of statically indeterminate beam/frame structures as well as for more general stochastic finite element systems. As in the case of classical variability functions, these integral expressions involve the spectral density function of a stochastic field modeling the uncertain material properties and so-called dynamic mean and variability response functions, recently established for linear stochastic statically determinate single degree of freedom oscillators. A finite element method-based fast Monte Carlo simulation procedure is used for the accurate and efficient numerical evaluation of these functions. Numerical examples are provided including a statically indeterminate beam/frame structure and a plane stress problem. The dynamic mean and variability response functions can be used consequently to perform sensitivity/parametric analyses with respect to various probabilistic characteristics involved in the problem (i.e., correlation distance, standard deviation) and to establish realizable upper bounds on the dynamic mean and variance of the response.

Keywords Dynamic variability response functions • Stochastic finite element analysis • Upper bounds • Stochastic dynamic systems

V. Papadopoulos (✉) • O. Kokkinos
Institute of Structural Analysis and Antiseismic Research, National Technical University of Athens, 9 Iroon Polytechneiou, Zografou Campus, Athens, 15780 Greece
e-mail: vpapado@central.ntua.gr; okokki@central.ntua.gr

M. Papadrakakis and G. Stefanou (eds.), *Multiscale Modeling and Uncertainty Quantification of Materials and Structures*, DOI 10.1007/978-3-319-06331-7__18,
© Springer International Publishing Switzerland 2014

1 Introduction

In stochastic finite element methodologies developed over the past decades, whether these are based on perturbation/expansion (Liu et al. 1986a, b), spectral Galerkin approximations (Ghanem and Spanos 1991) or computationally expensive Monte Carlo methods (Liu et al. 1986a; Grigoriu 2006; Matthies et al. 1997; Stefanou 2009), knowledge of the correlation structure and the marginal probability distribution function (*pdf*) of the stochastic fields describing the uncertain system parameters is a prerequisite for the prediction of the response variability of a stochastic static or dynamic system. As there is usually a lack of experimental data for the quantification of such probabilistic quantities, a sensitivity analysis with respect to various stochastic parameters is often implemented. In this case, however, the problems that arise are the increased computational effort, the lack of insight on how these parameters control the response variability of the system and the inability to determine bounds of the response variability. Furthermore, limited works are dealing with the dynamic propagation of system uncertainties, most of them reducing the stochastic dynamic PDE's to a linear random eigenvalue problem (Ghosh et al. 2005; Schueller 2011).

In this framework and to tackle the aforementioned issues, the concept of the Dynamic Variability Response Function (*DVRF*) has been proposed in (Papadopoulos and Kokkinos 2012), which is a function of deterministic parameters of the problem as well as of the standard deviation of the stochastic field modeling the uncertain system properties. In that work, closed form integral expressions, involving *DVRF* and the spectral density function of the stochastic field, were suggested for the computation of the dynamic variance of the response displacement as follows:

$$Var\left[u(t)\right] = \int_{-\infty}^{\infty} DVRF\left(t, \kappa, \sigma_{ff}\right) S_{ff}\left(\kappa\right) d\kappa \qquad (1)$$

A similar expression has also been proposed for the mean system response involving a Dynamic Mean Response Function (*DMRF*). This approach was formulated for linear statically determinate single degree of freedom stochastic oscillators under dynamic excitations and it was demonstrated that the integral form expressions for the dynamic mean and variance can be used to effectively compute the first and second order statistics of the transient system response with reasonable accuracy, together with time dependent spectral-distribution-free upper bounds. They also provide an insight into the mechanisms controlling the uncertainty propagation with respect to both space and time and in particular the mean and variability time histories of the stochastic system dynamic response. Furthermore, once the *DMRF* and *DVRF* were established, sensitivity analyses with respect to various probabilistic parameters such as correlation distances, standard deviation were performed at a very small additional computational cost.

Based on the aforementioned recent development, closed form integral expressions in the form of Eq. (1) are proposed in the present work for the mean and variance of the dynamic response of statically indeterminate beam/frame structures and then extended to more general stochastic finite element systems (i.e. plane stress problems) under dynamic excitations. In this case **DVRF** and **DMRF** are vectors comprised of a *DMRF* and *DVRF* for each degree of freedom of the FE system. A general so-called Dynamic FEM fast Monte Carlo simulation (DFEM-FMCS) is proposed for the accurate and efficient evaluation of **DVRF** and **DMRF** for stochastic FE systems. Numerical results are presented, demonstrating that as in the case of classical *VRFs* proposed in the late 1980s (Shinozuka 1987) along with different aspects and extensions (Wall and Deodatis 1994; Graham and Deodatis 1998), as well as in the case of *DMRF* and *DVRF* for single degree of freedom stochastic oscillators (Papadopoulos and Kokkinos 2012), the **DVRF** and **DMRF** matrices appear to be independent of the functional form of the power spectral density function $S_{ff}(\kappa)$ and appear to be marginally dependent on the *pdf* of the field modeling the uncertain system parameter. It is reminded that the existence of VRF has been proven only in the case of statically determinate structures under static loading (Shinozuka 1987; Papadopoulos and Deodatis 2006). In all other cases this existence had to be conjectured and the validity of this conjecture was demonstrated through comparisons of the results obtained from Eq. (1) with brute force MCS. It should be mentioned here that the *VRF* concept was recently extended in (Teferra and Deodatis 2012) for structures with non-linear material properties where a closed form analytic expression of *VRF* revealed the clear dependence of the integral form of Eq. (1) on the standard deviation as well as higher order power spectra of $f(x)$. Finally, realizable upper bounds of the mean and dynamic system response are evaluated.

2 Time-History Analysis of Stochastic Finite Element Systems

Without loss of generality consider the linear stochastic FE system of Fig. 1 which is a fixed-fixed beam/frame structure. The inverse of the elastic modulus is assumed to vary randomly along its length according to the following expression:

$$\frac{1}{E(x)} = F_0 \left(1 + f(x)\right), \tag{2}$$

where E is the elastic modulus, F_0 is the mean value of the inverse of E, and $f(x)$ is a zero-mean homogeneous stochastic field modeling the variation of $1/E$ around its mean value.

Fig. 1 Geometry and loading
of the fixed–fixed frame
discretized with 60 beam
elements

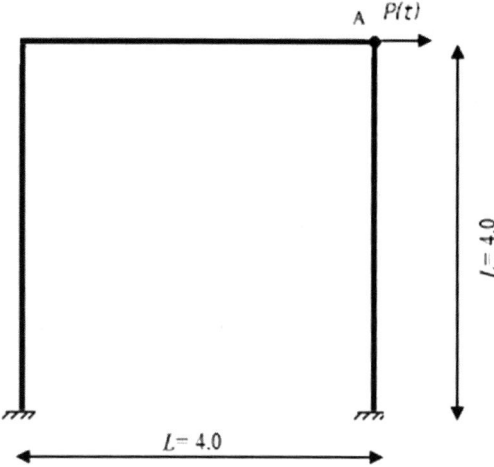

For the derivation of the deterministic system dynamic response the trivial second-order differential equation for the discretized FE dynamic system equilibrium neglecting damping is as follows:

$$\mathbf{M}\ddot{\mathbf{u}}(t) + \mathbf{K}\mathbf{u}(t) = \mathbf{P}(t) \tag{3}$$

where $\mathbf{M}$ is the mass matrix of the discretized FE system, $\mathbf{K}$ is its stiffness matrix and $\mathbf{P}(t)$ is its loading vector. At last, $\mathbf{u}(t)$ is the time-history of the displacement vector of the system, providing information about the response of each node of the FE mesh, and $\ddot{\mathbf{u}}(t)$ is the second order time-derivative of $\mathbf{u}(t)$.

Direct integration of Eq. (3) can be performed using i.e. a Newmark scheme of the following form:

$$^{t+\Delta t}\widehat{\mathbf{R}} = {}^{t+\Delta t}\mathbf{R} + \mathbf{M}\left(a_0{}^t\mathbf{U} + a_1{}^t\dot{\mathbf{U}} + a_2{}^t\ddot{\mathbf{U}}\right) \tag{4}$$

where $a_0 = \frac{1}{a\Delta t^2}$; $a_1 = \frac{1}{a\Delta t}$; $a_2 = \frac{1}{2a} - 1$; $a_6 = \Delta t\,(1 - \delta)$; $a_7 = \delta\Delta t$. After choosing a time step Δt parameters α and δ are selected under the limitations $\delta \geq 0.50$ and $a \geq 0.25(0.5 + \delta)^2$. After initialization of ${}^0\mathbf{U}, {}^0\dot{\mathbf{U}}$, and ${}^0\ddot{\mathbf{U}}$, the displacements at time $t + \Delta t$ are calculated solving the following linear system of equations

$$\widehat{\mathbf{K}}^{t+\Delta t}\mathbf{U} = {}^{t+\Delta t}\widehat{\mathbf{R}} \tag{5}$$

where $\widehat{\mathbf{K}}$ is the effective stiffness matrix given by

$$\widehat{\mathbf{K}} = \mathbf{K} + a_0\mathbf{M} \tag{6}$$

Finally accelerations and velocities at time $t + \Delta t$ accrue from the following equations:

$$^{t+\Delta t}\ddot{\mathbf{U}} = a_0 \left(^{t+\Delta t}\mathbf{U} - {}^t\mathbf{U}\right) - a_1{}^t\dot{\mathbf{U}} - a_2{}^t\ddot{\mathbf{U}} \tag{7}$$

$$^{t+\Delta t}\dot{\mathbf{U}} = {}^t\dot{\mathbf{U}} + a_6{}^t\ddot{\mathbf{U}} + a_7{}^{t+\Delta t}\ddot{\mathbf{U}} \tag{8}$$

Matrices $\widehat{\mathbf{R}}$ and $\widehat{\mathbf{K}}$ in Eqs. (5) and (6) and consequently vectors $\mathbf{U}, \dot{\mathbf{U}}$ and $\ddot{\mathbf{U}}$ are random due to the variation of $E(x)$ in Eq. (2). Thus, the solution of Eq. (5) requires the implementation of some stochastic methodology in order to invert the stochastic operator $\widehat{\mathbf{K}}$ at each time step and predict the stochastic dynamic response of the FE system.

3   Analysis of Mean and Variance of Dynamic System Response Using DMRF and DVRF

Following a procedure similar to the one presented in (Papadopoulos and Kokkinos 2012) for linear stochastic oscillators under dynamic loading, it is possible to express the variance of the dynamic response of a stochastic finite element system in the following integral form expression:

$$Var\left[\mathbf{u}(t)\right] = \int_{-\infty}^{\infty} \mathbf{DVRF}\left(t, \kappa, \sigma_{ff}\right) S_{ff}\left(\kappa\right) d\kappa \tag{9a}$$

where **DVRF** is the vectorized dynamic version of *DVRF*, assumed to be a function of deterministic parameters of the problem related to geometry, loading, (mean) material properties and the standard deviation σ_{ff} of the stochastic field modeling the system's flexibility. A similar integral expression can provide an estimate for the mean value of the dynamic response of the system (Papadopoulos et al. 2006):

$$\varepsilon\left[\mathbf{u}(t)\right] = \int_{-\infty}^{\infty} \mathbf{DMRF}\left(t, \kappa, \sigma_{ff}\right) S_{ff}\left(\kappa\right) d\kappa \tag{9b}$$

where again **DMRF** is the vectorized dynamic version of *DMRF*, which is a function similar to the **DVRF** in the sense that it also depends on deterministic parameters of the problem as well as σ_{ff}.

3.1   Numerical Estimation of the DVRF and the DMRF Using Fast Monte Carlo Simulation

The numerical estimation of DVRF and DMRF involves a dynamic FEM-based fast Monte Carlo simulation (DFEM-FMCS) whose idea is to consider the random field $f(x)$ in Eq. (2) as a random sinusoid (Papadopoulos et al. 2005; Papadopoulos and Deodatis 2006) and plug its monochromatic power spectrum into Eqs. (9a) and (9b), in order to compute the respective mean and variance response at various wave numbers as a function of time t. The steps of the FEM-FMCS approach are the following:

(i) Generate N (5–10) sample functions of the below random sinusoid with standard deviation σ_{ff} and wave number $\overline{\kappa}$ modeling the variation of the inverse of the elastic modulus $1/E$ around its mean F_0:

$$f_j(x) = \sqrt{2}\sigma_{ff} \cos\left(\overline{\kappa}x + \varphi_j\right) \qquad (10)$$

where $j = 1,2,\ldots,N$ and φ_j varies randomly under uniform distribution in the range $[0,2\pi]$. These samples are generated by dividing the range $[0,2\pi]$ at 5–10 equally spaced distances and selecting the centres of these distances as values of random phase angles φj's.

(ii) Using these N generated sample functions it is straightforward to compute their respective dynamic mean and response variance, $\varepsilon[\mathbf{u}(t)]_{\overline{\kappa}}$ and $Var[\mathbf{u}(t)]_{\overline{\kappa}}$, by solving the corresponding FEM system under the applied dynamic loading using Eqs. (5), (7) and (8). Random matrix $\widehat{\mathbf{K}}$ is constructed by assigning a different value of E at each FE, using e.g. the mid-point method.

(iii) The value of the **DMRF** at wave number $\overline{\kappa}$ can then be computed as follows:

$$\mathbf{DMRF}\left(t,\overline{\kappa},\sigma_{ff}\right) = \frac{\varepsilon[\mathbf{u}(t)]_{\overline{\kappa}}}{\sigma_{ff}^2} \qquad (11a)$$

and likewise the value of the **DVRF** at wave number $\overline{\kappa}$

$$\mathbf{DVRF}\left(t,\overline{\kappa},\sigma_{ff}\right) = \frac{Var[\mathbf{u}(t)]_{\overline{\kappa}}}{\sigma_{ff}^2} \qquad (11b)$$

Both previous equations are direct consequences of the integral expressions in Eqs. (9a) and (9b) in the case that the stochastic field becomes a random sinusoid.

(iv) Get **DMRF** and **DVRF** as a function of both time t and wave number κ by repeating previous steps for various wave numbers and different time steps. The entire procedure can be repeated for different values of the standard deviation σ_{ff} of the random sinusoid.

3.2   Bounds of the Mean and Variance of the Dynamic Response

Upper bounds on the mean and variance of the dynamic response of the stochastic system can be established directly from Eqs. (9a) and (9b), as follows:

$$\varepsilon\left[\mathbf{u}(t)\right] = \int_{-\infty}^{\infty} \mathbf{DMRF}\left(t, \kappa, \sigma_{ff}\right) S_{ff}\left(\kappa\right) d\kappa \leq \mathbf{DMRF}\left(t, \kappa^{\max}(t), \sigma_{ff}\right) \sigma_{ff}^2 \tag{12a}$$

$$Var\left[\mathbf{u}(t)\right] = \int_{-\infty}^{\infty} \mathbf{DVRF}\left(t, \kappa, \sigma_{ff}\right) S_{ff}\left(\kappa\right) d\kappa \leq \mathbf{DVRF}\left(t, \kappa^{\max}(t), \sigma_{ff}\right) \sigma_{ff}^2 \tag{12b}$$

where $\kappa^{\max}(t)$ is the wave number at which $\mathbf{DMRF}$ and $\mathbf{DVRF}$, corresponding to a given time step t and value of σ_{ff}, reach their maximum value. For the minimum, $\kappa^{\max}(t)$ is substituted with $\kappa^{\min}(t)$ and inequality signs switch direction. An envelope of time evolving upper and lower bounds on the mean and variance of the dynamic system response can be extracted from Eqs. (12a) and (12b). As in the case of linear stochastic systems under static loads (Papadopoulos et al. 2005, 2006; Papadopoulos and Deodatis 2006), this envelope is physically realizable since the form of the stochastic field that produces it is the random sinusoid of Eq. (10) with $\overline{\kappa} = \kappa^{\max}(t)$.

4   2D Formulation

In the case of a problem where the inverse elastic modulus is considered to vary randomly over a 2D domain, the following equation is adopted:

$$\frac{1}{E(x, y)} = F_0\left(1 + f(x, y)\right), \tag{13}$$

where E is the elastic modulus, F_0 is the mean value of the inverse of E, and $f(x,y)$ is now a two-dimensional, zero-mean homogeneous stochastic field modeling the variation of $1/E$ around its mean value F_0. Accordingly, the integral expressions for the variance and mean response displacement $u(t)$ become:

$$Var\left[\mathbf{u}(t)\right] = \int_{-\infty}^{\infty}\int_{-\infty}^{\infty} \mathbf{DVRF}\left(t, \kappa_x, \kappa_y, \sigma_{ff}\right) S_{ff}\left(\kappa_x, \kappa_y\right) d\kappa_x d\kappa_y \tag{14a}$$

$$\varepsilon\left[\mathbf{u}(t)\right] = \int_{-\infty}^{\infty}\int_{-\infty}^{\infty} \mathbf{DMRF}\left(t, \kappa_x, \kappa_y, \sigma_{ff}\right) S_{ff}\left(\kappa_x, \kappa_y\right) d\kappa_x d\kappa_y \tag{14b}$$

where $\mathbf{DVRF}(t,\kappa_x,\kappa_y,\sigma_{ff})$ and $\mathbf{DMRF}(t,\kappa_x,\kappa_y,\sigma_{ff})$ are in this case two-dimensional, possessing the following bi-quadrant symmetries:

$$\mathbf{DMRF}\left(\kappa_x,\kappa_y\right) = \mathbf{DMRF}\left(-\kappa_x,-\kappa_y\right) \tag{15}$$

$$\mathbf{DVRF}\left(\kappa_x,\kappa_y\right) = \mathbf{DVRF}\left(-\kappa_x,-\kappa_y\right) \tag{16}$$

$S_{ff}(\kappa_x,\kappa_y)$ is the spectral density function of the stochastic field $f(x,y)$ possessing the same symmetries as $\mathbf{DMRF}$ and $\mathbf{DVRF}$. The 1D random sinusoid in Eq. (10) now becomes a 2D one with the following form:

$$f_j(x) = \sqrt{2}\sigma_{ff} \cos\left(\overline{\kappa}_x x + \overline{\kappa}_y y + \varphi_j\right); j = 1,2,\ldots,N. \tag{17}$$

Upper bounds on the mean and variance of the response displacement for a given time instance t can be established for the 2D case as follows:

$$Var\left[\mathbf{u}(t)\right] \leq \mathbf{DVRF}\left(t,\kappa_x^{\max},\kappa_y^{\max},\sigma_{ff}\right)\sigma_{ff}^2 \tag{18a}$$

$$\varepsilon\left[\mathbf{u}(t)\right] \leq \mathbf{DMRF}\left(t,\kappa_x^{\max},\kappa_y^{\max},\sigma_{ff}\right)\sigma_{ff}^2 \tag{18b}$$

where $(\kappa_x^{\max},\kappa_y^{\max})$ is the wave number pair at which the $\mathbf{DMRF}$ or the $\mathbf{DVRF}$ take their maximum value (for a given value of σ_{ff} and a given location (x,y)), and σ_{ff}^2 is the variance of the stochastic field $f(x,y)$ modeling the inverse of the elastic modulus. Again, for the minimum, $\kappa_{x,y}^{\max}(t)$ is substituted with $\kappa_{x,y}^{\min}(t)$ and inequality signs switch direction. It should be emphasized that $(\kappa_x^{\max},\kappa_y^{\max})$ are not necessarily the same for the $\mathbf{DMRF}$ and the $\mathbf{DVRF}$.

5 Numerical Examples

Example 1 For the fixed-fixed frame shown in Fig. 1 with length and height equal to $L=4$ m, the inverse of the modulus of elasticity is assumed to vary randomly along its length according to Eq. (2) with $F_0 = (1.35 \times 10^8 \ KN/m)^{-1}$ and $I=0.1$ m^4. The total mass of the beam is assumed to be $m_{tot} = 6,000$ kg, distributed evenly among the finite element nodes of the model. For the analysis of the frame structure we used 60 beam elements, 20 for each column and the plateau, of equal length, resulting in 177 d.o.f.'s.

Two load cases are considered: LC1 consisting of a concentrated dynamic periodic load $P(t) = 100\sin(2t)$ at the right top corner of the frame (see Fig. 1) and LC2 consisting of a dynamic load $p_n(t) = -m_n\ddot{U}_g(t)$ acting on each node n of the beam with m_n being the corresponding nodal mass and $\ddot{U}_g(t)$ the acceleration time history of the 1940 El Centro earthquake. The stochastic field $f(x)$ in Eq. (2)

is considered to vary across the length of the two columns and the plateau of the frame running continuously from the left fixed edge to the right. The spectral density function (*SDF*) used for the modeling of the inverse of the elastic modulus stochastic field is given by the following formula:

$$S_{ff}(\kappa) = \frac{1}{4}\sigma^2 b^3 \kappa^2 e^{-b|\kappa|} \tag{19}$$

with $b = 1,2,10$ being three different values of the correlation length parameter examined.

For standard deviations σ_{ff} of the stochastic field $f(x)$ higher than 0.2 a truncated Gaussian and a lognormal *pdf* is used to model $f(x)$. For this purpose, an underlying Gaussian stochastic field denoted by $g(x)$ is generated using the spectral representation method (Shinozuka and Deodatis 1991) and the power spectrum of Eq. (19). The truncated Gaussian field $f_{TG}(x)$ is obtained by simply truncating $g(x)$ in the following way: $-0.9 \le g(x) \le 0.9$, while the lognormal $f_L(x)$ is obtained from the following transformation as a translation field (Grigoriu 1995):

$$f_L(x) = F_L^{-1}\{G[g(x)]\} \tag{20}$$

The *SDF* of the underlying Gaussian field in Eq. (20) and the corresponding spectral densities of the truncated Gaussian and the Lognormal fields denoted $S_{f_{TG}f_{TG}}(\kappa)$ and $S_{f_L f_L}(\kappa)$, respectively, are different from the one in Eq. (19) and are computed from the following formula

$$S_{f_i f_i}(\kappa) = \frac{1}{2\pi L_x} \left| \int_0^{L_x} f_i(x) e^{-i\kappa x} dx \right|^2 ; i = TG, L \tag{21}$$

where L_x is the length of the sample functions of the non-Gaussian fields modeling flexibility. As the sample functions of the non-Gaussian fields are non-ergodic, the estimation of power spectra in Eq. (21) is performed in an ensemble average sense (Grigoriu 1995).

Figure 2 presents 3D plots of *DMRF(u_A)* and *DVRF(u_A)* for the horizontal displacement u_A of point A of the frame as a function of time t and frequency κ for $\sigma_{ff} = 0.2$. In this figure it can be observed that *DMRF(u_A)* remains almost constant with respect to κ, while evolving substantially as a function of t. On the contrary *DVRF(u_A)* demonstrates a substantial volatility with respect to both κ and t. Therefore, in contrast to *DMRF(u_A)*, *DVRF(u_A)* accommodates the possibility of considerable variation of the variability response for different statistical parameters of the stochastic field. This is further demonstrated in Fig. 3 in which the upper and lower bounds of the dynamic mean and variability response are depicted containing minima and maxima respectively, in comparison to the estimated mean and variability responses for case of a Gaussian stochastic field with the power spectrum of Eq. (19) and $\sigma_{ff} = 0.2$. The aforementioned bounds are derived directly

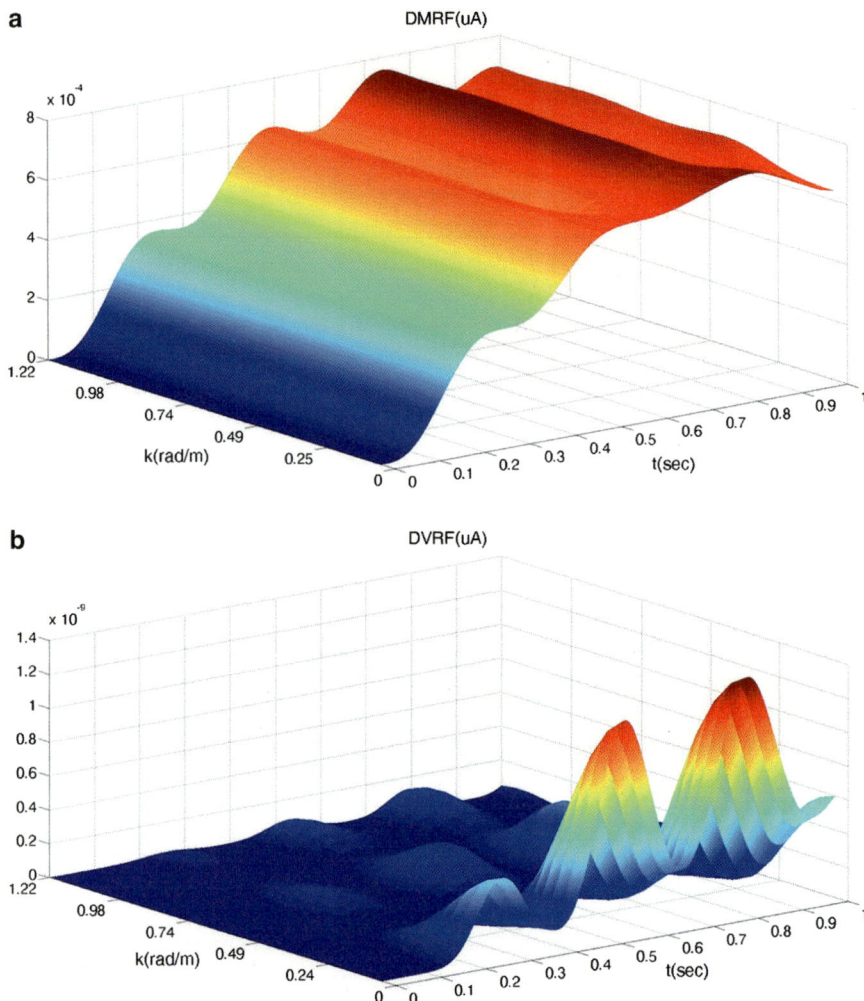

Fig. 2 3D plots of (**a**) *DMRF* and (**b**) *DVRF* of the horizontal displacement u_A, as a function of frequency κ (rad/m) and time t(sec) for LC1 and $\sigma_{ff} = 0.2$

from Eq. (12) having previously computed $DMRF(u_A)$ and $DVRF(u_A)$ with the computationally efficient DFEM-FMCS in Eq. (11), while in the case of the Gaussian field with $\sigma_{ff} = 0.2$, the mean and variance were obtained with the integral expression in Eq. (9). From this figure it can be seen that the upper mean dynamic response and the one estimated for the Gaussian field, are almost identical, while they differ significantly in the case of the response variability, reaching a maximum difference of more than *70 %* at *t = 0.8 s*. It should be pointed out here that bounds of each response do not necessarily need to coincide in the frequency number that they occur.

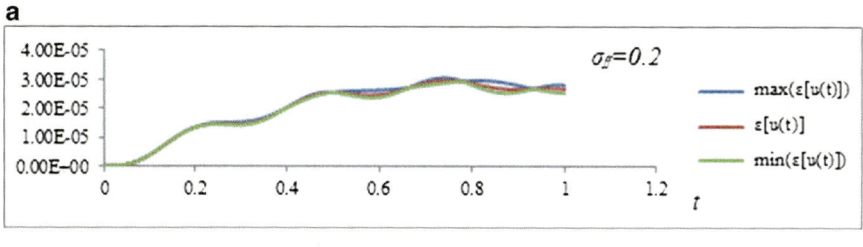

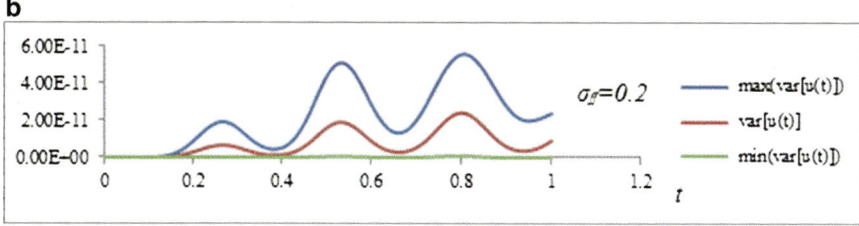

Fig. 3 Upper and lower bounds on the (a) mean and (b) variance of the response displacement for LC1 and $\sigma_{ff} = 0.2$

In order to demonstrate the validity of the proposed approach, the results obtained from the DFEM-FMCS procedure and Eq. (9) were compared with Brute Force Monte Carlo Simulation. In Fig. 4a–f the results of this comparison are presented for the dynamic mean and response variability of u_A (Fig. 1) using a Gaussian stochastic field and $\sigma_{ff} = 0.2$ for three different values of correlation length parameter b. In this manner the independence of **DMRF** and **DVRF** from the spectral density function is also showcased. Figure 5 presents the same comparison but for a truncated Gaussian field with $\sigma_{ff} = 0.3912$ while Fig. 6 examines a lognormal field case with $\sigma_{ff} = 0.399$. From all these figures it can be observed that the results of the DFEM-FMCS are in close agreement with the corresponding results of MCS. The prediction of the mean value is almost identical for the two methods in all cases considered, while the maximum error in the variance does not exceed 20 % and is attributed to a slight dependence of the *DVRF* on the *pdf* of the random field modeling $1/E(x)$. This error becomes negligible in the case of small standard deviations of the order of $\sigma_{ff} = 0.2$.

Example 2 Consider now the shear wall in Fig. 7 with length and height equal to $L = 4$ m, the inverse of the modulus of elasticity is assumed to vary randomly within its surface according to Eq. (13) with $F_0 = (1.35 \times 10^8 KN/m)^{-1}$, $v = 0.2$ and $t = 1.0$ The total mass of the beam is assumed to be $m_{tot} = 4,000$ kg, distributed evenly among the finite element nodes of the model. The wall is discretized into a total of 100 plain stress elements, 121 nodes and 242 d.o.f.'s. In this example the 2D version DFEM-FMCS procedure has been implemented, using Eq. (14) for the estimation of the dynamic mean and variability.

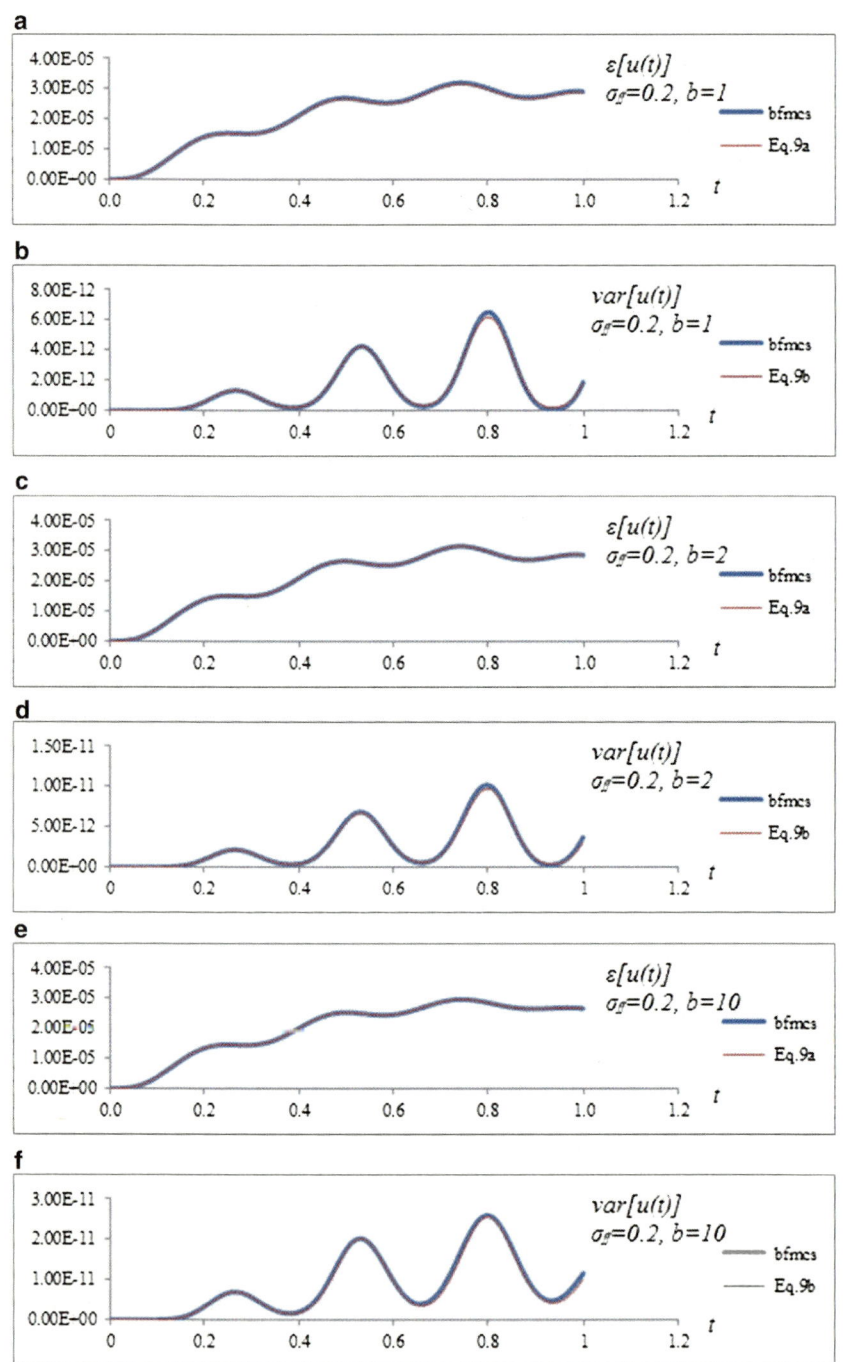

Fig. 4 Time histories of the (**a**), (**c**), (**e**) mean and (**b**), (**d**), (**f**) variance response displacement of the frame structure for a Gaussian field with $\sigma_{ff} = 0.2$ for LC1 and for three different correlation length parameter values b = 1,2 and 10.. Comparison of results obtained from Eqs. (9a and 9b) and MCS

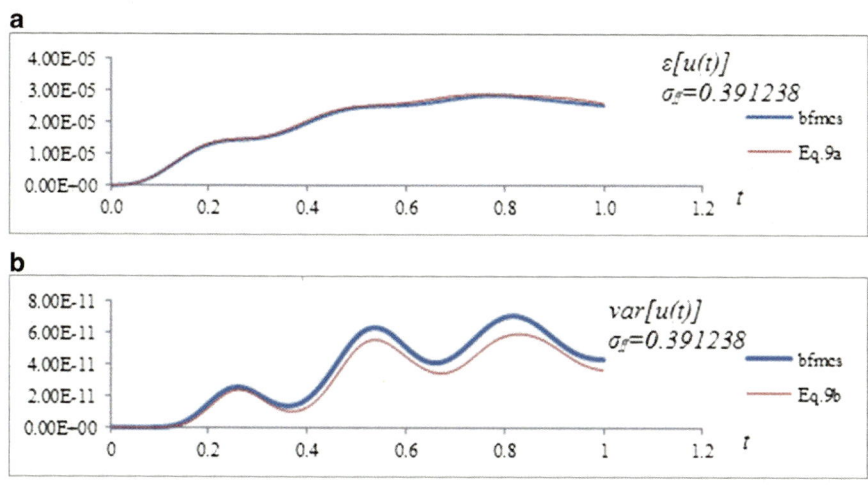

Fig. 5 Time histories of the (**a**) mean and (**b**) variance response displacement of the frame structure for a truncated Gaussian field with $\sigma_{ff} = 0.391238$ for LC1. Comparison of results obtained from Eqs. (9a and 9b) and MCS

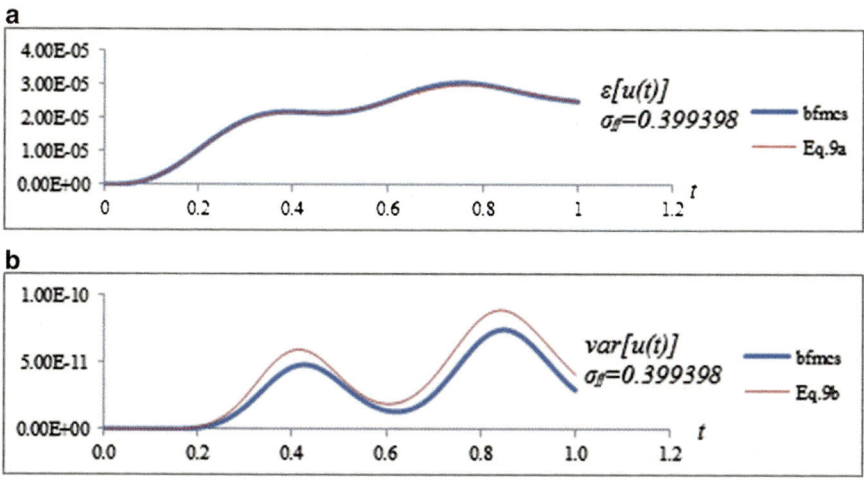

Fig. 6 Time histories of the (**a**) mean and (**b**) variance response displacement of the frame structure for a lognormal field with $\sigma_{ff} = 0.399398$ for LC1. Comparison of results obtained from Eqs. (9a and 9b) and MCS

The concentrated load is applied as shown in Fig. 3. For this case the following 2D spectrum is implemented:

$$S_{f_0 f_0}(\kappa_x, \kappa_x) = \frac{\sigma_f^2}{4\pi} b_x b_y \exp\left[-\frac{1}{4}\left(b_x^2 \kappa_y^2 + b_y^2 \kappa_y^2\right)\right] \quad (22)$$

with $b_x = b_y = 2.0$.

Fig. 7 Geometry, loading
and finite element mesh of the
shear wall

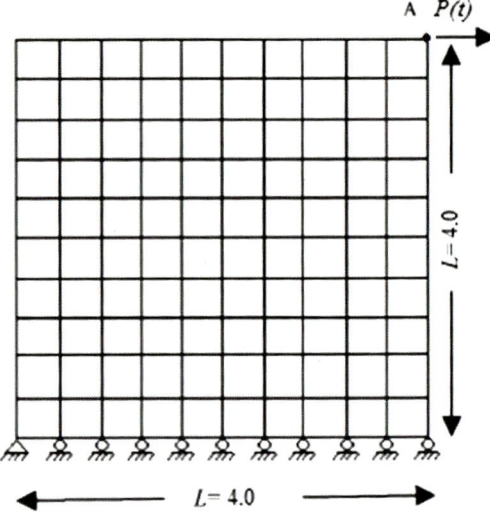

A *P(t)*

L = 4.0

L = 4.0

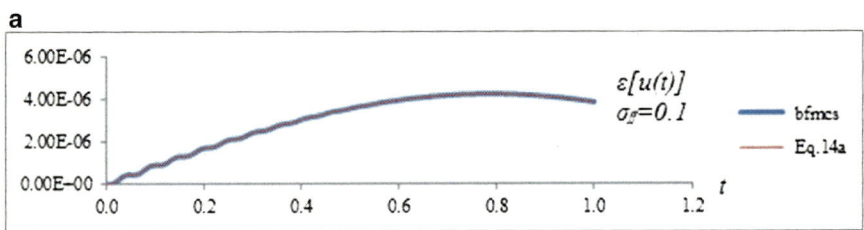

a

$\varepsilon[u(t)]$
$\sigma_{ff}=0.1$

bfmcs

Eq.14a

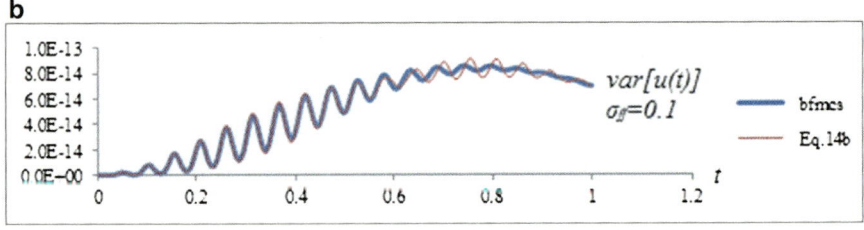

b

$var[u(t)]$
$\sigma_{ff}=0.1$

bfmcs

Eq.14b

Fig. 8 Time histories of the (**a**) mean and (**b**) variance response displacement of the shear wall for a Gaussian field with $\sigma_{ff} = 0.1$ for LC1. Comparison of results obtained from Eqs. (14a and 14b) and MCS

In Fig. 8, charts depict the comparison for the dynamic mean and variability response of the shear wall horizontal displacement at point A and LC1 for a Gaussian stochastic field with $\sigma_{ff} = 0.1$. The prediction of Eq. (14) in this case is very satisfactory with errors ranging up to *5–8* %. In Fig. 9 the upper bounds of the mean and variance response displacement are depicted in comparison to the corresponding responses for the case of a Gaussian field with $\sigma_{ff} = 0.1$. As expected the accruing upper bounds vary considerably from the respective mean and variance response obtained for $\sigma_{ff} = 0.1$.

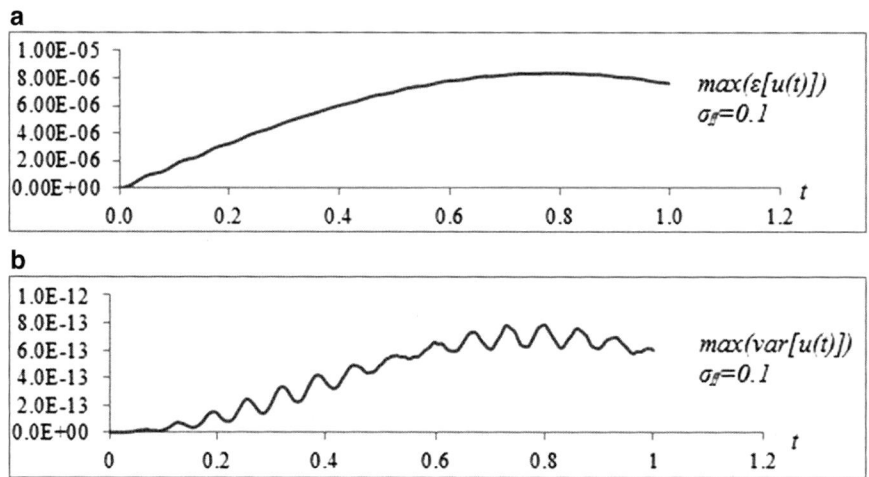

Fig. 9 Time histories of the (**a**) mean and (**b**) variance response displacement upper bounds of the shear wall for a Gaussian field with $\sigma_{ff} = 0.1$ for LC1. Results obtained from Eqs. (18a and 18b)

6 Concluding Remarks

In the present work, Dynamic Variability Response Functions and Dynamic Mean Response Functions are obtained for a statically indeterminate frame structure and a shear wall with random material properties under dynamic excitation using both beam and plain stress elements. The inverse of the modulus of elasticity was considered as the uncertain system parameter.

The *DVRF* and *DMRF* provide with an insight of the dynamic system sensitivity to the stochastic parameters and the mechanisms controlling the response mean and variability and their evolution in time.

Acknowledgements This work has been supported by the European Research Council Advanced Grant MASTER – Mastering the computational challenges in numerical modeling and optimum design of CNT reinforced composites (ERC-2011-ADG-20110209).

References

Ghanem R, Spanos PD (1991) Stochastic finite elements: a spectral approach. Springer, Berlin (2nd edn), Dover Publications, New York (2003)

Ghosh D, Ghanem R, Red-Horse J (2005) Analysis of eigenvalues and modal interaction of stochastic systems. AIAA J 43(10):2196–2201

Graham L, Deodatis G (1998) Weighted integral method and variability response functions for stochastic plate bending problems. Struct Saf 20:167–188

Grigoriu M (1995) Applied Non-Gaussian processes: examples, theory, simulation, linear random vibration, and MATLAB solutions. Prentice Hall, Englewood Cliffs

Grigoriu M (2006) Evaluation of Karhunen-Loève, spectral and sampling representations for stochastic processes. J Eng Mech (ASCE) 132:179–189

Liu WK, Belytschko T, Mani A (1986a) Probabilistic finite elements for nonlinear structural dynamics. Comput Methods Appl Mech Eng 56:61–86

Liu WK, Belytschko T, Mani A (1986b) Random field finite elements. Int J Numer Methods Eng 23:1831–1845

Matthies HG, Brenner CE, Bucher CG, Guedes Soares C (1997) Uncertainties in probabilistic numerical analysis of structures and solids – stochastic finite elements. Struct Saf 19:283–336

Papadopoulos V, Deodatis G (2006) Response variability of stochastic frame structures using evolutionary field theory. Comput Methods Appl Mech Eng 195(9–12):1050–1074

Papadopoulos V, Kokkinos O (2012) Variability response functions for stochastic systems under dynamic excitations. Probab Eng Mech 28:176–184

Papadopoulos V, Deodatis G, Papadrakakis, M (2005) Flexibility-based upper bounds on the response variability of simple beams. Comput Methods Appl Mech Eng 194(12–16): 1385–1404, 8, pp

Papadopoulos V, Papadrakakis M, Deodatis G (2006) Analysis of mean response and response variability of stochastic finite element systems. Comput Methods Appl Mech Eng 195(41–43):5454–5471

Schueller GI (2011) Model reduction and uncertainties in structural dynamics. In: Papadrakakis M, Stefanou G, Papadopoulos V (eds) Computational methods in stochastic dynamics. Springer, Dordrecht/New York

Shinozuka M (1987) Structural response variability. J Eng Mech 113(6):825–842

Shinozuka M, Deodatis G (1991) Simulation of stochastic processes by spectral representation. Appl Mech Rev 44(4):191–203

Stefanou G (2009) The stochastic finite element method: past, present and future. Comput Methods Appl Mech Eng 198(9–12):1031–1051

Teferra K, Deodatis G (2012) Variability response functions for beams with nonlinear constitutive laws. Probab Eng Mech 29:139–148

Wall FJ, Deodatis G (1994) Variability response functions of stochastic plane stress/strain problems. J Eng Mech 120(9):1963–1982

Stochastic Models of Defects in Wind Turbine Drivetrain Components

Hesam Mirzaei Rafsanjani and John Dalsgaard Sørensen

Abstract The drivetrain in a wind turbine nacelle typically consists of a variety of heavily loaded components, like the main shaft, bearings, gearbox and generator. The variations in environmental load challenge the performance of all the components of the drivetrain. Failure of each of these components of the drivetrain will lead to substantial economic losses such as cost of lost energy production, cost of repairs, cost of crew and cost of transportation. For offshore wind turbines, the marine environment affects the repair & maintenance process and in some case because of the rush environment, the maintenance team cannot operate properly and the wind turbine does not work for several days and consequently the cost of lost energy increases drastically. In this paper is presented stochastic models for fatigue failure based on test data and the accuracy of the models are compared.

Keywords Wind turbine • Reliability • Drivetrain • Defects • Stochastic model

1 Introduction

Reliability of wind turbine drivetrain components is very important for wind turbine manufacturers and owners. Offshore wind turbines are large structures exposed to wave excitations, highly dynamic wind loads influenced by the wind turbine control system and wakes from other wind turbines. Therefore, most components in a wind turbine experience highly dynamic and time-varying loads. These components may fail due to wear or fatigue and this can lead to unplanned shut down and repairs.

H. Mirzaei Rafsanjani (✉) • J.D. Sørensen
Department of Civil Engineering, Aalborg University, Sohngårdsholmsvej 57,
9000 Aalborg, Denmark
e-mail: hmr@civil.aau.dk; jds@civil.aau.dk

M. Papadrakakis and G. Stefanou (eds.), *Multiscale Modeling and Uncertainty
Quantification of Materials and Structures*, DOI 10.1007/978-3-319-06331-7_19,
© Springer International Publishing Switzerland 2014

The drivetrain consists of a variety of heavily loaded components, such as the main shaft, bearings, gearbox and generator. The variability of the loads challenges the performance of all the components of drivetrain. The failure of each component of the drivetrain will lead to economic losses such as cost of lost energy production, cost of repairs, cost of crew and cost of transportation. The environmental exposure affects the repair & maintenance of offshore wind turbine. Sometimes, because of the harsh environment, the maintenance team cannot operate properly and therefore the wind turbine cannot be accessed for several days. Consequently, the cost of lost energy increases drastically.

Due to fluctuating loads, fatigue is one of the main failure modes in wind turbine components. The current design of large wind turbines against fatigue is usually based on the life design approach (Campbell 2008). In the safe life design, fatigue testing is carried out on baseline materials to produce S-N curves. For many years it has been assumed in designs that all loads and strengths are deterministic. The strength of an element was determined in such a way that it exceeded the load with a certain margin and accounted for by a safety factor defined as the ratio between the strength and the load (Dong et al. 2013). Recently, safety factors are changed to partial safety factors in new codes. Hence, characteristic values of the uncertain loads and resistances are specified and partial safety factors are applied to the loads and strengths in order to ensure that the structure is safe enough. Hence, the uncertainties in the loads, strengths and the modeling can be accounted partially for in such a semi-probabilistic safety format.

This paper focuses on probabilistic methods for assessment of the reliability and stochastic modeling of the fatigue strength using structural reliability methods; see Entezami et al. (2012) allowing a rational modeling of all uncertainties. An important aspect in modeling fatigue failure of large cast steel components is to take into account scale effects. Two approaches are considered in this paper for stochastic modeling of the fatigue life including scale effects. One method is based on the classical Weibull approach and the other on application of a LogNormal distribution as done e.g. for the fatigue life of welded steel details.

2   Wind Turbine Drivetrain

Currently, most operating wind turbines use a modular configuration (Hau 2006). Typically, all individual components of the drivetrain are mounted onto a bedplate. The basic drivetrain components are the main bearing, shaft, gearbox, brakes, high-speed shaft and the generator, see Hau (2006) and Lindley (1976). A typical configuration of these components in the nacelle of a wind turbine is shown in Fig. 1.

Reliability of wind turbine gearboxes is studied in a number of research projects, e.g. the GRC project at National Renewable Energy Laboratory (NREL), (Oyague 2009). This include as important areas research on fault diagnosis and condition monitoring. Several methods have been considered, such as vibration and acoustic emission (Soua et al. 2013) and Local mean decomposition (Liu et al. 2012).

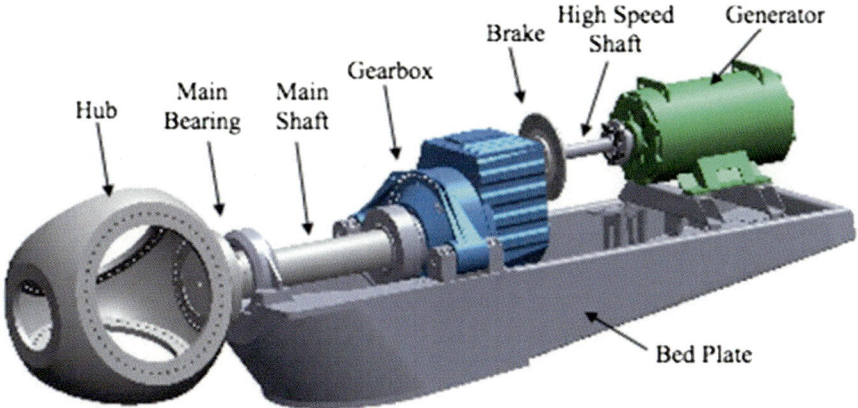

Fig. 1 Wind turbine drivetrain components (Oyague 2009)

Some studies on probabilistic modeling of failures in wind turbine drivetrain components have been carried out (Dong et al. 2013) but without a detailed stochastic modeling of the uncertainties related to the parameters in the limit state equations modeling each failure mode.

As mentioned above, most of the studies concentrated on gearbox failures. Moreover, in some studies failure of other parts like brake system (Entezami et al. 2012) is considered. By reviewing failure statistics of wind turbines, it is seen that focus is on reliability of blades, foundation and electrical parts whereas reliability of mechanical part such as bearing is only considered in few public studies.

Therefore, in this paper, the main bearing or main shaft are considered. Bearing and shaft of wind turbines are those having the highest downtimes in case of failure, see e.g. (Sheng and Veers 2011) and (Tavner et al. 2012). The current fatigue design is based on the life design approach (Shirani and Härkegård 2011a). In the safe life design S-N curves are based on tests as discussed above. However, the fatigue strength is typically highly uncertain and statistical uncertainties due to a limited number of tests can be important in modeling the fatigue strength. Moreover, model uncertainties related to e.g. application of the Miner rule for fatigue damage accumulation should be considered.

3 Fatigue Life Modeled by a LogNormal Distribution

The fatigue life can be modeled as the number of cycles to failure at a specified stress level. As the applied stress level decreases, the number of cycles to failure increases. The fatigue strength of metals is often assumed to follow the Basquin equation (Campbell 2008):

$$\sigma_a = \sigma_f (2N)^{-\frac{1}{m}} \tag{1}$$

where σ_a is the alternating stress amplitude, σ_f is the fatigue strength, $2\,N$ is the number of load reversals to failure, and $-1/m$ is the fatigue strength exponent. Equation (1) can also be written

$$N = \frac{1}{2}\left(\frac{\sigma_a}{\sigma_f}\right)^{-m} \tag{2}$$

In Eq. (2), σ_a is affected by geometrical size effects and can be estimated by the following equation (Shirani and Härkegård 2011a)

$$\frac{\sigma_a}{\sigma_{a0}} = \left(\frac{V}{V_0}\right)^{-\frac{1}{b_\sigma}} \Rightarrow \sigma_a = \sigma_{a0}\left(\frac{V}{V_0}\right)^{-\frac{1}{b_\sigma}} \tag{3}$$

where V_0 is the reference volume and σ_{a0} is the fatigue strength corresponding to the volume V_0. The stress exponent b_σ determines the effect of the specimen size on the fatigue life. By substitution of Eq. (3) in Eq. (2), the following equation is obtained:

$$N = \frac{1}{2}\left(\frac{\sigma_{a0}}{\sigma_f}\right)^{-m}\left(\frac{V}{V_0}\right)^{\frac{1}{b_n}} = \frac{1}{2}\left(\frac{\sigma_{a0}}{\sigma_f}\right)^{-m}\frac{1}{S_V^{b_n}} \tag{4}$$

The relative component volume influences the size effect and therefore the volume ratio is introduced by a scaling parameter s_V:

$$S_V = \frac{V}{V_0} \tag{5}$$

Moreover, the b_n in Eq. (4) is:

$$b_n = \frac{b_\sigma}{m} \tag{6}$$

The Eq. (4) can be rewritten in logarithmic format as follows

$$\log N = \log\left[\frac{1}{2}\left(\frac{\sigma_{a0}}{\sigma_f}\right)^{-m} S_V^{1/b_n}\right] \Rightarrow \log N = m\log\sigma_f - m\log\sigma_{a0} + \frac{1}{b_n}\log S_V - \log(2)$$

This equation is rewritten introducing an uncertainty term ε:

$$\log N = m\log\sigma_f - m\log\sigma_{a0} + \frac{1}{b_n}\log S_V - \log(2) + \varepsilon \tag{7}$$

where ε is assumed to be normally distributed with mean value $= 0$ and standard deviation $= \sigma_\varepsilon$. ε models the scatter in fatigue life and can be considered here to cover both physical and model uncertainties related to imperfect knowledge or idealizations of the mathematical models used or uncertainty related to the choice of probability distribution types for the stochastic variables. It is noted that the test data considered below do not allow a bilinear S-N curve to be fitted.

The parameters in Eq. (7) can be estimated using available test data. In this paper test data extracted from Shirani and Härkegård (2011b) are used to exemplify the procedure for the stochastic modeling. Assuming that the Shirani data are representative the results of the statistical analysis can also be used to assess the reliability level for drivetrain components and to calibrate safety factors, see below.

In the following, the Maximum Likelihood Method is used for the statistical analysis. The likelihood function as function of the statistical parameters σ_f, m, and σ_ε to be estimated is written as follows accounting both for tests results where failure occurs and for test results where failure does not occur (run-outs):

$$
L\left(\sigma_f, m, \sigma_\varepsilon\right) = \prod_{i=1}^{n_F} P\left[\log n_i = m\log\sigma_f - m\log\sigma_{a0,i} + \frac{1}{b_n}\log S_V + \varepsilon - \log 2\right]
$$

$$
\times \prod_{i=1}^{n_R} P\left[\log n_i > m\log\sigma_f - m\log\sigma_{a0,i} + \frac{1}{b_n}\log S_V + \varepsilon - \log 2\right]
$$

$$(8)$$

where n_i is the number of stress cycles to failure or run-out (no failure) with stress range equal to $\sigma_{a0,i}$ in test number i. n_F is the number of tests where failure occurs, and n_R is the number of tests where failure does not occur after n_i stress cycles (run-outs). $n = n_F + n_R$ is the total number of tests. σ_f, m, and σ_ε are estimated solving the optimization problem max $L(\sigma_f, m, \sigma_\varepsilon)$. This can be done using a standard nonlinear optimizer, e.g. the NLPQL algorithm, see Schittkowski (1986).

Since the parameters σ_f, m and σ_ε are estimated by the maximum-likelihood technique, they become asymptotically (number of data should be larger than 25–30) normally distributed stochastic variables with expected values equal to maximum-likelihood estimates and covariance matrix equal to, see Lindley (1976):

$$
C_{\sigma_f, m, \sigma_\varepsilon} = \left[-H_{\sigma_f, m, \sigma_\varepsilon}\right]^{-1} = \begin{bmatrix} \sigma^2_{\sigma_f} & \rho_{\sigma_f, m}\sigma_{\sigma_f}\sigma_m & \rho_{\sigma_f, \sigma_\varepsilon}\sigma_{\sigma_f}\sigma_{\sigma_\varepsilon} \\ \rho_{\sigma_f, m}\sigma_{\sigma_f}\sigma_m & \sigma_m^2 & \rho_{m, \sigma_\varepsilon}\sigma_m\sigma_{\sigma_\varepsilon} \\ \rho_{\sigma_f, \sigma_\varepsilon}\sigma_{\sigma_f}\sigma_{\sigma_\varepsilon} & \rho_{m, \sigma_\varepsilon}\sigma_m\sigma_{\sigma_\varepsilon} & \sigma^2_{\sigma_\varepsilon} \end{bmatrix} \quad (9)
$$

where $H_{\sigma_f, m, \sigma_\varepsilon}$ is the Hessian matrix with second-order derivatives of the log-likelihood function. σ_{σ_f}, σ_m, and $\sigma_{\sigma_\varepsilon}$ denote the standard deviation of σ_f, m and σ_ε respectively and ρ indicates correlation coefficients.

Alternatively to the LogNormal model for the S-N curve a Weibull model can be used, as described in the next section.

4   Fatigue Life Modeled by a Weibull Distribution

The influence of scale effects on damage modeling and fatigue life can from a theoretical basis be modeled by a Weibull model, see e.g. Madsen et al. (1986). Such a model is considered in this section assuming that the fatigue life can be modeled by a Weibull distribution. The distribution function for number of cycles to failure, N given stress range σ_{a0} is written:

$$F_N(n) = 1 - \exp\left[-\left(\frac{n}{N}\right)^{b_n}\right] \tag{10}$$

where b_n is a shape parameter. The corresponding density function becomes

$$f_N(n) = \frac{b_n}{N}\left(\frac{n}{N}\right)^{b_n-1}\exp\left[-\left(\frac{n}{N}\right)^{b_n}\right] \tag{11}$$

By substitution Eq. (4) and (6) in Eq. (11), the density function is written

$$f_N(n) = \frac{2b_n}{S_V^{1/b_n}}\left(\frac{\sigma_{a0}}{\sigma_f}\right)^m\left(\frac{2n}{S_V^{1/b_n}}\left(\frac{\sigma_{a0}}{\sigma_f}\right)^m\right)^{b_n-1}\exp\left[-\left(\frac{2n}{S_V^{1/b_n}}\left(\frac{\sigma_{a0}}{\sigma_f}\right)^m\right)^{b_n}\right] \tag{12}$$

The statistical parameters σ_f and m in Eq. (12) can be estimated by the Maximum Likelihood Method with the log-likelihood function:

$$\ln L\left(\sigma_f, m\right) =$$

$$\ln\left(\prod_{i=1}^{n} f_N(n_i)\right) = \sum_{i=1}^{n_F}\ln\left(\frac{2b_n}{S_V^{1/b_n}}\left(\frac{\sigma_{a0,i}}{\sigma_f}\right)^m\left(\frac{2n_i}{S_V^{1/b_n}}\left(\frac{\sigma_{a0,i}}{\sigma_f}\right)^m\right)^{b_n-1}\right.$$

$$\left.\exp\left[-\left(\frac{2n_i}{S_V^{1/b_n}}\left(\frac{\sigma_{a0,i}}{\sigma_f}\right)^m\right)^{b_n}\right]\right) + \sum_{i=1}^{n_R}\ln\left(\exp\left[-\left(\frac{2n_i}{S_V^{1/b_n}}\left(\frac{\sigma_{a0,i}}{\sigma_f}\right)^m\right)^{b_n}\right]\right) \tag{13}$$

where n_i is the number of stress cycles to fail or run-out (no failure) with stress range $\sigma_{a0,i}$ in test number i. n_F is the number of tests where failure occurs, and n_R is the number of tests where failure did not occur after n_i stress cycles (run-outs). $n = n_F + n_R$ is the total number of tests. σ_f and m are estimated solving the optimization problem max $L(\sigma_f, m)$, as described above.

5 Characteristic Values

In deterministic, code based design safety is introduced though application of deterministic values in terms of characteristic values and partial safety factors to obtain design values of both loads and strengths.

If statistical uncertainty is not taken into account then corresponding to a stress range, $\sigma_{a0,c}$ a characteristic value of the fatigue life, n_c defined as a 5 % quantile can be estimated directly from the distribution function of the fatigue life.

If statistical uncertainty is to be taken into account and the physical/model uncertainties for the fatigue life is modeled by a Lognormal distribution then a characteristic value for the fatigue life, n_c corresponding to the stress range, $\sigma_{a0,c}$ defined as a 5 % quantile can be obtained from

$$P\left(\log n_c > m \log \sigma_f - m \log \sigma_{a0,c} + \frac{1}{b_n} \log S_V + \varepsilon - \log 2\right) = 0.05 \quad (14)$$

with a corresponding limit state equation written as

$$g\left(\sigma_f, m, \varepsilon, \sigma_\varepsilon\right) = m \log \sigma_f - m \log \sigma_{a0,c} + \frac{1}{b_n} \log S_V + \varepsilon - \log 2 - \log n_c \quad (15)$$

Here the stochastic variables are ε, m, σ_ε and σ_f and they are introduced to model the physical/model and statistical uncertainties. For given $\sigma_{a0,c}$ (Eq. (15)) can be solved with respect to the characteristic fatigue life, n_c using e.g. FORM (First Order Reliability Methods), see Madsen et al. (1986).

Similarly if the fatigue life is modeled by a Weibull distribution and statistical uncertainty is accounted for then a limit state equation can be applied:

$$g\left(\sigma_f, m, \varepsilon, \sigma_\varepsilon\right) = \log n_c + \log 2 - \frac{1}{b_n} \log (S_V)$$

$$+ m \log (\sigma_{a0,c}) - m \log (\sigma_f) - \log (-1n(0.95))^{1/b_n - \varepsilon} \quad (16)$$

In Eq. (16), ε, σ_ε, m and σ_f model the physical/model and statistical uncertainties, respectively. As mentioned before, these parameters can be obtained from the test results.

6 Results

As mentioned above the data by Shirani and Härkegård (2011b) will be used to illustrate the above statistical analysis and reliability assessment for wind turbine components. The test data follows the specification listed in Table 1.

Table 1 The test plan (Shirani and Härkegård 2011b)

Material	Load ratio	Specimen	Number of specimen	Testing frequency
T95	0	Ø21	12	10
T95	−1	Ø21	12	10
T95	−1	Ø50	12	1
T150	−1	Ø21	18	10
T150	−1	120*140	9	40

Table 2 5 % quantile using LN distribution

Test	σ_f [MPA]	m	ε	σ_ε
D21 T95 R = 0	443.51	12.107	−1.1896	0.3220
D21 T95 R = −1	1,022.58	8.8793	−1.3270	0.3171
D50 T95 R = −1	1,003.92	8.3760	−1.2605	0.1652
D21 T150 R = −1	792.87	9.5477	−1.1181	0.2261
120*140 T150 R = −1	405.60	14.47	−1.5295	0.3524

Table 3 5 % quantile using Weibull distribution

Test	σ_f [MPA]	m	ε	σ_ε
D21 T95 R = 0	444.10	12.366	−1.579	0.3389
D21 T95 R = −1	974.68	9.1787	−1.4400	0.3528
D50 T95 R = −1	781.91	10.257	−1.6236	0.1657
D21 T150 R = −1	700.05	10.799	−1.5422	0.2571
120*140 T150 R = −1	412.59	14.39	−1.1564	0.3522

The material is EN-GJS-400-18-LT ductile cast iron with graphite nodules contained within a ferritic matrix (Sheng and Veers 2011). The specimens are extracted from two types of castings with 95 mm thickness, (95 mm × 200 mm × 750 mm) cast blocks, and 150 mm thickness, (150 mm × 300 mm × 1,550 mm) cast blocks. These blocks are illustrated in Table 1 by T95 and T150.

Two series of specimens were machined from T95 block, specimens with 21 mm and specimens with 50 mm diameter, see Sheng and Veers (2011). Specimens with 21 mm diameter were tested at load ratios $R = -1$ and $R = 0$, but specimens with 50 mm diameter were just tested at load ratio $R = -1$. Furthermore, two series of specimens were machined from T150 block, specimens with 21 mm and heavy section specimens with 120 × 140 mm cross section. All specimens were tested at load ratios R = −1 (Shirani and Härkegård 2011b).

The statistical analysis is performed following the methodology described in Sects. 3 and 4 for estimation of the parameters in the LogNormal and Weibull models. The results are shown in Tables 2 and 3 and 5 % quantiles are estimated as described above. The results of each test category are shown in Figs. 2, 3, 4, 5 and 6 showing test results for broken/failed and run-out specimens. Furthermore, the results of fit to LogNormal distribution and Weibull distribution are shown. Further, the figures show two types of 5 % quantiles for the LogNormal distribution, namely quantiles estimated when only failure data considered in calculating the 5 % quantile

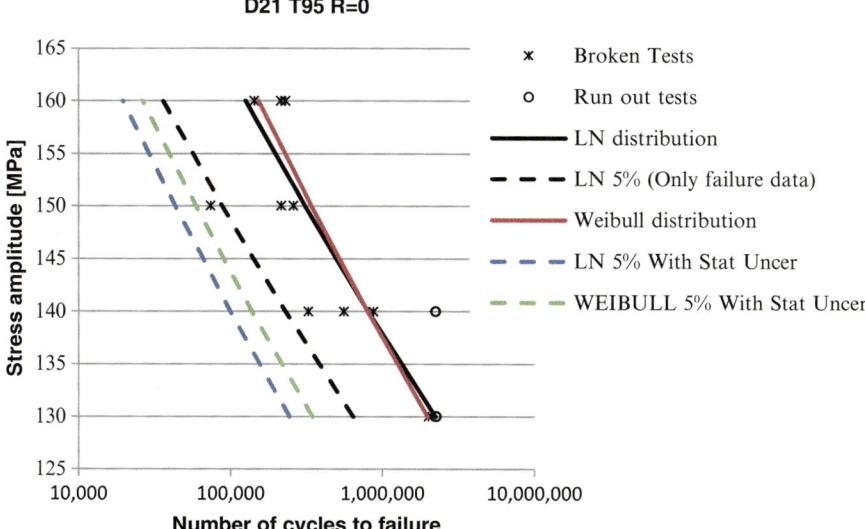

Fig. 2 Results for D21 T95 R0

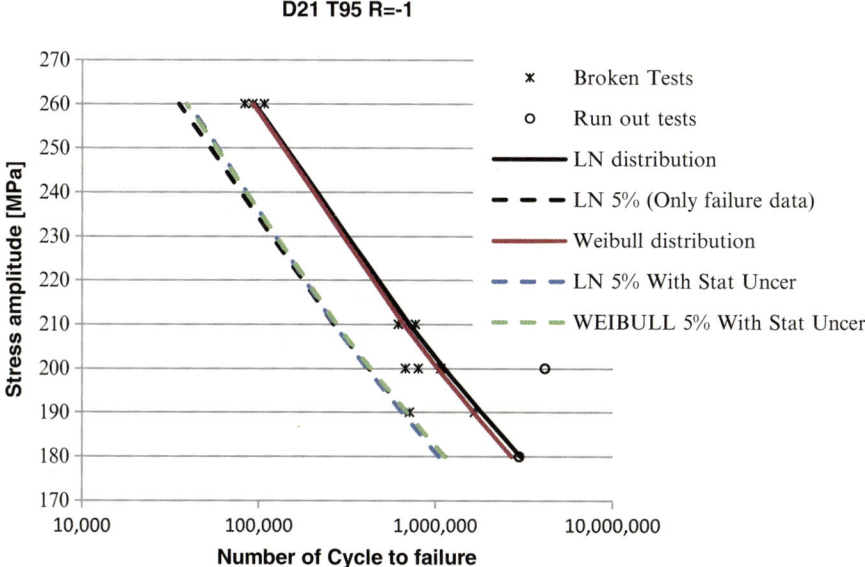

Fig. 3 Results for D21 T95 R-1

and the other quantile is estimated when the statistical uncertainties is taken into account. Moreover, the 5 % quantile of Weibull distribution is estimated when the statistical uncertainties are considered.

The results show that generally only a small difference is obtained between the mean (best fit) curves using Weibull and LogNormal distributions. Larger differences

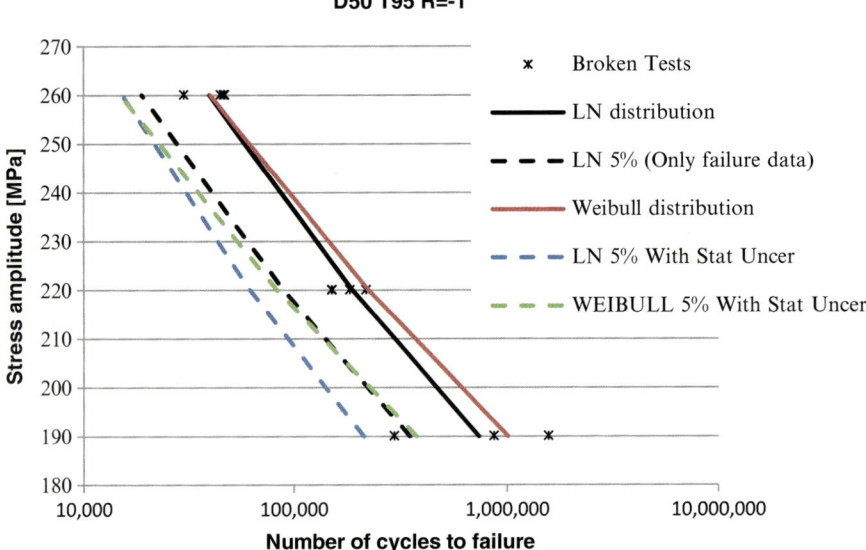

Fig. 4 Results for D50 T95 R-1

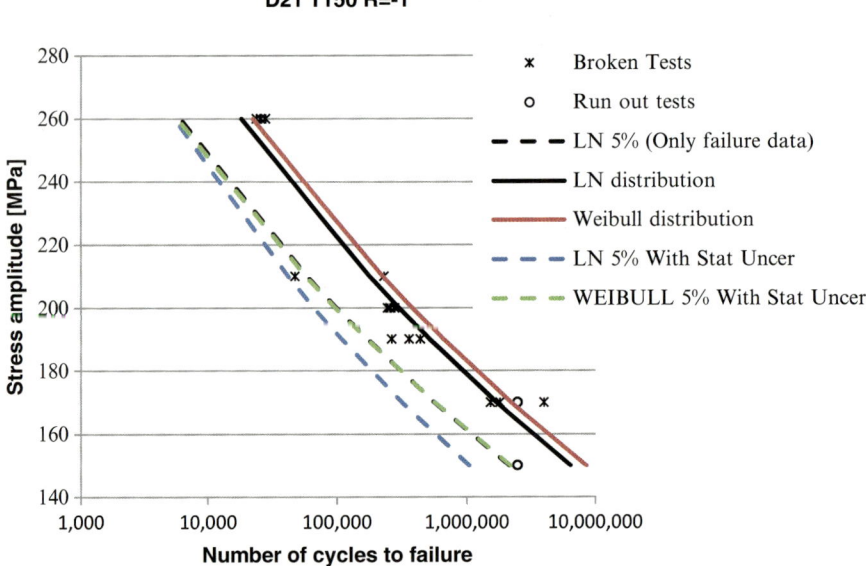

Fig. 5 Results for D21 T150 R-1

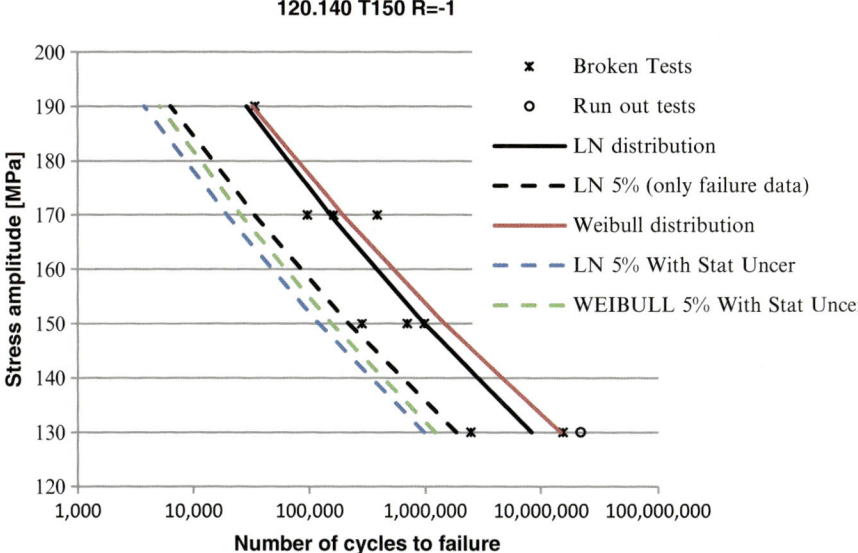

Fig. 6 Results for 120*140 T150 R-1

are seen in some cases when the 5 % quantiles are considered. Generally, the LogNormal distribution results in smaller number of cycles to failure than the Weibull distribution. Further, it is also seen that as expected smaller fatigue lives are obtained when statistical uncertainty is taken into account. Also it is seen in most cases to be important to take into account in the statistical analysis that some tests result in no-failure/run-out. As demonstrated in the examples this is easily accounted for using the Maximum Likelihood Method.

7 Conclusion

In this paper stochastic models for modeling fatigue failure in wind turbine drivetrain components are considered. Firstly, two stochastic models for uncertainties influencing fatigue failure are described based on a Weibull and a LogNormal distribution function. These uncertainties include model uncertainties, statistical uncertainties and size effects. It is described how the statistical parameters can be estimated using the Maximum Likelihood Method and how 5 % quantiles can be obtained taking into account statistical uncertainties though formulating limit state equations and applying FORM (First Order Reliability Methods).

In an illustrative example, statistical procedure is applied to a set of data to demonstrate the importance of taking into account both tests resulting in failure and in no-failure/run-out using the Maximum Likelihood Method. The results indicate that generally only a small difference is obtained between the mean (best fit) curves using Weibull and LogNormal distributions. When 5 % quantiles (characteristic values) are compared larger differences are seen with the LogNormal model resulting in smaller number of cycles to failure than the Weibull model. Further, it is also seen that as expected smaller fatigue lives are obtained when statistical uncertainty is taken into account.

Acknowledgments The work is supported by the Strategic Research Center "REWIND – Knowledge based engineering for improved reliability of critical wind turbine components", Danish Research Council for Strategic Research, grant no. 10–093966.

References

Campbell FC (2008) Elements of metallurgy and engineering alloys. ASM International, Materials Park, pp 243–264

Dong W, Xing Y, Moan T, Gao Z (2013) Time domain-based gear contact fatigue analysis of a wind turbine drivetrain under dynamic conditions. Int J Fatigue 48:133–146

Entezami M, Hillmansen S, Weston P, Papaelias M (2012) Fault detection and diagnosis within a wind turbine mechanical braking system using condition monitoring. Renew Energy 47:175–182

Hau E (2006) Wind turbines: fundamentals, technologies, application, economics, 2nd edn. Springer, New York

Lindley DV (1976) Introduction to probability and statistics from a Bayesian viewpoint, vol 1 + 2. Cambridge University Press, Cambridge

Liu WY, Zhang WH, Han JG, Wang GF (2012) A new wind turbine fault diagnosis method based on the local mean decomposition. Renew Energy 48:411–415

Madsen HO, Krenk S, Lind NC (1986) Methods of structural safety. Prentice-Hall, Inc., Englewood cliffs

Oyague F (2009) Gearbox modeling and load simulation of a baseline 750-kW wind turbine using state-of-the-art simulation codes. National Renewable Energy Laboratory. Available via NREL. http://www.nrel.gov/docs/fy09osti/41160.pdf

Schittkowski K (1986) NLPQL: a fortran subroutine solving constrained nonlinear programming problems. Ann Oper Res 5:485–500

Sheng S, Veers P (2011) Wind turbine drivetrain condition monitoring – an overview. In: Machinery Failure Prevention Technology (MFPT) Society conference proceedings, Virginia Beach, VA, USA, 10–12 May 2011

Shirani M, Härkegård G (2011a) Fatigue life distribution and size effect in ductile cast iron for wind turbine components. Eng Fail Anal 18:12–24

Shirani M, Härkegård G (2011b) Casting defects and fatigue behavior of ductile cast iron for wind turbine components: a comprehensive study. Materialwiss Werkst 42:1059–1074

Soua S, Lieshout PV, Perera A, Gan T, Bridge B (2013) Determination of the combined vibrational and acoustic emission signature of a wind turbine gearbox and generator shaft in service as a pre-requisite for effective condition monitoring. Renew Energy 51:175–181

Tavner PJ, Greenwood DM, Whittle MWG, Gindele R, Faulstich S, Hahn B (2012) Study of weather and location effects on wind turbine failure rates. Wind Energy 16:175–187

Stochastic FEM to Structural Vibration with Parametric Uncertainty

K. Sepahvand and S. Marburg

Abstract In this paper, we employ non–sampling techniques based on the generalized polynomial chaos (gPC) expansions to numerical simulation of damped vibration problems including random material and damping parameters. A general stochastic finite element method (SFEM) formulation is presented for damped linear structural vibration. Uncertainty involved in stiffness and damping matrices are represented by the gPC expansions. A hybrid SFEM and the gPC expansion is implemented to generate samples of the parameters for the FEM deterministic code from which the gPC expansions of natural frequencies and damping ratios are calculated. For that, experimental modal data are used to evaluate the coefficient of proportional uncertain damping matrix. The model is validated using experimental modal data for samples of composite plates.

Keywords Random damping • Stochastic vibration • Stochastic FEM • Polynomial chaos

1 Introduction

The behavior prediction of vibration problems with uncertain parameters is an important topic in novel engineering research with various applications. The deterministic numerical simulation of such problems leads to an approximate and nominal solution of reality. In such conditions, the powerful tool stochastic FEM (SFEM) is applied to achieve reliable results in numerical simulations.

K. Sepahvand (✉) • S. Marburg
Institute of Mechanics, Univerität der Bundeswehr, Munich, Germany
e-mail: sepahvand@unibw.de

M. Papadrakakis and G. Stefanou (eds.), *Multiscale Modeling and Uncertainty Quantification of Materials and Structures*, DOI 10.1007/978-3-319-06331-7__20,
© Springer International Publishing Switzerland 2014

Various SFEM procedures have been developed in past decades (Vanmarcke and Grigoriu 1983; Der Kiureghian and Ke 1988; Matthies et al. 1997; Ghanem and Abras 2003; Keese 2003; Baroth et al. 2007), most based on the sampling techniques as Monte Carlo (MC) methods (Papadrakakis and Papadopoulos 1996). A comprehensive review on the SFEM has been reported in Stefanou (2009). The application of the method to vibration problems has been reported in many works, i.e. Sarkar and Ghanem (2002), Schuëller and Pradlwarter (2009), Adhikari and Sarkar (2009), Sepahvand et al. (2012), and Soize (2013). In the most of these works, the impact of parameter uncertainties on structural damping has been ignored. In this paper, we investigate structural elastic parameter uncertainties and their effects on the natural eigenfrequencies and damping ratio. To this end, a SFEM formulation of structural free vibration is presented. It is assumed that the parameter uncertainties will appear in stiffness matrix.

2   Stochastic FE Modeling of Structural Damped Vibration

As established in many studies the general time dependent FE model of structure vibration can be presented as

$$\left[-\omega^2 \mathbf{M} + \mathbf{K}(j\omega)\right] \mathbf{U} = \mathbf{F} \tag{1}$$

where $\mathbf{M}$ and $\mathbf{K}$ are global mass and stiffness matrices, respectively, $\mathbf{F}$ and $\mathbf{U}$ are nodal force and displacement vectors and $j = sqrt(-1)$. It is assumed that the stiffness matrix can be decoupled as

$$\mathbf{K}(j\omega) = \mathbf{K}_e + \mathbf{K}_v(j\omega) \tag{2}$$

Elastic stiffness matrix $\mathbf{K}_e$ is assumed frequency independent, whereas $\mathbf{K}_v$ serves as the stiffness part which contribute in damping stiffness. It is assumed that uncertainties attribute only in $\mathbf{K}_v$ and $\mathbf{K}_e$ is considered deterministic. This is particularly useful when one deals with sandwich composite structures where the elastic faces can be modeled deterministically and the viscoelastic layer as uncertain part. In such conditions, the stochastic FE from of Eq. (1) can be written as

$$\left[-\omega^2(\boldsymbol{\xi})\mathbf{M} + \mathbf{K}_e + \mathbf{K}_v(j\omega, \boldsymbol{\xi})\right] \mathbf{U}(\boldsymbol{\xi}) = \mathbf{F} \tag{3}$$

where $\boldsymbol{\xi} = \{\xi_1, \xi_2, \ldots, \xi_M\}$ is the vector of independent random variables stand for uncertainties in material parameters, e.g. elastic and damping parameters of viscoelastic layer. The stochastic matrix $\mathbf{K}_v(j\omega, \boldsymbol{\xi})$, eigenfrequencies ω and the stochastic vector $\mathbf{U}(\boldsymbol{\xi})$ can be approximated using gPC expansions as

$$\mathbf{K}_v(j\omega, \boldsymbol{\xi}) = \sum_{i=0}^{N_1} \mathbf{k}_i(j\omega)\Psi_i(\boldsymbol{\xi}) = \mathbf{k}^T(j\omega)\boldsymbol{\Psi}(\boldsymbol{\xi}) \tag{4}$$

$$\omega^2(\boldsymbol{\xi}) = \sum_{i=0}^{N_2} a_j \Psi_j(\boldsymbol{\xi}) = \mathbf{a}^T \boldsymbol{\Psi}(\boldsymbol{\xi}) \tag{5}$$

$$\mathbf{U}(\boldsymbol{\xi}) = \sum_{k=0}^{N_3} \mathbf{u}_k \Psi_k(\boldsymbol{\xi}) = \mathbf{u}^T \boldsymbol{\Psi}(\boldsymbol{\xi}) \tag{6}$$

In these expansions Ψ_i, Ψ_j and Ψ_k denote the orthogonal polynomials of the vector of random variables. In a classical gPC problem, the deterministic coefficients $\mathbf{k}_i$ are calculated based on the information available on the statistical distribution of uncertain parameters and the type of random variables (Sepahvand et al. 2010). Substituting the gPC expansions in Eq. (3) leads to

$$\epsilon(\boldsymbol{\xi}) = \left[-\mathbf{a}^T\boldsymbol{\Psi}(\boldsymbol{\xi})\mathbf{M} + \mathbf{K}_e + \mathbf{k}^T(j\omega)\boldsymbol{\Psi}(\boldsymbol{\xi})\right]\mathbf{u}^T\boldsymbol{\Psi}(\boldsymbol{\xi}) - \mathbf{F} \tag{7}$$

This represents the error of approximate gPC solution of stochastic FE model in Eq. (1). These unknown deterministic coefficient vectors $\mathbf{a}^T$ and $\mathbf{u}^T$ are obtained by using Galerkin method, i.e. projecting the error $\epsilon(\boldsymbol{\xi})$ onto space of basis functions $\Psi_p(\boldsymbol{\xi})$. This yields to

$$\langle \epsilon(\boldsymbol{\xi}),\ \Psi_k(\boldsymbol{\xi})\rangle \to 0 \tag{8}$$

Once the stochastic basis functions, $\Psi(\boldsymbol{\xi})$s, are chosen, the solution process reduces to computation of the unknown coefficients by minimization of the error. Consequently, any optimization process to minimize stochastic error must be performed with respect to the random space discretization by the gPC, i.e.

$$\langle\left[-\mathbf{a}^T\boldsymbol{\Psi}(\boldsymbol{\xi})\mathbf{M} + \mathbf{K}_e + \mathbf{k}^T(j\omega)\boldsymbol{\Psi}(\boldsymbol{\xi})\right]\mathbf{u}^T\boldsymbol{\Psi}(\boldsymbol{\xi}),\ \Psi_k(\boldsymbol{\xi})\rangle - \langle\mathbf{F},\ \Psi_k(\boldsymbol{\xi})\rangle = 0 \tag{9}$$

There are two broad classes of methods that can be used to solve the above stochastic model: (i) intrusive and (ii) non–intrusive methods. Implementation of the intrusive method requires projection of the stochastic model into an equivalent deterministic model by using stochastic Galerkin projection, whereas in the non–intrusive method, the model is employed as third party solver or black–box, and the solution is investigated at specific collocation points of the stochastic basis function (Huang et al. 2007). In this work we employ the second method as described in next section.

3 Case Study

As a case study, the free vibration of orthotropic plates is investigated in this section. The E–moduli E_{11}, E_{22} and the shear modulus G_{12} are considered as uncertain parameters. The uncertain damping parameter is represented by means of proportional model as

$$C(\xi) = \alpha M + \beta K_v(\xi) \tag{10}$$

The orthogonal transformation of the damping matrix with respect to normalized eigenvectors $\{\Phi\}$ for deterministic case is used to calculate the pre–defined constants α and β, i.e.

$$\{\Phi\}^T C \{\Phi\} = [\alpha + \beta \omega_i^2] \tag{11}$$

As in most of the practical engineering vibration analysis, these constants are estimated form measured modal data for limited modes assuming that would be valid for the overall vibration modes of the plate. We used experimental modal damping and frequencies of the first few modes to estimate the constants. From Eq. (11) we can write

$$2\hat{\zeta}_i \hat{\omega}_i = \alpha + \beta \hat{\omega}_i^2, \qquad i = 1, 2, \ldots, m \tag{12}$$

where $\hat{\zeta}$ and $\hat{\omega}$ are measured damping and natural frequencies, respectively, and m is the number of first few modes, see Fig. 1. As $m > 2$, a least–square minimization is performed for evaluating the constants. The first order, 2–dimensional (2d) gPC expansions are served to represent the uncertainty in material parameters, for instance for G_{12}

$$G_{12}(\xi_1, \xi_2) = \bar{G}_{12} + \delta_1 \xi_1 + \delta_2 \xi_2, \qquad \xi_1, \xi_2 \in N[0, 1] \tag{13}$$

In which $\bar{G}$ is the mean value and $\delta_1 = \delta_2 = 0.15\bar{G}$. Accordingly, the uncertain eigenfrequency $\omega_i = 2\pi f_i$ and damping ratio η_i at ith–mode are approximated using second order, 2d–gPC as

$$\lambda_i = \omega_i^2 [1 + j\eta_i] = \sum_{i=0}^{2} a_i \Psi_i(\xi_1, \xi_2) \left[1 + j \sum_{i=0}^{2} b_i \Psi_i(\xi_1, \xi_2) \right] \tag{14}$$

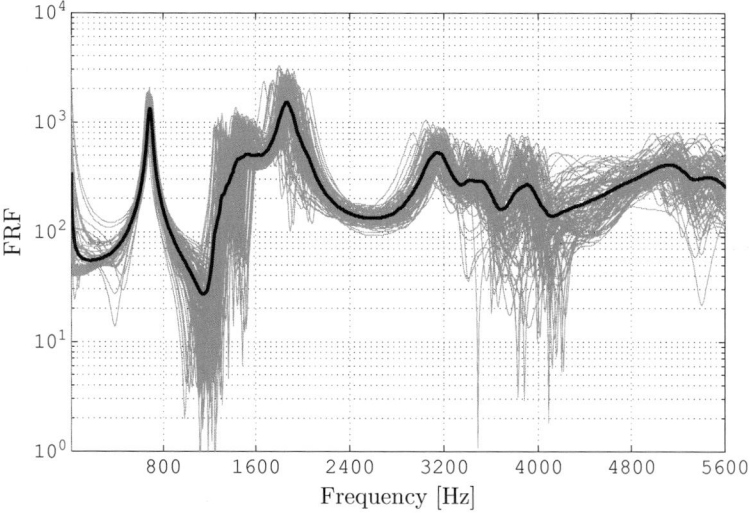

Fig. 1 Experimental FRF for 100 identical plate samples, the average FRF (*bold line*) is used to estimate the constants in Eq. (10)

Table 1 The gPC coefficients of natural frequencies (Hz) and damping ratios (%) in Eq. (14)

Mode	(a_0, b_0)	(a_1, b_1)	(a_2, b_2)	(a_3, b_3)	(a_4, b_4)	(a_5, b_5)
1	(660.8,1.78)	(49.4,0.95)	(49.4,0.95)	(−2.2,0.10)	(−2.2,0.10)	(−4.3,0.11)
2	(1336.1,0.83)	(101.4,1.42)	(101.4,1.42)	(−4.4,−2.35)	(−4.4,−2.35)	(−8.7,−2.70)
3	(1461.9,1.67)	(110.40,2.11)	(110.4, 2.11)	(−4.9,0.20)	(−4.9,0.20)	(−9.6,0.21)
4	(1879.2,1.21)	(138.30,2.64)	(138.3,2.64)	(−6.4,−9.91)	(−6.4,−9.91)	(−12.6,0.44)

In which a_i and b_i are deterministic unknown coefficients and $\Psi_i(\xi_1, \xi_2)$ are 2d orthogonal Hermite polynomials, cf. Sepahvand et al. (2010). Non–intrusive SFEM is employed to calculate these coefficients. To this end, nine samples of sparse points are generated from roots of third order Hermite polynomials, i.e. $(0, \pm 1.732)$, and deterministic FE model of the plate is solved on each sample point to estimate 2×6 unknown coefficients. The simulated natural frequencies and damping ratios, then, passed a nonlinear optimization process to evaluate the coefficients. The results are shown in Table 1. To evaluate the validity of the results obtained from the proposed method and to test the convergence property, the constructed probability density function (PDF) of the uncertain frequencies and damping ratios are compared with the constructed PDFs from the measured data for 100 samples. The results are shown in Figs. 2 and 3. A good agreement is observed between the second order gPC estimations (bold lines) and experimental results.

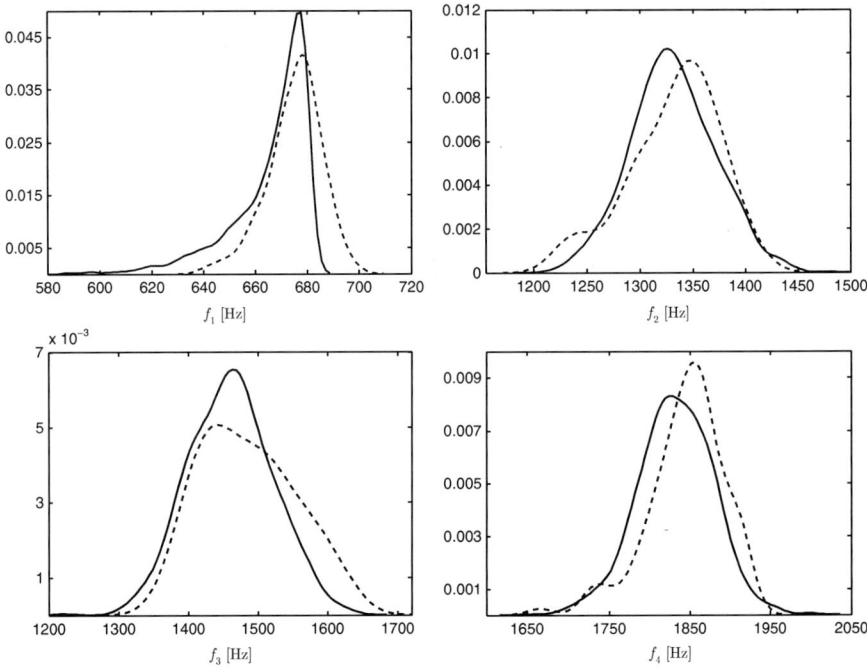

Fig. 2 PDFs of the first four eigenfrequencies constructed from the second order gPC (*bold lines*) in comparison with experimental results (*dashed lines*)

4   Conclusions

This paper presented application of stochastic FEM for structural damped vibration problems in which material and damping properties exhibit random variation, by using collocation points. The method appears to be efficient, requiring only several runs to accurately compute the solution statistics in comparison with Monte Carlo simulation which may require thousands of realizations. The paper contribution focuses on the experimental validation of the results. As a numerical example, the free vibration of damped orthotropic plate is investigated in which random stiffness matrix is employed to model uncertain proportional damping. The unknown deterministic coefficients of damping ratios and eigenfrequencies for the first four modes have been calculated from nine collocation points. This strategy helps us to use the available deterministic FEM code developed in any commercial software.

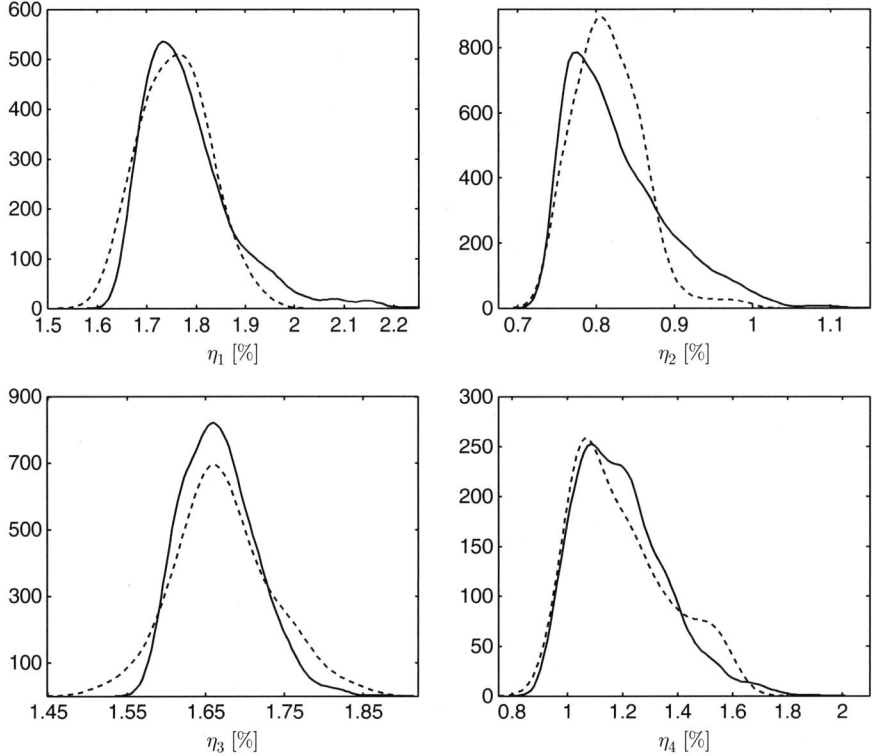

Fig. 3 PDF of the first four damping ratios represented by second order gPC (*bold lines*). The *dashed lines* show the experimentally measured values for 100 samples

References

Adhikari S, Sarkar A (2009) Uncertainty in structural dynamics: experimental validation of a wishart random matrix model. J Sound Vib 323(3–5):802–825

Baroth J, Bressolette Ph, Chauvière C, Fogli M (2007) An efficient {SFE} method using lagrange polynomials: application to nonlinear mechanical problems with uncertain parameters. Comput Methods Appl Mech Eng 196(45–48):4419–4429

Der Kiureghian A, Ke J-B (1988) The stochastic finite element method in structural reliability. Probab Eng Mech 3(2):83–91

Ghanem R, Abras J (2003) A general purpose library for stochastic finite element computations. In: Bathe KJ (ed) Computational fluid and solid mechanics 2003, pp 2278–2280. Elsevier Science Ltd, Oxford

Huang S, Mahadevan S, Rebba R (2007) Collocation–based stochastic finite element analysis for random field problems. Probab Eng Mech 22(2):194–205

Keese A (2003) Numerical solution of systems with stochastic uncertainties – a general purpose framework for stochastic finite elements. PhD thesis, Fachbereich Mathematik and Informatik, TU Braunschweig, Braunschweig

Matthies HG, Brenner CE, Bucher CG, Soares CG (1997) Uncertainties in probabilistic numerical analysis of structures and solids-stochastic finite elements. Struct Saf 19(3):283–336

Papadrakakis M, Papadopoulos V (1996) Robust and efficient methods for stochastic finite element analysis using Monte Carlo simulation. Comput Methods Appl Mech Eng 134 (3–4):325–340

Sarkar A, Ghanem R (2002) Mid–frequency structural dynamics with parameter uncertainty. Comput Methods Appl Mech Eng 191(47–48):5499–5513

Schuëller GI, Pradlwarter HJ (2009) Uncertain linear systems in dynamics: retrospective and recent developments by stochastic approaches. Eng Struct 31(11):2507–2517

Sepahvand K, Marburg S, Hardtke H-J (2010) Uncertainty quantification in stochastic systems using polynomial chaos expansion. Int J Appl Mech 2(2):305–353

Sepahvand K, Marburg S, Hardtke H-J (2012) Stochastic free vibration of orthotropic plates using generalized polynomial chaos expansion. J Sound Vib 331:167–179

Soize C (2013) Stochastic modeling of uncertainties in computational structural dynamics–recent theoretical advances. J Sound Vib 332(10):2379–2395

Stefanou G (2009) The stochastic finite element method: past, present and future. Comput Methods Appl Mech Eng 198:1031–1051

Vanmarcke E, Grigoriu M (1983) Stochastic finite element analysis of simple beams. J Eng Mech 109(5):1203–214